To
Louise

love

Dad . & Linda

Practical Fermentation Technology

Practical Fermentation Technology

BRIAN MCNEIL & LINDA M. HARVEY

Strathclyde Fermentation Centre, Strathclyde University, UK

John Wiley & Sons, Ltd

Other Wiley Editorial Offices

John Wiley & Sons Inc., 111 River Street, Hoboken, NJ 07030, USA

Jossey-Bass, 989 Market Street, San Francisco, CA 94103-1741, USA

Wiley-VCH Verlag GmbH, Boschstr. 12, D-69469 Weinheim, Germany

John Wiley & Sons Australia Ltd, 42 McDougall Street, Milton, Queensland 4064, Australia

John Wiley & Sons (Asia) Pte Ltd, 2 Clementi Loop #02-01, Jin Xing Distripark, Singapore 129809

John Wiley & Sons Ltd, 6045 Freemont Blvd, Mississauga, Ontario L5R 4J3, Canada

Wiley also publishes its books in a variety of electronic formats. Some content that appears in print may not be available in
electronic books.

Library of Congress Cataloging-in-Publication Data

McNeil, B. Practical fermentation technology / Brian McNeil & Linda M. Harvey.
   p. cm.
Includes bibliographical references and index.
ISBN 978-0-470-01434-9 (cloth)
1. Fermentation. I. Harvey, L. M. II. Title.
TP156.F4M36 2008
660′.28449 – dc22

                                                                          2007041702

British Library Cataloguing in Publication Data

A catalogue record for this book is available from the British Library

ISBN 978-0470-014349

Typeset in 10/12pt Times by SNP Best-set Typesetter Ltd., Hong Kong
Printed and bound in Great Britain by Antony Rowe Ltd, Chippenham, Wiltshire

For David & Louise

Contents

List of Contributors

Stephen A. Baldwin, Astbury Centre for Structural Molecular Biology, Institute of Membrane and Systems Biology, University of Leeds, Leeds LS2 9JT

Roslyn M. Bill, School of Health and Life Sciences, Aston University, Aston Triangle, Birmingham B4 7ET

Frances Burke, Eli Lilly, Speke Operations, Fleming Rd, Speke, Liverpool, L24 9LN

Beverley Finn, Strathclyde Fermentation Centre, Strathclyde Institute of Pharmacy and Biomedical Sciences, Strathclyde University, Royal College Building, 204 George Street, Glasgow, G1 1XW

Ger Fleming, Department of Microbiology, National University of Ireland, Galway, Ireland

Linda M. Harvey, Institute of Pharmacy and Biomedical Sciences, Strathclyde University, Royal College Building, 204 George Street, Glasgow, G1 1XW

Andrew Hayward, Director of European Operations, Broadley Technologies Ltd, Wrest Park, Silsoe, Beds, MK45 4HS

Peter J.F. Henderson, Astbury Centre for Structural Molecular Biology, Institute for Membrane and Systems Biology, University of Leeds, Leeds, LS2 9JT

Richard B. Herbert, School of Chemistry, University of Leeds, Leeds LS2 9JT

Ryan J. Hope, Astbury Centre for Structural Molecular Biology, Institute of Membrane and Systems Biology, University of Leeds, Leeds LS2 9JT

Mohammed Jamshad, School of Health and Life Sciences, Aston University, Aston Triangle, Birmingham B4 7ET

Erik Kakes, Applikon Biotechnology, De Brauwweg 13, 3125 AE Schiedam, The Netherlands

Robert Kinley, Lilly UK, Erl Wood Manor, Windlesham, Surrey, GU20 6PH

Sue Macauley-Patrick, Sartorius Stedim UK Limited, Longmead Business Centre, Blenheim Road, Epsom, Surrey, KT19 9QQ

Guy Matthews, Applikon Biotechnology, Deer Park Business Centre, Eckington, Pershore, WR10 3DN

Ferda Mavituna, School of Chemical Engineering and Analytical Science, University of Manchester, Sackville Street, Manchester, M60 1QD

Brian McNeil, Institute of Pharmacy and Biomedical Sciences, Strathclyde University, 204 George Street, Royal College Building, Glasgow, G1 1XW

James R. Moldenhauer, Eli Lilly and Company, Lilly Corporate Center, Indianapolis, IN 46285

Halina T. Norbertczak, Astbury Centre for Structural Molecular Biology, Institute of Membrane and Systems Biology, University of Leeds, Leeds LS2 9JT

John O'Reilly, Astbury Centre for Structural Molecular Biology, Institute of Membrane and Systems Biology, University of Leeds, Leeds LS2 9JT

John Patching, Department of Microbiology, National University of Ireland, Galway, Ireland

Simon G. Patching, Astbury Centre for Structural Molecular Biology, Institute of Membrane and Systems Biology, University of Leeds, Leeds LS2 9JT

Peter C.J. Roach, Astbury Centre for Structural Molecular Biology, Institute of Membrane and Systems Biology, University of Leeds, Leeds LS2 9JT

Nicholas G. Rutherford, Astbury Centre for Structural Molecular Biology, Institute of Membrane and Cellular Biology, University of Leeds, Leeds LS2 9JT

Charles G. Sinclair, School of Chemical Engineering and Analytical Science, University of Manchester, Sackville Street, Manchester, M60 1QD

Peter G. Stockley, Astbury Centre for Structural Molecular Biology, Institute of Molecular and Cellular Biology, University of Leeds, Leeds LS2 9JT

Henrietta Venter, Department of Pharmacology, University of Cambridge, Cambridge CB2 1PD

Stewart White, Devro plc, Gartferry Road, Moodiesburn, Glasgow, G69 0JE

Acknowledgements

The editors would like to thank all involved in the production of *Practical Fermentation Technology*.

Preface

Fermentation is a very ancient practice indeed, dating back several millennia. More recently, fermentation processes have been developed for the manufacture of a vast range of materials from chemically simple feedstocks, such as ethanol, right up to highly complex protein structures. The advent of this latter range of products and processes has revolutionised the practice of clinical medicine and many areas of fundamental research, and has also significantly increased the need for skilled individuals in the fermentation area.

The key question is 'How can we deliver these skills to those who need them?' In essence, this question hints at the potential difficulties: fermentation itself is an applied science, an underpinning technology. Many of the new entrants have no track record in the area, and being an applied science, in this context at least, publicly supported training in this area has been subject to the usual neglect in funding terms in many countries.

Typically, the acquisition of a set of practical skills might involve the skills being passed on or down from an experienced practitioner to a relative newcomer via demonstration, explanation and repetition. However, given the expansion in the use of fermentation techniques, this bespoke one to one approach is not always possible, especially for the many new scientists, engineers and technicians entering the fermentation area, in labs which have no previous experience of this area to draw upon.

This book is aimed at helping these relative newcomers to fermentation. It is not intended as a substitute for the type of training described above, but it may help the newcomer avoid some of the more obvious mistakes and pitfalls we have all made, and especially, to prevent them from 're-inventing the wheel'.

The contributors to this book are academic and industrial scientists and engineers with many years practical experience of actually carrying out fermentation processes of different types, at a range of scales from bench-top, right up to the largest industrial production scales. This book is intended to help the beginner or less experienced fermentation scientist, by bringing together and setting down our practical experiences in fermentation technology. It is not intended to cover fundamental or theoretical aspects underpinning fermentation, which are, in any respect, already well covered by a range of accessible books and reviews. Instead, we have focused on the practical skills and associated problems, cross referencing to appropriate reading material dealing with the underpinning science or engineering where relevant.

This book proceeds from a brief background on the development of fermentation, through the criteria for selection of lab based fermenters, with a chapter on the more specialised needs and challenges of equipping a lab for membrane protein expression (a very common raison d' être for the setting up of small fermentation labs or suites nowadays) before moving on to practical aspects of cultivation modes, medium preparation and sterilisation, and culture preservation and inoculum work up techniques. The modelling of fermentation processes is then discussed, followed by a discussion of practical aspects of scaling up or scaling down fermentations. The typical sensors used to monitor fermentations are then described and associated practical challenges discussed. Finally, rather more specialised, and less frequently discussed areas relating to fermentation are described, including the selection and use of Supervisory Control and Data Acquisition systems (SCADA). The penultimate chapter focuses on a brief discussion of the variability inherent in fermentation processes, its consequences and how this can be described and quantified. The final chapter deals with a description of the various continuous culture systems (e.g. chemostats), and of how powerful these technologies can be in helping us understand better the physiology of microbes, or cultured cells (animal or plant) via definition and control of their environment in the fermenter.

This book essentially moves from the pre-fermentation stages (equipment selection, lab set up, culture techniques, medium and inoculum preparation) to discussion of how we describe, scale up, and monitor fermentations. Finally ending with some rather neglected areas post fermentation relating to what we do with data arising from the fermentation, how we assess acceptable degrees of variability in our process, and finally how we can use fermentations as powerful tools in microbiological research.

This work is not intended to be a handbook, but we hope that by reading how we deal with some of the challenging aspects of the practice of fermentation, the learning curve of the newcomer might be accelerated, and their path to competence smoothed a little.

Brian McNeil

1

Fermentation: An Art from the Past, a Skill for the Future

Brian McNeil and Linda M. Harvey

The origins of fermentation are lost in ancient history, perhaps even in prehistory. We know that the ancient Egyptians and Sumerians both had knowledge of the techniques used to convert starchy grains into alcohol. For most of history these processes, or similar ones based on fruit juice conversion, have represented the most commonly accepted interpretation of the word 'fermentation'.

However, 'fermentation' has many different and distinct meanings for differing groups of individuals. In the present context we intend it to mean the use of submerged liquid culture of selected strains of microorganisms, plant or animal cells, for the manufacture of some useful product or products, or to gain insights into the physiology of these cell types. This is a relatively narrow definition, but would include the 'traditional' fermentations described above. By contrast, the modern fermentation industry, which is largely a product of the Twentieth century, is dominated by aerobic cultivations intended to make a range of higher value products than simple ethanol.

In recent years there has been a tremendous expansion in the use of fermentation technology by individuals with less training in the subject than previous exponents of these techniques. This book is aimed at scientists and engineers relatively new to the subject of fermentation technology, and is intended to be the text equivalent of the briefings and chats that mentors in this area would have with newcomers. It is meant to be a means of passing on the experiences we have had in many years of fermentation to relative newcomers to the subject, so that perhaps, you will be able to avoid some of the more obvious pitfalls we fell into. It is specifically not intended as a reference text to the principles underlying fermentation science and engineering, as such volumes already exist. Each

Practical Fermentation Technology Edited by Brian McNeil and Linda M. Harvey
© 2008 John Wiley & Sons, Ltd

chapter in this book is accompanied by a short 'further reading section' or supporting reference section, which generally contain a list of a few book chapters, or relevant reviews supporting the material in the chapter.

The fermentation industry today is very much in a state of flux, with rapid changes in location, product spectrum, and scale of processes occurring. To a large extent this has been brought about by macroeconomic forces compelling the relocation of large scale bioprocesses outside high labour cost regions, but also by the successful deciphering of the human genome with its myriad of new therapeutic targets , and the significant advances in the construction of advanced fermentation expression systems for making novel proteins and antibodies.

Thus, fermentation skills and knowledge are now essential to driving forward systematic research into drug/receptor interactions, function of membrane proteins in health and disease, and are powering an unparalleled expansion in our capability to combat serious diseases in the human population, including cancers, degenerative illnesses such as Alzheimer's, and increasingly common complaints of developed societies such as asthma. The new fermentation-derived medicines, including biopharmaceuticals, hold out the prospect of improved specificity of treatment, and decreased side effects. It is truly a revolutionary period in clinical medicine as these new agents manufactured by fermentation routes enter the market. The 'new' fermentation products, therapeutic proteins, antibodies(simple and conjugated) are more complex and costly than previous products, but, in essence, the need to focus upon the fermentation step is now clearer than ever. Basically, the 'quality' of these products(the potency, efficacy, stability and immunogenicity) is determined by the upstream or fermentation stage, so the need for a clear understanding of what happens in that stage, how it can best be monitored, controlled, and carried out in a reproducible fashion, is greater than ever. It is in exactly these areas that the many often highly capable individuals entering fermentation are unwittingly deficient in background.

The long heralded era of personalized medicine may well be imminent due to recent advances in cultivation and replication of stem cells. This will make the need for scientists and engineers who understand culture techniques even greater in coming years. Thus, the demand for fermentation skills is likely to increase in the immediate future. Fermentation has contributed much to the well-being and wealth of human populations over millennia; it will continue to do so to an even greater extent in the future We hope this book will help those coming recently to this field to contribute more effectively to that process.

2

Fermentation Equipment Selection: Laboratory Scale Bioreactor Design Considerations

Guy Matthews

2.1 Introduction

This chapter will cover the design considerations of small-scale bioreactors, and illustrate the kind of logic involved in selecting fermentation equipment for a small scale 'general purpose' laboratory where the fermentation equipment will typically be usable for several fermentation processes. Equipping a research scale laboratory for, specialist purposes, for example, for expression of recombinant proteins, particularly, membrane proteins, is covered in detail in Chapter 3.

Bioreactors in some form or another have been around for thousands of years, although up until the 1900s their use was limited to the production of potable alcohol. It was really from the 1940s onwards that fermentation as it is known today began to appear with the need to produce antibiotics during World War Two. At this point the need for process development to improve yields drove research. As it would not be practical to carry out this research on production scale equipment, small scale bioreactors became common place. Small scale for the purposes of this chapter is defined as having a total volume of between 1 litre and 20 litres.

Bioreactors at this volume can be used for a number of purposes, scale up/down studies, clone selection, medium development, process development and, in some cases, production.

Practical Fermentation Technology Edited by Brian McNeil and Linda M. Harvey
© 2008 John Wiley & Sons, Ltd

The bioreactor should be capable of the following as a minimum:

- aseptic production for extended periods of time;
- meeting the local containment regulations;
- monitoring and controlling the following parameters:
  pH by either acid/base addition or CO_2/base addition;
  mixing such that the culture remains in suspension and DO_2 is maintained; all this
      should be achieved without damage to the organisms;
  temperature regulation;
  sterile sampling capability.

Thought should also be given to the future in terms of spare head plate ports, so that the configuration of the system is not fixed.

Scale up and scale down should also be considered such that the aspect ratio of the vessels being used is consistent with any larger systems, which the small scale system is designed to mimic. This is discussed in much greater detail in Chapter 8.

Other issues that should be considered are the available utilities, the turn around time of the vessel from finishing one run to starting another, and the manual handling issues especially at the 10-L plus scale where the weight of the system may require two people for safe manual handling.

2.2   Types of Bioreactor

By far the most common type of bioreactor in use today, and the focus of this chapter, is the STR (stirred tank reactor). This essentially consists of a vessel with an aspect ratio of around 3 : 1, and a mixing system typically driven through the headplate, although with some steam in-place systems the mixing will be driven though the base. The head plate will have ports that allow for the addition of probes, reagents and gas as well as the removal of samples.

Numerous other types of bioreactor are in existence, the main alternatives to the STR are discussed below.

Tower fermenters, as the name suggests, are vessels characterised by a high height-to-diameter ratio, anywhere from 6 : 1 to 15 : 1. They are aerated by gas sparging via a simple sparger usually located near the fermenter base. These systems can be operated continuously by the creation of settling zones by using baffles, which allow the product to be taken off and the cells returned to the main body of the vessel.

In *airlift fermenters* the mixing system (Motor, driveshaft and impellors) is replaced by a constant flow of gas introduced into a riser tube. Within the vessel flows develop that, as the air rises and then the medium containing cells falls, ensure thorough mixing. The airlift vessel may be baffled to improve mixing. These vessels provide very gentle mixing, and so are particularly suited to cells that are too shear sensitive to be mixed by an impeller.

Hollow fibre chambers are used to grow anchorage-dependent cells. The system consists of a bundle of fibres and the cells grow within the extra capillary spaces (ECS) with in a cartridge. Medium and gas perfuse through the capillary lumea to the ECS. where they are available to the cells. The size of the lumea can be selected such that any product is retained in the ECS or passes through the lumea such that the system acts as a *perfu-*

sion bioreactor. An alternative to hollow fibres would be to use microcarriers in a stirred tank reactor. Microcarriers are typically chromatographic grade DEAE sepadex beads. The beads are positively charged, and so are attracted to negatively charged animal cells and provide a surface on which the cells can grow.

Small scale stirred tank bioreactors may be divided in to three basic groups.

2.2.1 Autoclavable Systems

These represent perhaps the most commonly utilised systems, perhaps due to their relative simplicity, cost and basic utility requirements. Full functionality of such glass systems may be via an actuation panel, which houses all the devices for the control of liquid or gas flow, namely pumps, solenoid valves, and rotameters or mass flow controllers. All of this is controlled by a local controller potentially connected to a SCADA (Supervisory Control and Data Acquisition package), a specific software package that will log all data generated by the system and control the system. The use of a SCADA allows a greater level of functionality than that delivered by just the local controller, and is discussed in more detail in Chapter 10.

Figure 2.1(a) represents a typical small-scale autoclavable system. The vessel is made from glass, with a stainless steel head plate, and is serviced by an actuation panel with a local controller that is connected to a SCADA (Supervisory Control and Data Acquisition package).

2.2.2 Stainless Steel Steam in Place Systems

Figure 2.1(b) shows a vessel of approximately the same volume as that in Figure 2.1(a), but in stainless steel. The key difference between this and the glass system is that the steel vessel has a steam supply piped to it so sterilisation will occur *in-situ*, thus removing the need for a large on site autoclave. Its main advantage over the glass system is the ability to apply over pressure. The application of over pressure (running the system at a higher than atmospheric pressure) is used to enhance the oxygen transfer rate. For obvious safety reasons it would not be acceptable to apply an overpressure to a glass vessel. The removal of manual handling considerations, as the stainless steel system is sterilised *in-situ* rather than in an autoclave, would be another reason to move to a stainless steel system at these small volumes. Some operators prefer an autoclavable system to SIP bioreactors as they feel that the former suffer less contamination than the latter. The evidence for this largely anecdotal and subjective.

The steam supplied to such a system needs to be particle and oil free as either of these can affect the product being made by leaving a residue in the vessel or damage the system by blocking valves. The steam also needs to be dry, as water will condense which could potentially lead to a cold spot that does not achieve the sterilisation temperature, usually 121 degrees centigrade.

2.2.3 Single Use Bioreactors

Figure 2.1(c) shows a single use system where the vessel has been replaced by a bag. This bag is multi-layered and meets all the criteria for a product in contact with a biological process, namely that it should be nonshedding, leaching, or chelating. The bag construction is discussed in the next section covering materials of construction. The bag is

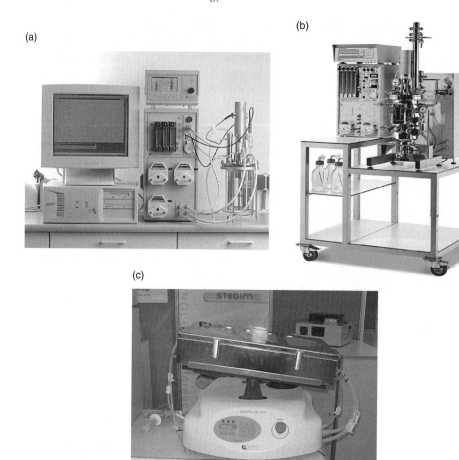

Figure 2.1 *(a) Typical small scale bench mounted glass autoclavable bioreactor. (b) Small scale steam in place bioreactor. (c) Bench top single use bioreactor*

supplied presterilised, so does not require an autoclave or steam supply. The benefit being that the turn around time (time between runs) is minimised. At the moment these systems tend to be limited to mammalian and insect cell culture.

Before any of the above systems can be considered there are a number of fundamental points to decide upon and these are discussed below.

2.3   Construction Aspects

2.3.1   Materials of Construction

The materials from which a bioreactor is constructed have been widely discussed, for example, by Cowan and Thomas (1988). The material of construction is selected on the basis that it displays the following physical properties: it must be chemically inert such that it does not leach elements into the medium or chelate elements (principally metal ions) from the medium. So long as the materials of construction meet these requirements then the choice of material for the bioreactor is dictated by the scale at which the process is to be operated, the process itself and economic considerations. This requirement for the material of construction to be chemically inert, as well as being common sense, is also a regulatory requirement, as laid down by the FDA (Food and Drug Administration) and MHRA (Medicine and Healthcare Products Regulatory Agency) where the equipment is being used for the manufacture of a medicinal or diagnostic product.

Larger scale bioreactors, 20 litres and above, have their material of construction dictated primarily by the pressure codes and manual handling considerations. Bioreactors at the 20-L plus scale will almost exclusively be made from stainless steel and primarily 316L grade stainless steel or equivalent. Exceptions to this are systems used in corrosive environments, when such materials such as Hasteloy will be used, or in the single use technologies where a plastic will be the material of construction.

At the smaller scale the systems will be made from borosilicate glass, with a head plate made from 316-L stainless steel and O-rings from PTFE. However systems can be and are made of 316-L stainless steel at this scale. Generally the only reasons for choosing stainless steel at this scale are the lack of a large enough autoclave, the need to apply an overpressure or a culture that is prone to adhere strongly to borosilicate glass. The production of 316L stainless steel is very closely regulated and can be defined as a chromium-containing steel alloy. 316L contains other elements all of which contribute to its properties (Table 2.1).

The resulting metal is then electropolished to give an ultrasmooth finish. These finishes are described as having an ra (roughness average) value. This means that when something is quoted as have a 0.4 ra finish, on average no point will be higher than 0.4 μm above the surface of the metal. It is possible by using mechanical polishing instead of electropolishing (which dissolves peaks on the metal surface) to achieve a higher ra finish. However mechanical polishing will potentially lead to areas where contaminants can gather as the surface of the metal is folded over rather than dissolved away. The electropolishing process leads to an open structure on the surface, which can be easily cleaned.

The benefits of this are improved cleanability as there are no surfaces on which debris can settle. It also enhances corrosion resistance.

Borosilicate glass is chemically inert, easy to clean and robust. The glass is made up of SiO_2 81%, B_2O_3 13%, $Na_2O + K_2O$ 4% and Al_2O_3 2%. The glass itself will have to meet criteria for the thickness of the glass, the number of air bubbles within the glass,

Table 2.1 *The make up and function of the components of 316 L stainless steel*

Component	Content (%)	Function
Steel	65	
Chromium	16–18	Corrosion resistance
Nickel	12–14	Improves formability and ductility
Molybdenum	2–3	Improves chlorine resistance
Carbon	0.03 maximum	Reduces sensitisation, which is carbon precipitation during welding.

and how many of these bubbles are on the surface. The glass thickness and number of air bubbles determine the strength of the glass, while a bubble that has formed and burst on the surface of a vessel will leave an area that is difficult to clean and be a potential source of contamination.

Until recently the two main materials of construction were glass and steel. Approximately 4 years ago a single use technology was introduced by Stedim (www.stedim.com) that added a third material, plastic, in the form of bags. To describe these as plastic bags would be to misinterpret them to the same degree as describing a stainless steel bioreactor as being made of a steel drum. The single use fermenter is a 300-μm thick six-layered gas impermeable bag that meets all the validation criteria for materials compatibility as discussed above. This has made major inroads into the traditional STR (stirred tank reactor) market, because (a) the turn around time on single use systems is minimal B) the utility requirements are greatly reduced C) the validation route is very simple for such systems thus, they are used in validated production environments. The only drawback with these systems is the lack of an impellor and baffles to achieve higher OTR's, and, as such, there use has so far been limited to mammalian and insect cell systems with relatively modest oxygen demands.

Any type of system, be it glass, steel or plastic, should be validatable, that is, any manufacturer's claims in relation to pH, temperature control, mixing and O_2 transfer capability must be supported by a documented validation protocol. In such a protocol, all the claimed performance criteria are assessed against independent measuring devices, which themselves are calibrated against recognised international standards. All this work is documented and the records stored. Only when a system is fully validated can it be used in the production of any product that will be used either as a medicine or as part of a clinical trial.

2.3.2   Location Considerations

The system should be placed in an environment that is suited to its use. By this it is meant that the room chosen as the fermenter's location should have suitable containment facilities such that the operation of the system poses no threat to the health of those working with it or those working in the same building.

2.3.3   Utilities

The location should have suitably specified utilities, typically electrical power, chilled water, air, oxygen, nitrogen or carbon dioxide, and possibly, in the case of small scale steam in place (SIP) systems, steam at the correct volume, pressure and temperature.

The power requirements of a low volume system are typically based around 13-amp power, although in some cases 3-phase power may be required. The gas requirements are detailed later on, but typically air is supplied via a compressor with the other gases supplied from cylinders. All gases should be supplied as particulate, oil and moisture free. They are routinely required to be supplied at 1–3 bar as process gases. If pneumatic actuation is used on valves within the system then the air supply will need to be at 6 bar, but the use of pneumatic valves at this volume is unusual.

The water supplied to the fermenter, although never in product contact, should be of a quality that meets the requirements of the process in terms of temperature, particulates, hardness, pH and organic carbon content. If an integrated heating element is used the water source should be checked for hardness (dissolved mineral content) as hard water (defined as greater than 120 mg/L) will cause the system to fur up over time.

2.3.4 Autoclave Size

An important consideration and one that can be easily overlooked when specifying a system, is the size of the available autoclave. Consideration should also be given to the type of autoclave, and the manual handling implications of a front- or top-loading system, especially when dealing with the top end of the volumes under discussion here. A way round this is the use of a pulley or lift system to load the vessel into the autoclave. The restrictions imposed by the need to use an autoclave are one reason to use a steam in place system. However, typically on a cost basis the investment required for a 20-L glass system and a suitably sized autoclave, will be less than that required for 20-L steam in place system.

2.3.5 Risk and Containment

Advice should be sought on the local containment requirements given the classification of the organism with which you are working. East *et al.* (1984) give useful information on this.

The European Federation of Biotechnology classifies microorganisms into five groups based on the perception of risk to the human population.

European Federation of Biotechnology – Risk Classes of Microorganisms

Harmless microorganisms (EFB Class 1).   Microorganisms that have never been identified as causative agents of disease in man, and that offer no threat to the environment.

Low-risk microorganisms (EFB Class 2).   Microorganisms that may cause disease in man and might, therefore, offer a hazard to laboratory workers. They are unlikely to spread in the environment. Prophylactics are available and treatment is effective.

Medium-risk microorganisms (EFB Class 3).   Microorganisms that offer a severe threat to the health of laboratory workers but a comparatively small risk to the population at large. Prophylactics are available and treatment is effective.

High-risk microorganisms (EFB Class 4).   Microorganisms that cause severe illness in man, and offer a serious hazard to laboratory workers and people at large. In general effective prophylactics are not available and no effective treatment is known.

Environmental-risk microorganisms.   Microorganisms that offer a more severe threat to the environment than to man. They may be responsible for heavy economic losses. This group includes several classes, Ep 1, Ep 2, Ep 3, to accommodate plant pathogens.

In assessing the risk involved in cultivation of a chosen strain in a fermenter there may be a requirement to justify the volume at which you intend working if the organism is in a sufficiently high category. It is usually held that the cultivation of a given microorganism at large volumes (this can be as low as one litre) raises the risk level. The reasons for this are fairly obvious, fermenters are vigorously aerated and stirred thus there is a potential, if leakage occurs, to cause serious environmental contamination via aerosols.

Depending upon the type of organism you are working with the room in which the bioreactor will be located will require certain features to ensure containment. Along with this all personnel should be familiar with good laboratory practices (Table 2.2).

2.4   Vessel Design

2.4.1   Basic Design

Nearly all vessels at the small scale regardless of whether they are glass or steel, for mammalian or bacterial use, share a basic vessel design (Figure 2.2). They will consist of a head plate that is flat, straight vessel walls and a hemispherical base. This gives good mixing characteristics without creating excessive shear forces. Where the stainless steel and glass systems start to differ is in construction of the walls of the vessel. Two options exist for the glass systems, nonjacketed and jacketed, and both relate to how temperature is controlled. The simplest and cheapest system, nonjacketed, relies on a wrap-around heater tape or equivalent to supply heat into the system; heat can be removed by having a heat exchanger sited within the vessel. The vessel wall is constructed of one single layer of glass.

The second option when using glass vessels is to have a jacketed vessel (Figure 2.3) whereby the vessel has a water-filled outer jacket. The water performs the temperature regulation by both heat input and removal. The water temperature is regulated by an external thermal circulator.

The jacketed vessel was considered the industry standard as the temperature control was regarded as more accurate and stable when compared to the nonjacketed system. The drawback of the jacketed system was the higher investment cost and the fragility of the water connections. As control technology has developed, the stability and the accuracy of the nonjacketed system has reached a level where it now matches the jacketed system. The industry norm is now the nonjacketed system; the only reason a jacket system would be required is for elevated temperatures (+70 °C) that cannot be achieved by direct heating, or if the process requires a rapid temperature shift, which can only be achieved by a water-jacketed vessel due to its large heat transfer area.

Flat bottomed vessels may also be bought, but characterisation of mixing is most clearly understood in hemispherical-based vessels, and they represent more closely what will be done in the larger scale, so it is the hemispherical-based system that is most widely used.

A small scale stainless steel vessel (Figure 2.4) in terms of its heat input/removal works in much the same way as the jacketed glass system, with an externally regulated water supply to control heat input/removal. The reason why wrap-around jackets have not been developed for these systems is that they are sterilised *in situ* and for sterilisation steam is supplied to the jacket.

2.4.2   Aspect Ratio

Convention states that the aspect ratio of a vessel (the ratio between its height and its diameter) should be 1 to 1 at the working volume for cell culture, and 2.2 to 1 at the working volume for microbial systems. The reasoning behind this is simply that the aspect

Table 2.2 Summary of containment requirements

Biosafety level	Agents	Practices	Safety equipment	Facilities
1	*Class 1* Harmless microorganisms	Standard microbiological practices	Personal protective equipment typically laboratory coat, safety glasses and gloves	Wash hand basin
2	*Class 2* Low risk microorganisms	BSL 1 plus • Limited access • Biohazard warning signs • Standard operating procedures for decontamination • Sharps precautions	BSL 1 plus Class 1 or better biological safety cabinet	BSL 1 Plus autoclave
3	*Class 3* Medium risk microorganisms	BSL 2 plus • Controlled access • Waste decontamination • Decontamination of laboratory clothing prior to laundering	BSL 2 plus respiratory protection as required	BSL 2 Plus partial laboratory isolation Laboratory capable of fumigation Negative air flow in laboratory Nonrecirculated exhausted air
4	*Class 4* High risk microorganisms	BSL 3 plus • Clothing change prior to entry • Shower on exit • All material subject to decontamination procedures before it leaves the laboratory	BSL 3 plus Class 3 biological safety cabinet or Class 1 or 2 in conjunction with the use of full body suit at positive pressure with personal air supply	BSL 3 Plus separate building / isolated area Dedicated supply, exhaust and decontamination system Effluent decontamination Double ended autoclave

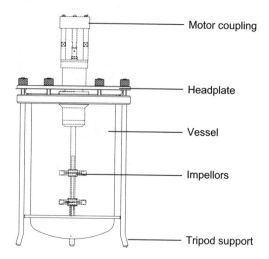

Figure 2.2   *Typical small scale nonjacketed glass bioreactor*

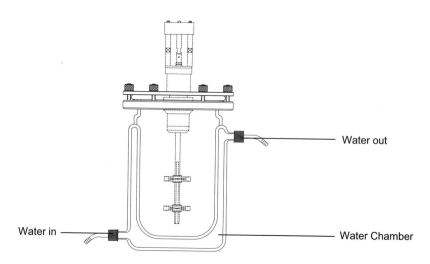

Figure 2.3   *Typical small scale water jacketed glass bioreactor*

ratio of the cell culture vessel gives a relatively large surface area across which gas can diffuse. In the microbial system the aspect ratio is such that the gas remains in the liquid longer, thus giving more opportunity for oxygen transfer to occur. Close attention should be paid to the aspect ratio of any vessel that is going to be used for scale-up studies to ensure that it matches the final production vessel as closely as possible. By doing this developing a process at small scale that does not work at the larger scale because the OTR (oxygen transfer rate) is insufficient for the process will be avoided.

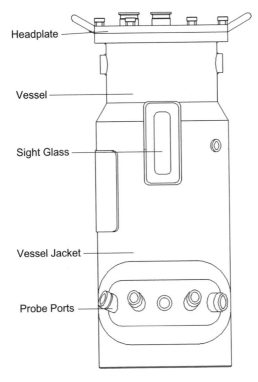

Headplate

Vessel

Sight Glass

Vessel Jacket

Probe Ports

Figure 2.4   *Small scale stainless steel bioreactor*

2.4.3   Vessel Configuration

There are a number of different processes that can be run in a typical small-scale bioreactor, and although these use fundamentally the same equipment, there may be some differences in terms of vessel fittings. The differences are driven by the physiology of the cells being grown. Mammalian culture processes are typically of longer duration (due to the slow growing nature of the cells) and are far more sensitive to shear forces than are bacterial systems. Due to its nature the mammalian system will require low shear mixing. The motor used to drive the impellors will be a low speed design, so that the signal and therefore the speed is very stable. Whereas a bacterial system will have high speed mixing using Rushton impellors and baffles, to ensure the highest possible level of oxygen transfer. However, in many respects a basic fermenter system, especially the very flexible STR, can be adapted to suit a number of culture types, for example, the addition of a covering of tinfoil will enable the cultivation of a light-sensitive organism, while the addition of a light source will facilitate the growth of an organism that requires light at a particular wavelength.

Work being carried out to culture bacteria from smoke stacks deep beneath the ocean surface has led to the development of bioreactors that can function at high temperatures and pressures. While systems to accommodate halophiles and acidophiles have also been successfully constructed. The design of such a system will need to take account of the

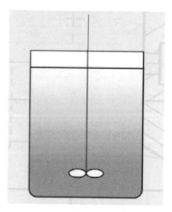

Figure 2.5   *Representation of a batch fermenter set up*

environmental condition that is trying to be mimicked, for example, the growth of halophiles or acidophiles may require that the vessel be constructed from a higher (more chemically resistant) grade of stainless steel such as hasteloy.

2.4.4   Process Configuration

The exact configuration of a bioreactor is usually process driven. These processes can be broken down in to three main groups. These are discussed in much more detail in Chapter 4, but, briefly, in terms of fermenter requirements.

Batch Process

Simplest of all (Figure 2.5), the vessel is prepared, inoculated and the process left to run. Typically a batch fermentation of mammalian cell line would last a few days or up to a week and cell densities of 2×10^6 cells/mL can be achieved. (1×10^8–1×10^9 can be achieved in an *E.coli* fermentation). The process will stop when a key nutrient runs out or metabolic waste products accumulate. This process will require, in terms of hardware, the minimum amount. The system will usually require dissolved oxygen, pH, temperature and mixing control and monitoring. The capacity to take samples should also be available without compromising the process sterility. This type of fermenter will be the simplest and usually the cheapest to purchase. Most sellers of fermenters make simple basic nonjacketed glass systems that are suitable for straightforward batch cultures. Such systems may have a limited number of addition ports, which might make it more difficult/ costly to adapt them to either fed-batch or continuous cultures at a later date.

Fed Batch

Fed batch is a slightly more sophisticated process (Figure 2.6) where the vessel is prepared inoculated, but instead of leaving the process to run until a key nutrient has been fully utilised, concentrated nutrients are added over time. A mammalian cell culture process may run for 7 to 14 days, slightly higher cell densities will be achieved 3–4×10^6 although the end viability of the cells will be typically <50% as there will be a build up of cytotoxic

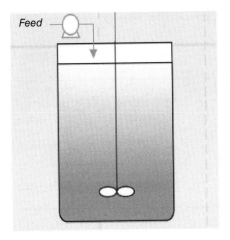

Figure 2.6 *Representation of a fed-batch fermenter set up*

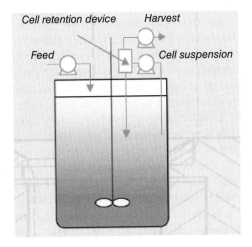

Figure 2.7 *Representation of a perfusion fermenter set up*

material in the vessel from cells that have lysed. In terms of hardware, the fed-batch process requires everything that a batch process requires along with a feed pump, which may or may not be controlled by a software package. If you are pumping medium into a vessel it would be advisable to ensure a method of level detection to ensure that the vessel is never over filled. The fed-batch stage of these processes is usually preceded by a classic batch phase with transition to the fed batch before nutrient exhaustion occurs.

Perfusion/Chemostat

The difference between a chemostat and a perfusion system (Figure 2.7) is simply that in a chemostat the cells are removed as part of the process while in a perfusion system

Table 2.3 *The hardware and medium requirements to produce 1 kg of a mouse –IgG antibody*

	Batch	Fed batch	Perfusion At 3 volumes/day
Bioreactor volume (active)	500 L	350 L	7 L
Cell concentration	2×10^6/ml (peak)	2×10^6/ml (average)	20×10^6/ml (steady state)
Runs per year	40	20	4
Duration per run	1 week	2 weeks	10 weeks
Consumption of medium per year	20 000 L	7000 L	7000 L
Mab concentration in harvest	50 μg/ml	150 μg/ml	150 μg/ml

the cells are retained within the vessel. Fast growing organisms such as bacteria are suited to a chemostat set up while slower growing cell types like mammalian cells are used in a perfusion set up.

A step up again in sophistication is a perfusion process. (Bierau *et al.*, 1998). The vessel is again prepared inoculated and fed over time. The key difference in a perfusion process is that medium is also removed from the vessel. To prevent the loss of cells, a perfusion device is placed on the medium take off. This cell retention device will prevent the loss of cells but allow the removal of media such that fresh media can be added. The duration of the process can be anywhere from 1 month to 3 months. In theory a perfusion process could go on indefinitely. Considerations of product quality and consistency need to be taken into account on very long perfusion runs, namely, is the product (protein, DNA, virus) the same at the end of a process as it was at the beginning? This is discussed in more detail in Chapter 12. Cell densities as high as 20×10^6 can be achieved in a perfusion process with viabilities of +95%. See the example in Table 2.3.

So if a perfusion system is offering much higher cell densities with higher viability, why does anybody run a batch/fed batch process? Perfusion can be complex requiring feed pumps and harvest pumps all working in synchronicity. With a constant feeding process there is also a great risk of introducing contamination. All these issues can be, and have been, addressed by equipment manufacturers to the point that very large perfusion processes are being run in validated environments. The cost of set up of a perfusion system, as it is the most hardware intensive, can also be off putting. Medium costs are also a factor, as feeding in up to 4 although more typically 1.5–2 vessel volumes per day of medium can be cost prohibitive. In the highly regulated environment of GMP (good manufacturing practice, a set of guide lines that govern pharmaceutical production standards) the governing bodies like to see clearly where a batch of material begins and ends and this in a perfusion system is not always easy to define. The other consideration is the impact on downstream processing (DSP), as a perfusion process will be delivering on a daily basis a large amount of material to DSP, which need to be equipped to deal with this so increasing the capital investment, whereas the more common and therefore familiar fed-batch process will deliver a fixed amount every 2–3 weeks.

2.4.5   Head Plate Fittings

The items attached to the head plate of a bioreactor are what gives it its functionality. These fittings perform two basic tasks – one is to get information out of the vessel, such as temperature, pH or DO, and the other is to get material into or out of the vessel, such as pH correction fluid, gases to maintain DO_2, nutrients, while samples of the fermentation fluid will usually need to be removed aseptically from the vessel. A head plate will typically have a series of ports varying in size from 6 mm up to 27 mm. There will be one larger port, but this is always in the centre and used for the drive shaft to connect the motor to the impellors. A greater number of ports gives greater versatility, but care should be taken to ensure that there is no compromise of the head plate strength, as any weakening of the head plate may cause it to warp and offer a potential route for a contaminant to get into the vessel. While a system needs to be as versatile as possible it is also important that the head plate is not over crowded such that it becomes difficult to work with. Any ports that are unused are blanked off with blind stoppers and any fittings that are not required are removed and replaced with blind stoppers.

2.4.6   Sampling

As much monitoring as possible should be done online in the vessel, as is the case with pH, DO and temperature, without the need to remove a sample, as every time a sample is taken the system is exposed to a possible source of contamination; for some tests this is unavoidable and so taking a sample is required. The sorts of tests that are routinely performed are cell counts, cell viability and checks on medium constituents that cannot be performed online. A device that allows for the aseptic removal of samples is a vital part of any small-scale or large-scale system. A typical sampling device (Figure 2.8) on a glass system will consist of a sample hood externally on the head plate and a sample pipe that sits within the vessel. A piece of short tubing is used to connect from the bioreactor sample pipe to the sample hood. A syringe is connected after a filter that is then used to pull a sample from the vessel. By using good aseptic technique it is possible to attach a bottle to the sample hood and take a sterile sample from the vessel.

This sort of set up would not be possible with a small scale stainless steel system as it would not be possible to sterilise such a system with steam. The more complex system (Figure 2.9) is required.

To take a sample, valve 3 is closed and is removed and replaced by the spool piece. Valve 2 is then attached to a sample bottle via tubing and the assembly is autoclaved. After autoclaving the spool piece is removed from between valves 4 and 2 and replaced with valve 3. Valves 1 and 4 are opened valves 2 and 3 are shut to allow steam to sterilise the line from the bioreactor with out steam going to the sample bottle or the vessel. After sterilization, valves 1 and 4 are shut. Then valves 2 and 3 are opened and the sample is taken. Once the sample has been removed the sterilisation process can be repeated to protect the operator.

Such a system could be incorporated on to a glass vessel, but would normally only be required when working with high risk microorganisms.

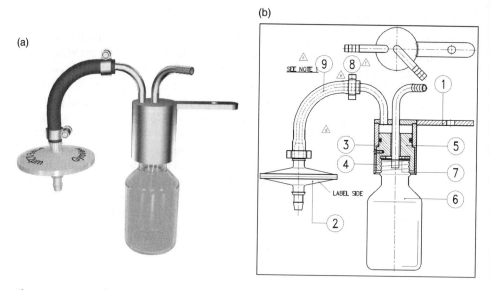

Figure 2.8 *Sampling systems. (a) Sample system for an autoclavable bioreactor. (b) Sample system for an autoclavable bioreactor. 1, Connector; 2, 0.22 µm filter; 3, O-ring; 4, grub screw; 5, sampling system insert; 6, bottle; 7, sample pipe lid; 8, clamp; 9, tubing*

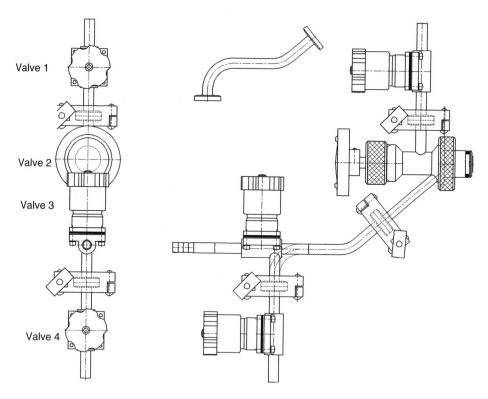

Figure 2.9   *Sample system for a steam in place bioreactor*

2.4.7   Air Inlet Spargers

Gases for the maintenance of dissolved oxygen, and in the case of mammalian culture, pH, need to be introduced into the vessel, and this is done in two ways – either by sparging into the liquid phase or by overlaying gas into the headspace. When sparging into the vessel this happens just below the impellors. All the gas is filtered prior to entry into the bioreactor by passing it through a 0.22-μm filter. Gas requirements are illustrated in Table 2.4.

Gas that is sparged in to the vessel will enter the vessel though several different types of sparge line, again dependent on the type of organism being grown. To minimize the risk of contamination a filter is always placed prior to the sparge line. The first illustrated (Figure 2.10a) is the conventional 'L' design, for high gas flow rates, typically up to 2 vessel volumes per minute. Gas is dispersed through precision drill holes in the base of the pipe, and the resulting gas bubbles driven in to the liquid phase by the agitation and baffling systems.

The alternative illustrated (Figure 2.10b) is the 'microporous' design, usually for low gas flow rates, typically up to 0.1 vessel volumes per minute. Gas is pushed through a fine sintered tip. Each pore is approx. 15 μm and so gas enters the liquid phase in a very

Table 2.4   *General guide as to the gas requirements of both microbial and mammalian systems*

Gas	Cell culture	Microbial culture	Cell culture	Microbial culture
	Sparging		Overlay	
Air	Approx. 0.1 vvm	1–2 vvm	0.1 vvm	10% of 'air to sparger'
O_2	10% of 'air'	20–30% of 'air'	NA	NA
CO_2	10–25% of 'air'	20–30% of 'air'	10% of 'air to sparger'	NA
N_2	10–25% of 'air'	20–30% of 'air'	NA	NA

vvm = volume per (working) volume per minute.

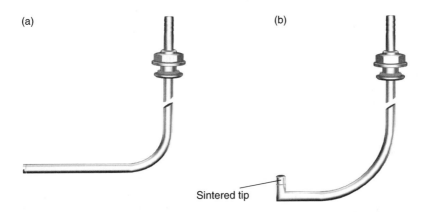

(a)

(b)

Sintered tip

Figure 2.10   *Example of spargers. (a) L sparger; (b) sintered sparger*

Figure 2.11   *Condenser*

fine dispersion of high surface area, thus improving mass transfer. This sparge design is sometimes favoured for cell culture as the bubbles are smaller and so contain less energy.

2.4.8   Exhaust Gas Condenser

The gas that has not been utilised by the process will pass through and exit the vessel. This gas will collect moisture as it goes through the medium so to prevent excessive evaporation and the blocking of the off gas filter a condenser (Figure 2.11) is placed prior to the filter. The condenser is of baffled design, thus increasing the surface area and condensate return to the vessel. Chilled water is circulated round the condenser to maximise the condensing of gas moisture so that it is returned to the vessel. The cold water source should be a consideration as is it no longer environmentally or economically acceptable to leave a tap running to drain. A chiller unit should be used in this way, the environmental impact in minimised and the temperature of the water is constant. The same chiller could be used to supply a heat exchanger to remove the excess heat generated during an exothermic fermentation.

2.4.9   Fluid Additions

Given the limitations, especially at smaller volumes, of the availability of ports in the head plate, one device that saves a lot of space is a triple inlet (Figure 2.12a). This port provides three narrow entry points into the vessel and can be used for liquids of low viscosity required at low volumes.

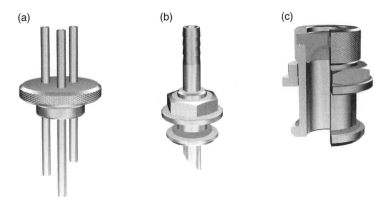

(a)   (b)   (c)

Figure 2.12   *Methods of introducing fluids. (a) Triple inlet; (b) single inlet; (c) septa*

Single inlets (Figure 2.12b) can also be used where either a liquid needs to go in a large volume, is viscous or contains particulate matter.

Septa (Figure 2.12c) can be used as a quick method of getting liquid into a vessel. The septum consists of a rubber plug housed within a fitting that goes through the head plate. The liquid to be added is placed in a syringe with a needle. The septum surface is wiped with a disinfectant and the needle simply pushed through. On health and safety grounds there is a drive to move away from having sharps in the laboratory to avoid needle stick injuries, especially around GMO and pathogenic organisms; for that reason septa are becoming less widely used.

2.4.10   Heat Exchangers

The maintenance of the correct temperature is critical to any process: too cold and the cells will not grow at an optimum rate and the process will take longer than it should; too hot and the cells may be killed or inhibited. Likewise, temperature gradients and changes may impact upon the process, and the system should prevent or minimise these.

To maintain ambient or lower temperatures in the bioreactor, either a jacketed glass vessel could be selected, or an internal heat exchanger (Figure 2.13) fitted to the head plate through which cold/tap water can be introduced, thus dropping the vessel tempera-ture. 'Water out' can either be piped to drain although this is becoming more economically and environmentally unacceptable, or returned in a closed loop installation, to a chiller bath (thus saving water). For bacterial applications, a heat exchanger is vital as these processes tend to be exothermic and if left without heat removal tend to over heat.

2.4.11   Pressure Relief Valves

The application of over pressure, anything over +0.5 bar, is used to enhance oxygen transfer rates, and in scale up studies to make the small system more representative of what will happen at the large scale.

Common sense should dictate that an overpressure should not be applied to a glass vessel. Although a new vessel would be structurally strong and capable of withstanding

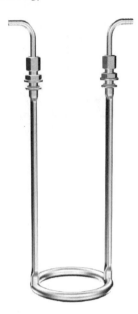

Figure 2.13   *Internal heat exchanger*

a level of pressure, as soon as a vessel has been autoclaved or handled in such a way that it is no longer possible to guarantee it has not been knocked or damaged in anyway, it is not safe to pressurise such a system. The consequences of a weakened vessel being over pressurised, that is, vessel fracturing and releasing the contents, mean that for safety reasons these vessels should never be operated with a deliberate overpressure.

However, it is possible to fit pressure release valves (Figure 2.14) onto small-scale bioreactors. These valves will release pressure at a fixed level, usually +0.5 bar. The valve opens and then reseats itself, while this may appear as a potential contamination source this is unlikely due to the outflow of gas. Of greater concern is the release of uncontrolled aerosols containing the organism being cultured, to the laboratory environment. For that reason overpressure tends not to be applied to glass vessels. One of the main reasons for utilising a steel vessel at lower volumes is that overpressure can be safely applied.

2.4.12   Baffles

Baffles (Figure 2.15) are flat paddle like structures, typically 10% of the vessel internal diameter. As so often in fermentation, the question arises as to 'Why the magic number? – 10 in this case'. Winkler (1990) showed that wider baffles have little effect, while with narrower baffles the baffling effect decreases rapidly. Baffles sit inside the vessel next to the wall and are used to break the rotational flow of liquid in the vessel that may occur at high agitation rates. The use of baffles eliminates laminar flow, creates a turbulent flow and enhances mass transfer. Baffles tend only to be used in bacterial/yeast systems. It should be remembered that other vessel fittings, such as pH / DO probes and heat exchangers that go into the vessel, will have a baffling effect. Mammalian systems tend not to use baffles, as the shear forces created behind a baffle may be sufficient to damage the cells.

Figure 2.14   *Pressure release valve for an autoclavable bioreactor*

Figure 2.15   *Baffle*

2.5   Drives/Coupling

2.5.1   Impellor Types

All bioreactors require mixing to some extent. The mixing achieved by the impellors will be required to perform a number of different functions. These will include maintaining a uniform environment in the vessel in terms of heat, nutrient, and, if present, solid particle distribution, fluid and gas mixing, and air/oxygen dispersion. All these things must be achieved efficiently, but without damage to the cells being grown.

To achieve these goals, several different types of impellor may be used. The choice of impellor is dictated by the physical robustness of the cells being grown, and the need for oxygen transfer.

A typical mammalian culture will use a marine impellor (Figure 2.16) that will be half the diameter of the vessel. This impellor will be either scoping, which means it creates an upward current in the vessel, or vortexing which means it creates a downward flow in the vessel. These impellors are used specifically due to the low sheer forces created when they are rotating thus minimising damage to the cells while, at the same time, maintaining a homogeneous mix with in the vessel.

Rushton impellors (Figure 2.17) are typically used in bacterial and yeast fermentations. These impellors will be one-third of the diameter of the vessel, although for viscous cultures (filamentous fungi or Streptomycete cultures) a 0.5 diameter Rushton turbine might be used. The stirring speeds that Rushton impellors are used at can be up to ten-times those of the marines. This is possible as microbial cells are far more robust than mammalian cells. This speed of mixing is essential in a bacterial fermentation, because as well as mixing, the Rushton impellors are also helping to maintain the dissolved oxygen, and so meet the oxygen demands of the bacterial fermentation in a far more effective way than do the marine impellors. They do this by breaking up the gas bubbles as they hit the impellors thus increasing the surface area of the bubbles leading to greater gas transfer. The maintenance of dissolved oxygen in a bacterial system is linked to the speed of the

Figure 2.16   *Marine impellor*

Figure 2.17   *Rushton impellor*

impellors, such that the speed may be automatically controlled to try to raise the DO_2 by creating smaller bubbles via faster mixing before a greater rate of gassing is applied. This is vital in most systems as there is a limit to how much gas can be supplied to a system. Foam may occur if too much gas is added to a fermentation, which leads potentially to blocked filters as medium gets past the condenser and on to the exhaust filter. The other reason that the gassing rate is limited to a maximum of 2 vessel volumes per minute is that beyond this point cavitation may occur; this is when the impellor stops breaking up the gas bubbles and instead the impellor is surrounded by a gas pocket and any new gas is simply diverted around the impellors.

Combinations of impellors may be utilised such that a Scoping marine impellor may be used to create a downward force while a Rushton is used at the liquid gas interface to reduce foaming but also to create a larger surface area over which gas can be pulled into the system.

Many other impellor types do exist, for example, the Lightnin range. Regardless of type and design, the requirement for good mixing without the creation of zones is the key consideration. The power that turns the impellor must be relayed from the motor to the impellor via a drive shaft, which raises the question below.

The need to seal the stirrer shaft on a bioreactor has always been an issue. Namely, how do you maintain a seal such that the process can run free from contamination yet allow the shaft to rotate so that mixing can be properly performed? To answer this question two main seal types are used on small-scale bioreactors, mechanical or magnetic.

2.5.2   Mechanical Seal

The Mechanical seal (Figure 2.18) is commonly used in both small- and large-scale bioreactors. The systems work by the action of a compression spring on two parts: one static within the bearing housing and the other rotating. This creates the seal, which is lubricated by the culture. Wear will occur which, if left unserviced, will result at some point, in the system failing. The results of such a failure in a base-mounted system would be the total loss of the culture, or in a top-mounted system, in the release of aerosols to the laboratory. If a higher category risk organism is being worked with there may be a requirement to

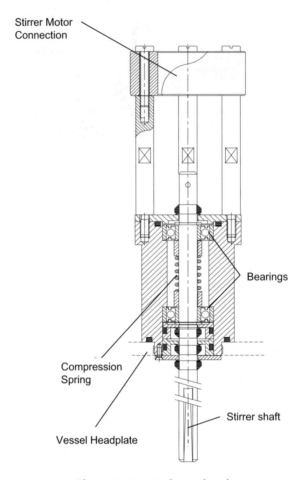

Figure 2.18 *Mechanical seal*

move to a double mechanical seal (DMS). DMSs utilise a second seal to provide a greater level of biosecurity. The first seal is lubricated by the culture while the second is lubricated by steam condensate, therefore DMSs are used only on steam in place systems. If the condensate is kept at a negative pressure such that any leakage would result in the ingression of medium into the seal the monitoring of the condensate may give and indication of the seal condition.

The other type of seal that is available is a magnetically coupled one, which is detailed below.

Both the mechanical and magnetic systems have advantages and drawbacks. The magnetic coupling offers absolute containment with no risk of aerosols, ideal for high value, high risk or long running processes. However, it may not be suitable for very high viscosity fermentations. The life expectancy and running costs of each system is about the same. Seals and bearings on both units do need to be replaced, typically on a 12-month cycle. A mechanical seal is lower cost when compared with a magnetically coupled system. The magnetic drive will add approximately 5% to the purchase price of a glass system and approximately 1% to a small SIP system.

2.5.3 Magnetic Seal

The issue of how to seal a rotating drive shaft has been addressed by the use of magnetically coupled drives (Figure 2.19) where the energy to turn the drive shaft is transferred via magnets. By using magnets is possible to fully seal the drive shaft. This is especially useful when dealing with organisms of high classification, long-term cultures (for example cell culture perfusion), or high cost systems.

The driving magnets are held in a housing connected to a motor by a drive shaft. Multiple ceramic magnets then transfer power to the driven magnets that are contained within a housing that is on the inner side of the head plate. This housing is in turn connected to the impellor shaft. There are limitations with magnetically coupled drive shafts in relation to the amount of energy that can be transferred. However, it only becomes a limiting factor at larger volumes. This volume range is typically 500 L for bacterial and 3000 L for mammalian cultures. The limitations are due to the viscosity of the culture. Given that a critical part of the system is the magnets, the life expectancy of these could be an issue. At least 20 years would be expected.

2.6 Probes and Sampling

The taking of a sample is both time consuming and a potential source of contamination, and in very low volume or slow growing cultures the loss may distort the process. Typically the total volume removed as a sample should be kept to less than 10% of the process volume. So if a parameter is to be measured in a bioreactor it is better if it can be measured on line, in the bioreactor, without the need for the taking of a sample. As standard in a bioreactor there is a pH probe, a dissolved oxygen probe and a temperature probe.

2.6.1 pH

In any type of fermentation the pH, the balance between H^+ and OH^- ions, will not remain static for long, due to metabolic activity in the vessel. If the highest levels of growth and production are to be achieved, the vessel contents will need to remain at the optimum set point. So that pH can be monitored and controlled a pH probe is always included. This probe is on a feed back loop that will, in the case of a bacterial system, add more acid or base, or in the case of mammalian cell culture, CO_2 gas or base to correct any pH drift. Details of the construction and function of these probes are described fully in Chapter 9.

2.6.2 DO₂

Dissolved oxygen is an important parameter in any fermentation (except anaerobic processes). Oxygen is not especially soluble in culture media (only 7 ppm at 37 °C) but can be consumed in large amounts by rapidly growing microbial cultures. For that reason is important to know the DO_2 level in the vessel and control it to the required set point. A

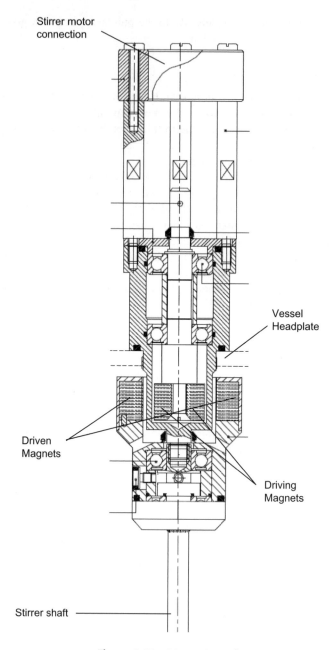

Figure 2.19   *Magnetic seal*

DO_2 sensor consists of two metal electrodes, a cathode and an anode. Both the cathode and the anode produce an electrical current that is proportional to the dissolved oxygen concentration. (Lee and Tsao, 1979). This signal is sent to a transmitter that converts it into the units of measurement that are then displayed on the local controller and/or recorded in the SCADA software. More detail on these probe types is available in Chapter 9.

2.6.3 Others

Various other probes can be fitted into a vessel, so long as there are sufficient head plate ports and the placing of probes does not alter too drastically the mixing profile of the vessel. These could include optical density probes. OD probes work on the basis that it is possible to measure the amount of light lost (scattered) over a set path length. OD probes can be used as an indicator of biomass. The draw back with OD is that it cannot distinguish between living and dead cells.

A system that utilises impedance has been developed (Aber Instruments: www.aber-instruments.co.uk) as a way of measuring viable cells online. A viable cell will hold up an electrical charge in direct relation to the number of viable cells present in a vessel.

High concentrations of dissolved CO_2 can inhibit cell growth and alter the glycosylation pattern of recombinant proteins. For that reason it may be worthwhile monitoring dissolved CO_2. A number of systems are available for this purpose, for example, Mettler Toledo (uk.mt.com) or YSI (www.YSI.com).

2.7 Control and Actuation

Any bioreactor system be it bacterial, fungal, insect or mammalian, is more than just a vessel with a motor to mix the contents. It will always be sited next to some sort of control system that will monitor the basic parameters inside the vessel: pH, DOT, temperature, level/foam and mixing speed. All these parameters are controlled by the input of a gas, liquid or energy. These elements have to be supplied in a controlled manner by pumps, rotameters, solenoids and mass flow controllers housed in an actuation panel (Figure 2.20). It is these actuation points working on the signals from the controller that make decisions based on input from the probes in the vessel.

2.7.1 Gas Flow and Delivery Control Devices

The variable area flowmeter (rotameter in this system) (Figure 2.21a) assembly is used to measure and adjust gas flow rate to the bioreactor. The assembly consists of two main parts, the flow tube and the float. Gas enters the flow tube through the smaller opening at the bottom, and exits through the upper end. Upward pressure causes the float to rise. Flow takes place through the circular area between the float and the inside surface of the flow tube. As the float rises the flow area increases due to the tapered bore of the flow tube. Dynamic equilibrium results when the buoyant force, due to the float and the upward force, due to the gas pressure, balance the weight of the float. The vertical position of the float at equilibrium corresponds exclusively to one particular flow rate. The flow rate is obtained by determining the height of the float on a scale, which is etched on to the flow tube. When applying gas to a system via rotameter the gas supply is constant and variable by the operator. It is possible to measure the gas flow rate to the system by this means in volumetric terms.

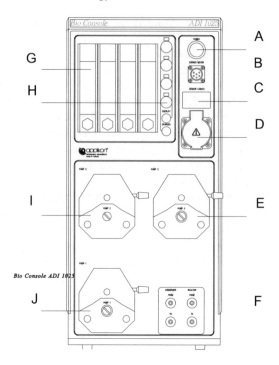

Figure 2.20   *Actuation panel. A, main power switch; B, motor cable connector; C, sensor cable tunnel; D, supply for heating blanket; E, tubing pump (level); F, thermo circulator connections; G, rotameters; H, gas selection block; I, tubing pump (base); J, tubing pump (acid)*

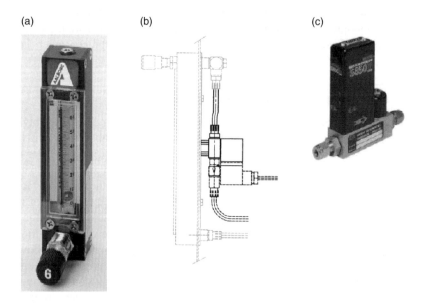

Figure 2.21   *(a) Rotameter; (b) solenoid valve; (c) mass flow controller*

The next step in terms of the supply of gas is to use a solenoid valve (Figure 2.21b). A solenoid valve can be installed on the outlet side of the rotameter. The solenoid valve is controlled by a signal from an external controller operating to control, for example, dissolved oxygen or pH. By incorporating a solenoid valve into a gas line, the supply of gas is controlled by the local control system in response to events within the bioreactor, such that a drop in DOT could cause the valve on an O_2 line to open for a predetermined period of time and release, at a flow rate determined by the rotameter, a set amount of gas into the system. It is possible to measure the amount of gas going into a system by knowing the pressure at which it is supplied, the diameter of the tubing and how long the solenoids have been open for. This set up essentially delivers a simple means of controlling DO_2 in the fermenter.

The most accurate, but the most expensive way (by a factor of 10 over a rotameter) of controlling gas flow to a bioreactor is the use of mass flow controllers (Figure 2.21c). They allow for the accurate measurement and recording of gas flow into the bioreactor, and unlike the rotameter, can be used to regulate the mass flow of gas to the reactor accurately.

Pumps

Most pumps (Figure 2.22) used to push liquid into a vessel are peristaltic and work in the same basic way. A peristaltic pump head consists of only two parts: the rotor and the housing. The tubing is placed in the tubing bed – between the rotor and housing – where it is occluded (squeezed). The rollers on the rotor move across the tubing, pushing the fluid. The tubing behind the rollers recovers its shape, creates a vacuum, and draws fluid in behind it. A 'pillow' of fluid is formed between the rollers. This is specific to the internal diameter of the tubing and the geometry of the rotor. Flow rate is determined by multiplying speed by the size of the pillow. This pillow stays fairly constant except with very viscous fluids.

Peristaltic pumps are usually robust in operation due to their simplicity. However, some key features are worth noting. Think very carefully about the flow rate range needed before ordering. Each variable speed pump will typically have a rotational range from 1 to 100%, and a range of tubing diameters that are suitable. Peristaltic pumps operating at less than 10% of their rotational range will often show marked fluctuation in delivered volume, so the tubing/pump rate combination should be chosen to avoid this range.

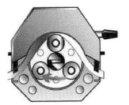

Figure 2.22   *Pumps*

Peristaltic pumps are not self priming, so a large air void in the tube may prevent the pump delivering. Similarly, the tubing used in the pumps typically stretches over time, so the volume delivered will vary. The pump output should be measured and regularly checked. For applications where the volume added to the fermenter is critical, the feed/titrant reservoir will be usually located on top of a balance (laboratory scales) and the output from this linked to the fermenter control system.

The type of tubing used is also important. Typical surgical grade silicone tubing, as used on most addition lines and air lines on a fermenter, is an excellent general purpose material, but can wear out quickly if used in a peristaltic pump. Instead a more physically tough tubing type such as Marprene (Watson-Marlow) or equivalent will show superior performance, leading to much extended life cycles, reduced risk of bursting and more accurate dosing of liquid into the reactor.

One very important consideration with peristaltic pumps is the system backpressure. Care must be taken with the build up of backpressure in the fermenter, although this may improve the driving force for O_2 transfer (see above); it may also make the addition of fluids (acid, base, nutrients, antifoam) very difficult. Typically if the filter on the gas exhaust line gets wet (condenser fails to remove liquid from the gas line or foaming out occurs) back pressure may rise very quickly, leading to a failure of the addition pumps to add liquid to the vessel leading to pH excursions and or nutrient depletion.

One major difference between peristaltic pumps comes in the actuation of the pump, either fixed speed or variable speed, or control system activated. A fixed speed pump responding to environmental changes in the bioreactor (usually foam, level or pH) and will pump at a fixed speed into the vessel for a predetermined amount of time either an antifoam, media or pH correction fluid. These are the simplest and cheapest, but least flexible pumps in terms of use. Pumps can also be of variable speeds, with the user setting the feed rate by a dial on the front, or they may be under the control of a software package such that they are activated in response to either time or events.

2.7.3   Temperature

Heat input will come form a heater blanket (Figure 2.23a) or a thermal circulator (in the case of a water jacketed vessel. It is the actuation panel under the control of a local controller that activates the power supply, or the opening and closing of solenoid valves for hot/cold water supply.

The temperature is controlled on a feed back loop with the information coming from a temperature probe (Figure 2.23b) in the vessel. The probe actually sits in the closed end well such that it is not actually in physical contact with the media but is situated such that it will be able to give accurate temperature measurement and control.

2.7.4   Foam

The cause of foaming in bioreactors is a combination of factors. In microbial fermentation the volume of air going though the system to maintain the required dissolved oxygen, combined with the level of agitation, can cause foaming. While in mammalian systems a much lower airflow into a medium rich in proteins can result in foaming.

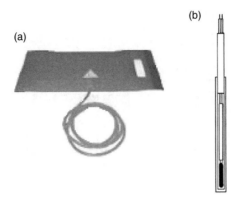

Figure 2.23 *(a) Heating blanket; (b) temperature probe*

Foam formation is an issue as it can, if it is allowed to, pass through the condenser, blocking the exit filters and causing an overpressure situation to occur. Foam can also affect the dissolved oxygen level by causing an increase in the residence time of air bubbles in the reactor, which in turn leads to their depletion. Foam is detected by the use of a foam probe. The presence of foam is detected when the foam that has formed touches the tip of the probe, thus completing the circuit within the vessel, which, in turn, actuates a pump to add a chemical antifoam agent. Normally there will be a built-in delay in the circuit to help ensure that the foam signal detected is a true one, not just a splash. So there will be an in-built requirement for the signal to last a set period of time before antifoam is added. This way antifoam addition is kept to a minimum.

The choice of an anti foam agent should be carefully made based on the following criteria:

- rapid dispersal within the vessel;
- rapid action on any foam present and ability to suppress any new foaming that occurs;
- have no direct metabolic impact on the culture, the product or the process (this is clearly an ideal! See above);
- have no or minimal impact on down stream processing;
- be nontoxic to humans/animals/the culture;
- have a minimal effect on oxygen transfer;
- be heat sterilisable.

2.8   Control Local and Remote (SCADA)

A typical bioreactor system will consist of three parts, the vessel, the actuation panel and a locally mounted control unit (Figure 2.24) (LMCU), which could also be described as a HMI (human–machine interface). The LMCU can operate anywhere between one and four vessels, but as the name suggests should be local to where the vessel is situated.

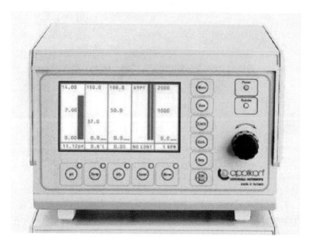

Figure 2.24   *Locally mounted control unit*

Clearly one way of keeping the costs of a multireactor set-up is to multiplex fermenters to a single LMCU, but at the cost of limiting operational flexibility and of all fermenters being unavailable if the single LMCU is out of action for any reason. A basic level of control and monitoring is performed from these units sufficient to run batch fermentation, typically pH, DOT, temperature, mixing speed, and level and/or foam detection are monitored and controlled. From this unit probes will be calibrated and alarm conditions noted. The local controller detects a drift from set point and corrects it simply by supplying, for example, more power to the heating system. Here the potential to over compensate for the drift exists, and a situation could arise where the controller is forever chasing its tail trying to get to the set point. For that reason PID (proportional, integral and derivative) control is used, the aim of which is to ensure that the system stays as close as possible to the set point and any deviation, when it is detected and corrected, takes place in such a way that any over shoot is minimized.

P   Proportional control (controller output relates to current deviation from set point)
I   Integral control (controller output relates to integrated deviation from set point)
D   Derivative control (controller output relates to the deviation trend)

If control beyond the functions of the LMCU are required, such as feeding strategies (activation of a pump) due to time-based or event-based stimuli or preprogrammed temperature shift, then the utilisation of a SCADA package may be required. SCADA (supervisory control and data acquisition) software sits on a computer that is connected to the LMCU or to many LMCUs. The SCADA, as it's name suggests, performs two main functions. Supervisory control in the form of control of set points, which may be a straightforward controlling to a fixed set point or more complex profiled settings, which may for example be temperature set to 37 °C for 18 hours after inoculation followed by a temperature shift down to induce protein production. This type of operation can be performed manually through the LMCU, but consideration would need to be given to the timing. The use of a SCADA removes these concerns.

The SCADA also has a data acquisition function. That is, to collect all the recordable data generated from a LCMU, so that there appears a record of the pH, DO, etc., profiles. The interval of data collection can be as quick as every second, which may be appropriate for DOT control on a bacterial fermentation, or less frequent in slower moving processes such as animal cell culture. It comes down to the amount of data that is manageable. SCADA software is discussed further in Chapter 10.

2.9 Single Use Bioreactors

Single use bioreactors, where a plastic bag replaces the vessel first appeared on the market in the late 1990s. There are now a number of suppliers (www.applikon.com; www.wavebiotech.com; www.xcellerex.com) and as such innovation is being driven by the need to be at the leading edge of technology. The attraction of single-use systems is a rapid turn-around time. When one fermentation is finished a new bag can be placed onto the system, filled with medium, brought up to temperature, probes calibrated and inoculated. This can be done far more rapidly than a conventional stirred tank that needs to be cleaned and then placed in an autoclave for sterilisation. The bags are supplied gamma irradiated, thus removing the need for the autoclaving step. The other major attraction is the short validation path. There is no cleaning or sterilisation validation to be performed as each batch gets a new bag each time.

The working volumes of these systems range in size from 1 litre to 500 litres. At all scales there is still the requirement of the original bioreactor and these have been addressed, the major difference being that instead of an impellor performing the mixing, the mixing is done by a rocking platform. As this is not as efficient as an impellor it is not possible to grow high density bacterial cultures in these systems, as it has not been possible to get sufficient oxygen transfer, so generally these systems have been limited to mammalian and insect cell culture. Whether these systems scan be scaled beyond 1000 L remains to be seen; the challenges faced are the containment of a large volume of moving liquid and providing robust enough hardware, both bags and platforms, that can cope with the rocking motion. If these systems are to overtake stirred tanks then they will need to combine the convenience of single use (quick turn around, low validation) with the performance of stirred tanks. The biggest challenge they face is to increase the OTR and build sufficient heat removal capacity.

References and Additional Reading

Cowan, C. T. and Thomas, C. R. (1988) Materials of construction in the Biological process industries. *Process Biochem.* **2391**, 5–11.
East, D., Stinnett, T. and Thomas, R. W. (1984) Reduction of biological risk in fermentation process by physical containment. *Dev. Ind. Microbiol.* **25**, 89–105.
Bierau, H., Perani, A., Al-Rubeai, M. and Emery, A. N. (1998) A comparison of intensive cell culture bioreactors operating with Hybridomas modified for inhibited apoptotic response. *Biotechnol.* **62(3)**, 195–207.
Lee, Y. H. and Tsao, G. T. (1979) Dissolved oxygen electrodes. *Adv. Biochem. Eng.* **13**, 36–86.
Winkler, M. A. (1990) Problems in fermenter design and operation. In *Chemical Engineering Problems in Biotechnology*, Winkler, M. A. (ed.), Springer pp. 215–350.

Further General Reading

Basic Bioreactor Design. (1991) Klaass Van Triet, J. Tramper, EEES. Marcel Dekker Ltd, New York.

Bioreactor System Design. (1994) Juan A. Asenjo, Jose C. Merchuk (eds), Bioprocess Technology Society. Marcel Dekker Ltd, New York.

Bioreactors in Biotechnology. (1991) A. H. Scragg (ed.), Ellis Horwood, Chichester.

Biochemical Engineering Fundamentals (1986). Bailey, J. and Oliss D. (eds), Chapter 9, Design and analysis of biological reactors. McGraw-Hill Higher Education, Berkshire.

3

Equipping a Research Scale Fermentation Laboratory for Production of Membrane Proteins

Peter C.J. Roach, John O'Reilly, Halina T. Norbertczak, Ryan J. Hope,
Henrietta Venter, Simon G. Patching, Mohammed Jamshad, Peter G. Stockley,
Stephen A. Baldwin, Richard B. Herbert, Nicholas G. Rutherford,
Roslyn M. Bill and Peter J.F. Henderson

3.1 Introduction

Fermentation has been used for millennia for the production of foodstuffs, but more recently for the manufacture of biological and medicinal products. Many pharmaceuticals and proteins are produced using fermentation processes, and it has become useful as a tool for the development of novel drugs and drug targets. In the last ten years, the DNA in genomes of numerous organisms has been sequenced. This explosion in biological information has revealed galaxies of new information, the exploration of which has scarcely started. One seminal observation is that genes encoding membrane proteins comprise about 30% of all the genome complement. Currently, the structures of approximately 30000 soluble proteins are known, whereas for membrane proteins the number is only around 100. In humans about 60% of identified targets for the development of pharmaceuticals are membrane proteins. Obtaining structural information is consequently important for elucidating their mechanism of action, manipulating proteins for commercial and biomedical purposes, and developing new drugs. A strong scientific focus on determining structures of membrane proteins, in particular, has resulted. There are several bottlenecks in achieving this, and one of the first is producing sufficient protein for

Practical Fermentation Technology Edited by Brian McNeil and Linda M. Harvey
© 2008 John Wiley & Sons, Ltd

structural studies by X-ray crystallography, nuclear magnetic resonance (NMR) and other technologies. The problem exists because individual membrane proteins tend to be of low abundance in the cell, a problem that can be overcome by genetic manipulation to achieve amplified expression in bacteria, especially *Escherichia coli*, or in yeasts such as *Saccharomyces cerevisiae* and *Pichia pastons*. Even with the best methods, the amplified production of a protein properly folded in the inner membrane *of E. coli* is likely to achieve a level of only about 3% of all the proteins in the cell. In order to produce enough *E. coli* to isolate enough protein for research (3–30 mg per batch), fermenters become a vehicle of choice.

Our laboratories are mainly concerned with the amplified production of membrane transport proteins for structural analysis. This is performed by using their recombinant genes carried on plasmid expression vectors in *E. coli* host cells. Our fermentation equipment has been chosen so that it can also be used for the growth of yeast strains. The equipment and growth conditions we use have also been successful on occasion for the production of soluble proteins.

The fermentation facility has expanded over a 12-year period from the use of a single 25 L maximum working volume (MWV)[1] MBR Braun fermenter to one 3–10 L autoclavable New Brunswick Scientific Bioflo 3000, two sterilise-in-place (SIP) 30 L Applikon fermenters, and one INFORS 100 L fermenter. Cell harvesting is performed using an Heraeus continuous flow Contifuge 17RS centrifuge for smaller scale work, and a CEPA Z41 tubular bowl centrifuge for the pilot scale vessels.

In this article we review the fermenters used, the ancillary equipment required, and the disposition of the equipment in restricted space. We are biased towards describing problems encountered in an academic environment, in which the rooms were converted to suit the new use rather than being purpose built. We will initially outline the scientific aims of the facility, since they dictate the fermentation requirements, and subsequently explore the facets of its operation, sometimes as a series of questions and answers. Finally we will describe ongoing strategies to improve productivity and to make proteins labelled with stable isotopes for NMR studies.

3.2   Special Considerations that Apply to the Production of Membrane Proteins

The needs in our laboratory fall into the following categories:

(i)   production of sufficient bacteria to provide four to eight individual cultures per month, each yielding at least 10 mg of purified membrane protein per batch for crystallisation trials, which must be performed for four to eight clones per month;

(ii) production of membrane preparations containing overexpressed wild-type or mutated membrane proteins for activity/binding assays, usually on a scale of one-third to one-quarter that for crystallisation studies (i);

---

[1] Manufacturers often use maximum vessel volume to describe their fermenters, but we find it important to define the maximum working volume. This can often be increased by a small extent, depending mostly on the presence or absence of foaming during the culture.

(iii) production of membrane proteins labelled with expensive stable isotopes for NMR experiments;

(iv) production of membrane proteins labelled with selenomethionine for X-ray crystallography.

The first and second categories are relatively cheap, but the ideal growth conditions (see below) must be determined by a series of small scale studies. Optimisation generally examines the construct efficiency, growth medium, pH, temperature, inducer concentration, time of inducer addition and the expression time. Generally, carbon and nitrogen sources are provided in excess to ensure that growth limitations do not occur. These experiments are usually carried out in flask culture, with the probability that scale up to the fermenter will result in different optimal conditions. Ideally these trials should be repeated in parallel minifermenters such as the 'Multifors' manufactured by INFORS, or an available bank of parallel 0.5–2.0 L fermenters using shaker flask conditions as a starting point. Optimisation can then be finally checked, and possibly modified, in the fermenter itself, not least to ensure reproducibility.

The production of materials for NMR studies, and incorporation of selenomethionine, are much more expensive and are preceded by numerous small scale pilot studies to ensure efficient utilisation of costly starting materials and minimise their wastage. Usually the study includes determining the minimum amount of these expensive materials that is required to achieve the desired outcome, prior to scale up in pilot vessels.

Our target proteins are generally located in the inner membrane of the *E. coli* host cell, which comprises 5–10% of the total cell protein, of which 10–60% is the protein desired. Consequently, to provide sufficient quantities of protein, a large volume of cells at high cell density is required. Overexpressed membrane proteins are often cytotoxic and high culture absorbances are difficult to achieve whilst maintaining good expression. Isopropyl-β-D-thiogalactoside (IPTG)-induced cultures rarely reach absorbances (680 nm) above 3.0, usually only growing to 0.8–1.5 absorbance units [we use the conversion factor that 1 mL of cells of $A_{680} = 1$ contain 0.68 mg dry cell mass]. Therefore large culture volumes of at least 30 L and, better, 100 L are needed for the crystallisation work. The situation is different for yeast cultures, in which very high cell densities can be routinely obtained and culture volumes of 3–10 L may be sufficient.

The production of cells on the scale required is too labour intensive and time consuming to be done in batches of flasks. The advantage of flask culture is its simplicity, where only agitation rate (equating to aeration) and temperature can be controlled. However, as discussed in Chapter 2, in a fermenter agitation rate, gas supply, temperature, foaming, pH, and additions to the medium (e.g., inducer or supplemental nutrient), can all be both monitored and controlled. Additionally, the culture absorbance, and exit gas concentrations of oxygen and carbon dioxide can be monitored.

Amplified expression of membrane proteins requires particular genetic constructs in which expression is often induced in exponential phase growth by the addition of IPTG. This minimises the cytotoxic effects of the expression and enables sufficient protein to be obtained, but results in cultures rarely reaching densities greater than an A_{680} of 3.0. The low cell density cultures obtained are a consequence of the inherent toxicity of the membrane proteins to the *E. coli* host. In cases of high protein expression the membrane becomes weakened, and prone to lysis and more susceptible to shear forces and chemicals, such as the IPTG inducer. The advent of autoinduction by lactose may enable the production of cultures of higher density while maintaining production of the desired protein. It will also be cheaper than using IPTG. This strategy is still being investigated for the production of membrane proteins.

3.2.1   What Size of Fermenter is Needed?

In our current research all proteins are produced during batch growth, because of its simplicity, and also the simplicity of scale up from shaker flask culture. Fed-batch culture (Chapter 4) is a mode of operation also suitable for production of membrane proteins in high density culture. Continuous culture is not feasible due to the cell lysis, which occurs in cases of high over expression.

The required working volume of a fermenter is determined by the quantity of protein needed, the yield of protein per cell mass and the biomass yield per litre of medium. In the case of soluble proteins the protein yield is typically very high. However, the yield of membrane proteins is relatively low, and hence either a small volume batch of very high cell density or large volume batch of low cell density is required to ensure that sufficient protein is obtained.

During initial research into a specific membrane protein, low quantities of protein will suffice. In these instances small-scale fermenters (e.g., 10 L working volume) are suitable. On reaching the stage of purification and crystallisation, thousands of different conditions require examination. Consequently, much greater quantities of protein are required, more than can be supplied by even a single 100 L culture. As a result, large volume and high yield is necessary.

In cases where several tests are to be performed on the cells or proteins, it is ideal to be able to use those obtained from the same batch to ensure comparability. The ability to produce sufficient cells in a single batch is therefore desirable.

With respect to autoclavable fermenters capacity is again important, but must be chosen according to the purpose. Optimisation trials are initially performed in small scale vessels, often less than 1L volume, as this saves on the cost of medium components and laboratory space, and also allows several vessels to be sterilised and run with relative ease. Small scale production may be performed on any suitable scale, which in our experience ranges from 5 to 10 L. Above this size large autoclaves are required, which are not always available. Also, manual handling of the heavy equipment may become prohibitory. Should autoclave space be unavailable, small scale SIP should be considered, although the costs associated with this, and the increased utilities, are much greater.

There is also an ongoing debate amongst those operating research scale fermenters about the reliability of autoclavable versus SIP fermenters, with many favouring the former on the grounds that failures in sterility are much less frequent. It is also worth noting that there is a large cost differential between the two types of system (See Chapter 2).

When deciding upon a suitable fermenter, the laboratory location and available space must be thoroughly examined. In situations of cramped bench space, or in rooms with unusual geometry, manufacturers are generally amenable to altering the fermenters to ensure that they fit whilst remaining accessible. Placement of control units under benches and the ability to buy fermenter 'banks' of several vessels on one platform are common methods of space saving. It should be noted, however, that in situations of restricted space the purchase options available may be limited.

Other important factors in deciding the scale of equipment are the running and maintenance costs and the cost of the growth medium. What appears to be relatively cheap on a small scale can often amount to a significant outlay per batch on a large scale. For this reason, it is again important that optimisation has been performed and proven on a small scale in order to minimise the associated costs and maximise output.

Provision must also be made for the safe disposal of waste. In our case the spent media is transferred to containers, space for which must be available in the facility, and treated with antiseptic detergent (cost implication) before disposal. For our Category 1 organisms we collect fermenter waste in 30 L polyethylene containers and disinfect with a 1% Virkon solution (as per the manufacturer's instructions). For higher containment category organisms an autoclavable kill tank near the harvesting equipment will be necessary.

3.3 Factors that Affect the Choice of Equipment

3.3.1 Who Will Operate the Fermenters?

In a typical academic research laboratory, fermenters can be operated by postgraduate or postdoctoral staff. However, the process is sophisticated and time consuming, especially if a continuous supply of materials is required. These individuals also have other responsibilities and activities as the needs of a research project develop and change. In our experience full-time skilled technical staff are essential if a fermenter facility is to be operated efficiently. They will generally service all aspects of the operation from setting up starter cultures to storage of cells (discussed in Chapter 6) and downstream processing to produce membranes (Figure 3.1).

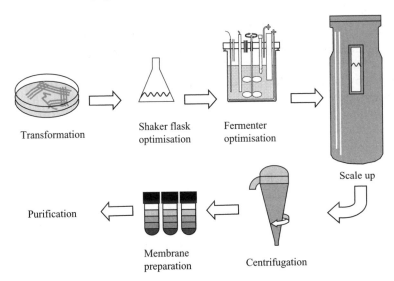

Transformation

Shaker flask optimisation

Fermenter optimisation

Scale up

Purification

Membrane preparation

Centrifugation

Figure 3.1   *Illustration of the key unit operations in producing* E. coli *cultures for membrane protein production*

3.3.2   What Kind of Fermenter is Needed?

Fermentation technology in batches is very well established (see Chapters 2 and 4). We use tanks with, ideally, a 3 : 1 height : diameter 'aspect ratio', driven by a powerful central shaft stirrer equipped with 'Rushton' impellers (two or more in parallel along the shaft, depending on fermenter size) and also baffles in the vessel. The engineering is designed to break up incoming forced air at the bottom into small, slowly rising bubbles. This enhances their area/volume ratio and retention time to maximise the transfer of oxygen into the liquid medium (Figure 3.2). It is important to note that solubility of oxygen decreases with increasing temperature; the growth rate of *E. coli* at temperatures below 37 °C may therefore be slower but better energised because of increased aeration and slower respiration. For agitation we prefer to avoid a top-driven stirrer shaft, which has a heavy motor that is difficult to manoeuvre in our confined space, and instead use a bottom-driven motor magnetically coupled to the stirrer shaft. Selecting a magnetic drive can minimise the servicing of seals, the risk of leakage, and contamination. Magnetic coupling may be inadequate if very high cell densities are achieved with viscous cultures, in which case double mechanical seals are the common alternative. However, the mechanical seals may need regular servicing and/or replacement.

When installing a fermenter, a bund under the machine (ideally of a volume equivalent to that of the culture) is desirable to contain spills. The floor should also be sealed to cover this eventuality. When using organisms above hazard Category 1, adequate containment is mandatory.

Fermenters may be operated in batch, continuous or fed-batch mode. Details of these modes of operation can be found in Chapter 4. At this time we have only used batch mode for production of membrane proteins. An idealised culture growth profile is shown in Figure 3.3(a), and an example from a real culture with monitoring of various parameters is in Figure 3.3(b).

Some points to note are as follows:

(i)   The aeration rate eventually cannot keep up with the culture oxygen consumption rate, so the concentration of dissolved oxygen drops (Figure 3.4). This can be ameliorated by increasing the rate of airflow, the stirring rate and the headspace (back) pressure (Figure 3.2). Dissolved oxygen control can be accomplished automatically by coupling the probe response to these control parameters.

(ii) The carbon source, or any other nutrient, may be consumed to the point where its availability limits growth (Figure 3.3a).

(iii) Toxins may accumulate and inhibit growth.

(iv) pH may change as a result of accumulation of metabolic end products. This can be avoided by titrating with acid or alkali, which can again be controlled automatically (Figure 3.4).

(v)   Foam may arise. Foaming is influenced by many factors including salts, pH, temperature, medium composition, air flow, agitation and even the overdosing of antifoam. The presence of foam may reduce aeration and compromise the efficiency of, or even block, the exit filter. Control can be performed by drip-feeding antifoam in response to a signal from the internal sensor. Antifoam is known to diminish dissolved O_2 (DO_2) slightly (Figure 3.2d), and may also damage cells and proteins.

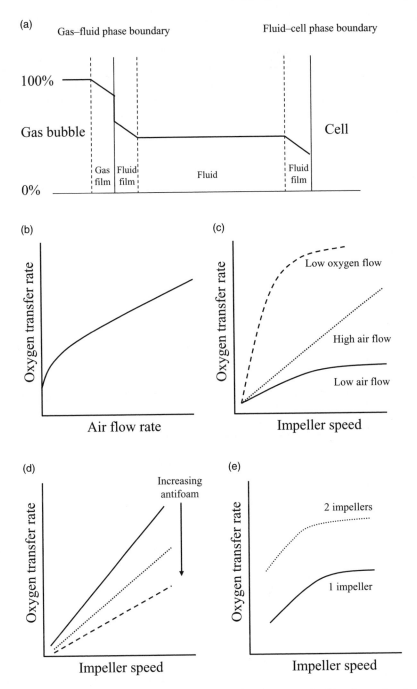

Figure 3.2 *Factors affecting aeration of a bacterial culture. (a) Profile of aeration between the boundary layers of a bubble and a cell: from Butterworth-Heinemann Ltd., (1992). Biotol Series. Operational modes of bioreach p. 16; (b) dependence of oxygen transfer rate on rate of air flow; (c) dependence of oxygen transfer rate on impeller speed; (d) effect of increasing antifoam on the oxygen transfer rate; (e) Influence of the number of impellers on the oxygen transfer rate*

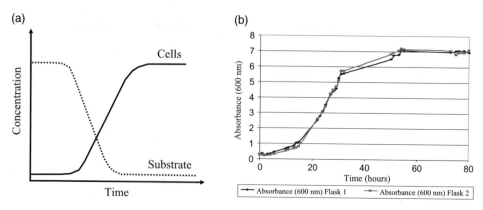

Figure 3.3   *Growth and nutrient depletion in batch culture. (a) Theoretically, the stationary phase is caused by depletion of nutrients, change in pH, accumulation of toxins or lack of oxygen; (b) an illustration of similar cessation of growth in an actual shaker flask culture of bacteria*

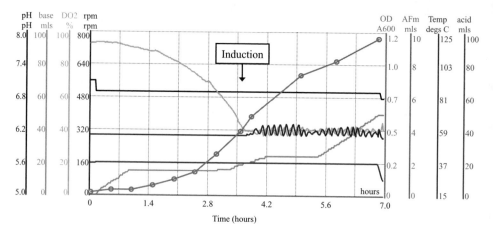

Figure 3.4   *Example growth profile of a fermenter culture. An example of the monitoring capability of one fermenter over-producing the zinc transporter protein, ZitB, in* E. coli. *Note that the cell densities are measured on 'off-line' samples taken from the vessel. The arrow indicates addition of the IPTG inducer. It is the induction of expression, rather than lack of nutrient, that diminishes the rate of growth*

Antifoams that can be metabolised may also have a physiological effect on the cells and cause diauxic growth. Some antifoam compounds may be ineffective against foams produced by different organisms since the foam may be caused by numerous different nutrients and proteins. The type of antifoam to be used needs to be tested to see if any change in physiological behaviour or protein quality results, and whether downstream processing, such as protein purification, is affected.

Table 3.1 *Flow rates and pressures of utilities required for operation of fermenters of varying scale*

	Fermenter type			
	5–10 L autoclavable	25 L *in situ* sterilisation	30 L SIP	100 L SIP
Steam pressure	n/a	n/a	2.5 bar	1.8 bar
Water pressure	1.4 bar	3 bar	2 bar	1.9–2.5 bar
Air pressure (pneumatics)	n/a	n/a[a]	5 bar	6 bar
Air pressure (culture)	1 bar	1.3 bar	1.8 bar	1.8–2 bar
Air flow rate	Up to 10 L/min	Up to 25 L/min	Up to 30 L/min	Up to 100 L/min

[a] Solenoid valves only.

The fermenters that we currently, or have previously, operated are briefly described in the following paragraphs. Information on gas and liquid flow rates and pressures are detailed in Table 3.1.

Low Volume Autoclavable Fermenter (3–10 L MWV)

We operate a New Brunswick Bioflo 3000 fermenter with interchangeable vessels of different volumes enabling us to grow expensive cultures containing, for example, stable isotopes or selenomethionine on scales from 3–10 L. Autoclaving of the equipment is cumbersome, especially when larger volumes are used. Ideally an autoclave of a suitable size is located next to such a fermenter, and a hoist may be a welcome accessory to transfer the vessel into and out of the autoclave safely. Failing this, a trolley for the movement between is essential and stairs must be avoided. The machine has a top-driven stirrer motor, which is heavy but can conveniently be removed for autoclaving. Forced air is taken from the building supply at approximately 1.5 bar, with water traps and 0.2 μm filters (with either PTFE or hydrophobic PVDF membranes) in place to remove any airborne contaminants.

One of the advantages of autoclavable fermenters is the reduced quantity of equipment required. Typically only a suitable autoclave, air pump, water and power supply are the minimum requirements. For small scale SIP systems (i.e., 10 L) the cost of the ancillary equipment may be prohibitive, and increases with greater capacity and supply rates.

Medium Volume Sterilisation in situ *(15–25 L MWV)*

We have operated a 25 L MWV MBR fermenter fitted with an integrated steam generator. Although this fermenter is sterilised *in situ*, no external steam supply is required, making it suitable for restricted laboratory space and avoiding the additional cost of a separate steam supply. A supply of water of highly regulated pressure is required for this fermenter. Forced air is taken from the building supply at 1.3 bar.

Larger Volume Steam-in-place (30 and 100 L MWV)

We have two 30 L MWV pilot scale fermenters manufactured by Applikon and one 100 L Techfors fermenter manufactured by INFORS. The steam supply to the first two is adequately supported by a Firbimatic TES Geyser Modular steam generator, but a Camptel Depositivo H2 was required to support sterilisation of 100 L in the larger machine. A Jun-air 2xOF1202 with a 150 L air receiver is used to supply compressed air to all the pilot scale fermenters.

3.3.3   Advantages of a Clean-in-place System?

On any large vessel it is advantageous to have a clean-in-place (CIP) system. Vessels above approximately 30 L in size can be awkward to clean due to the need for manual handling of the heavy top plate and an inability to access the lower part of the vessel. CIP systems work by the insertion of spray balls into the vessel, and the circulation of a cleaning solution, often caustic, via a pump. These systems are often costly but allow cleaning without the need to dismantle the head plate and associated fittings after each batch. The CIP also negates the requirement for steam sterilisation after a fermentation run, saving not only time but power, and also increasing the working lives of the pH and DO_2 probes.

3.4   What Supporting Facilities are Needed?

3.4.1   Electrical Outlets

The number and type of electrical outlets required is entirely dependent on the equipment. Close location of these outlets is also obviously important. Small scale fermenters, with their lower power requirements usually need only standard single phase sockets. In many cases two plugs are required for each unit, one for the control console and one for the control modules (heating, stirring, pH, etc.). A separate socket for an external air pump may also be necessary. Adequate provision of power for the available cell harvesting equipment must also be made.

Larger scale fermentation facilities obviously have higher power requirements. This not only includes power for the fermenters themselves, but also for the associated equipment. Examples include steam generators, air compressors and large scale centrifuges, which will often require three-phase power supplies.

In the case of our pilot scale fermenters, we require a three-phase supply for each fermenter (and control unit), steam generator, and tubular bowl centrifuge. A single-phase supply is used for the air compressor, air drier, water purifier, water pump, centrifuge pump and computer. Other peripheral equipment such as gas analysers, liquid analysers, feed pumps and calorimeters may also be used, each needing an electrical supply. Although much of this equipment may be used simultaneously for several vessels, it can be seen that the adequate provision of power is an important factor and must be addressed in advance.

3.4.2   Cooling Water

For almost all bacterial and yeast fermenter cultures, cooling water will be required. This may be to compensate for the exothermal nature of cell growth, individually small but collectively significant in high density broths, or to correct for overheating by the temperature control. The ability to cool the culture rapidly is also of use upon cell harvesting, both to prevent overgrowth and also to reduce protease activity.

Unlike heating, in which hot water may be recycled, cooling water is generally continuously flowing mains water, the temperature of which will vary depending upon the source and the season. The minimum temperature of mains water in our laboratory is approximately 15 °C. Instances in which this remains too high may require the addition of a chiller circulator, although this becomes less feasible as the scale increases as a result of capacity and cost. Nowadays, increasing environmental awareness often means that fermenter cooling or chilling water is recycled in a closed loop system, not simply sent to drain. The reasons for this are obvious when considering the average loss of potable standard water to drain from even a small lab fermenter in a 1h period. This mode of cooling water usage is clearly likely to become mandatory in more countries in the near future.

An adequate drain is essential. Usually the exit hose from the fermenter is linked directly to drain via wide bore pipework. It is absolutely essential that coolant lines (flexible tubing, reinforced tubing or piping) be checked fully before a fermenter is used. Particular attention should be focused on areas where piping meets flexible tubing, and care exercised to ensure that suitable clamps and fixtures are in use. Given the high flow rates of cooling systems, a leak can be serious.

3.4.3   Provision of Steam

Whereas in the past SIP fermenters may have been sterilised using an internal heater, nowadays SIP systems generally require an external steam source to sterilise, and in some cases also heat, the vessel. The generator capacity (in kg steam produced per hour) should usually be at least 50% of the maximum working volume of the vessel, i.e. for a 100 L working volume fermenter a capacity of 50 kg/h should suffice for both heating and sterilisation.

Higher capacity of the steam generator corresponds to higher cost, not only with respect to the equipment, but also from the power requirement. Costs may be reduced in some cases by using a single large generator for several small vessels, or in other cases by coupling together two smaller generators for a large vessel. Having a lower capacity generator does not necessarily correspond to an inability to sterilise a vessel effectively, but may simply increase the time required to reach the desired temperature. However, potential expansion capability should never be discounted when planning the laboratory set up.

A separate steam supply is not required in the case of small scale autoclavable fermenters.

In all cases in which it is required, the steam must be clean with all pipework made from grade 316 stainless steel. Otherwise it is possible for rust to enter the jacket and exterior, potentially damaging the fermenter. This should be avoided at all costs. As is the case with compressed air (see below), it is highly advisable to equip the laboratory with at least two steam generators, each capable of individually meeting the lab peak demand. This allows for alternation of use, greatly extending the equipment life cycle. It also allows for regular maintenance without disruption to laboratory work.

3.4.4   Compressed Air

Compressed air serves several functions in large scale fermenters. The most obvious of these is to provide oxygen to the broth, but it is also used to activate pneumatic valves. Application of air flow through an overlay valve, where gas is introduced into the headspace, can be used to increase headspace pressure for either foam suppression or for increasing dissolved oxygen, although this should never be performed in glass vessels. Compressed air may also be used to increase the rate of vessel emptying. As a consequence of all these considerations, a reliable airflow of sufficient capacity is of utmost importance.

Preferably the air compressor will be capable of supplying well-regulated, oil free, dry air. If this is the case, medical filters can also be installed to improve cleanliness prior to reaching the equipment. The compressor rating (the percentage of time in which the compressor can be run without overheating) is also important. When using a compressor with a rating less than 100%, downtime is required to cool the compressor units. In these cases it is necessary to ensure that the capacity is greater than the air requirement, i.e. if the rating is 50%, a capacity of at least double that required is necessary. Those with a rating of 100% can theoretically be operated indefinitely, although it is especially important that the air volume required for growth (e.g., 1 vvm) is not the compressor capacity if high pressure air is also required for the pneumatics.

3.4.5   Mass Flow Control

Gas flow control is performed using either a variable area flowmeter (VAF) or an automatic mass flow controller. Of the variable area flowmeters, the most common is the rotameter. Rotameters consist of a float situated inside a vertical tube, with the measurement relying on the force of the air flow to overcome the effect of gravity on the float. The application of an air flow forces the float to rise proportionately to the flow rate, until dynamic equilibrium is reached between the two forces. Small vessels are often fitted with the VAF since it is relatively cheap and robust, and can be easily replaced. Thermal mass flow controllers (MFC) are more complex devices that regulate gas flow electronically. Thermal MFCs determine the gas flow rate by the rate at which heat is absorbed from a heat source. Calculation of either the dissipated heat from the source, or of the heat absorbed by the gas, enables accurate determination of the mass of the gas flow over time. These are more expensive than the VAF and require more care, as they are prone to losing calibration if exposed to excess water. We prefer to utilise both systems, ideally in series, to allow manual readings of the mass flow controller output. This also enables checks on the MFC calibration to be performed.

3.4.6 Oxygen/Nitrogen

Under our current modes of operation, the requirement for enriched oxygen and/or nitrogen is not necessary. Typically in a high cell density culture, oxygen enriched process gas would be used, with oxygen being supplied either from a cylinder for small fermenters, or from an externally located generator for larger vessels. In many anaerobic cultures, the use of a 'blanket' of inert nitrogen gas (usually high purity, oxygen-free, nitrogen) aids development of anaerobic conditions in the liquid phase by the exclusion of oxygen. Anaerobic growth is not performed at this time, and as a consequence of the low density cell suspensions obtained from our *E. coli* membrane protein hosts under current conditions, the air supply and agitation rates provide sufficient oxygen without the need for a supplemental oxygen feed.

Provision for future use of these gases should nevertheless be made. This typically requires the installation of additional regulated gas lines and the associated bottle manifolds. Gas mixing devices may be incorporated into large scale fermenter equipment, or may be subsequently added. For small scale vessels, particularly autoclavable systems, these gases can directly replace the air supply line (especially in the case of nitrogen) or can be bled into the air stream (especially in the case of oxygen) by use of automatic controllers supplied by fermenter manufacturers.

3.4.7 Off-gas Analysis

Off-gas analysis can be performed by mass spectrometry, which is expensive but highly accurate, or by cheaper, less accurate, gas analysers such as the TanDem (Magellan Instruments), EGAS-L (Sartorius) or FerMac (Electrolab) analysers. These analysers measure both oxygen and carbon dioxide in the off-gas by electrochemical galvanic action (for oxygen) and infrared absorption (for carbon dioxide). Analyser multiplexes enable simultaneous sampling from multiple fermenters. Data is used for the calculation of the respiratory quotient, carbon dioxide evolution rate and oxygen uptake rate, and can also be integrated into the fermenter control software to enable control of growth parameters and feed pumps, should this be desired.

We currently use a TanDem for off-gas monitoring. The standard TanDem analyser is suitable for measuring O_2 and CO_2 between the ranges of 0–30% and 0–5% respectively. The analyser has an accuracy of 0.02% and a resolution of 0.01%. The maximum flow rate for this machine is 1 L/min, for which down-regulation is often required. Due to the method of analysis, the gas flow must also first be dried with a desiccant column. Failure to dry the gas can result in inaccurate measurements as a consequence of condensation on the CO_2 sensor.

3.4.8 Calorimetry

Virtually all chemical and physical processes result in either heat production or heat absorption. Microcalorimetry is the biophysical method of choice for the continuous measurement of such heat generation or absorption as a function of time. The method has proved to be particularly useful for research on metabolic activities of cellular systems, since it is noninvasive, very sensitive and enables metabolic activities to be monitored in real time and under real biological conditions. Although not as widely used for the analysis of microbiological cultures as off-gas analysis (which depends on metabolism being related to CO_2 production), microcalorimetry may be superior since it provides a route to obtain thermodynamic quantities such as ΔH, ΔG and hence ΔS. Such information has been helpful for the quantitative understanding of bioenergetics and the efficiency of metabolism and growth. For example, data from microcalorimetry can record growth, product formation and substrate utilisation; can optimise the control of biomass production and describe how much energy is required for biomass production or product formation; can monitor the metabolism of bacteria, yeast and other host cells to detect when new nutrient and energy sources are utilized and establish when and how a culture is sensitive to external parameters. The studies outlined below exemplify such applications.

Antoce and colleagues have examined the effects of methanol and ethanol on the growth of standard laboratory yeast strains using microcalorimetry. The heat evolved during incubation of yeast cultures at 30 °C was detected in the form of thermal profiles or 'growth thermograms'. Ethanol and methanol added to the culture medium produced changes in the growth thermograms that could be analysed to calculate the 50% inhibitory concentration and minimum inhibitory concentration. Correlation of the heat evolution curves with the number of cells and the turbidity of the culture was found to be very good. It was found that addition of ethanol and methanol up to 7.7% had clear effects of inhibition on growth of all yeast strains studied, reducing the growth rate constant and delaying growth.

Hans Westerhoff and colleagues have used microcalorimetry to measure total fluxes in growing yeast cells. Techniques such as NADH fluorimetry and the sampling of intracellular metabolites have shown that oscillations in the concentrations of glycolytic intermediates in yeast cells (also found in other organisms) depend on the growth phase in which they are harvested. The use of microcalorimetry complemented these techniques by demonstrating oscillations in the heat production rate accompanying glycolysis.

Cells grown on glucose and harvested at the diauxic shift oscillate until glucose is consumed, whereas cells harvested before or after the diauxic shift stop oscillating before the glucose is depleted. Microcalorimetric flux measurements of the differently-harvested cells showed differences in their specific average heat fluxes. These differences in specific heat production rates could be explained by differences in the average flux, differences in the products formed or a combination of the two. Measurement of the rate of glucose consumption and that of ethanol and glycerol production indicated that cells harvested during respiro-fermentative growth did not consume glucose any faster, but produced more ethanol than did cells harvested later. The glycerol production rates were comparable. When glycolysis takes place in a buffered medium the standard enthalpy of the conversion of 1 mol of glucose to 2 mol of ethanol and CO_2 is $-126\,kJ/mol$. Using this value for ΔH, 85% of the heat flux difference was accounted for by the difference in ethanol formation between the cells harvested at exponential growth or later, demonstrating the additional value of microcalorimetry in providing insight into the metabolic basis of the observed differences.

The microcalorimetric measurements also revealed a gradual decrease with time of the heat flux in exponential-phase cells. Under these conditions the glucose transport system has a low affinity component [K_m 28 mM (Walsh *et al.* 1994)]. Consequently, the transport rate will be significantly affected by the decrease in extracellular glucose, and this then affects the glycolytic flux. Cells harvested at the transition phase have a glucose transport system with only a high affinity for glucose ($K_m = 1.7\,\text{mM}$), and are therefore much less affected by a decrease in extracellular glucose. Indeed, only when the glucose concentration becomes very low at the end of the experiment, does the heat flux drop. These subtle changes in flux are suggestive of the importance of the glucose transport step in controlling the overall flux through glycolysis. Conventional flux determinations by following glucose consumption or ethanol production would not be sufficiently accurate to reveal such subtleties, and thus the application of microcalorimetry as a method to measure flux with high precision is very promising in the study of metabolic regulation and control.

3.4.9   Absorbance Monitoring

One of the most important parameters for our inducible proteins is that of cell density, by which we mean the concentration of cells in the liquid phase. Optimisation of the induction absorbance, amongst other parameters, allows maximum target-protein yield to be achieved. Since cells initially exhibit exponential growth when in an excess of essential nutrients, off-line absorbance readings can be used to predict the time at which a specific absorbance will be achieved. However, changes to the growth rate, caused by fluctuations in temperature, pH or medium composition will render such predictions inaccurate. Cell density probes can be used to measure the turbidity of the cell broth in real time, which can then be logged to a computer. If necessary these data can then be used to activate pumps for addition of either nutrient or inducer, and can be easily incorporated into programs to alter growth parameters (e.g., changing the temperature upon reaching induction absorbance). On-line monitoring of parameters such as this can not only increase automation of the fermenter, making the systems easier to run, but also improves reproducibility of batches. This is of great importance in any production facility.

It is also possible to follow growth by calorimetry and by off-gas analysis. Since these indirect measurements may not be related to cell increase if a change in the cells' metabolic pathways occurs, a calibration against cell mass should be incorporated into pilot studies.

3.4.10   Distilled Water Supply

Our fermenter cultures generally require the use of chemically-defined medium and hence all of our solutions and growth media are prepared using de-ionised or reverse osmosis water. We prefer to use purified water to ensure that the batches are reproducible, and that the defined medium is not contaminated with chemicals and organisms from the poorly treated laboratory water supply. Since this water, if left standing, will go 'off', there are two different methods which can be used to ensure a good supply.

(i)   Slow production of water, which is then stored in large quantities and recycled through the purification cartridges.
(ii) Rapid production of large quantities of water at the time of use.

As a consequence of this, water production facilities are purchased on the basis of the required production rate and storage volume. Rapid water production allows the choice of a small (or even no) storage tank, ideal when in a compact environment. However, large storage tanks may be wall mounted or placed under benches allowing greater freedom of choice.

The volume of water required is obviously dependent on the fermenter working volume. Since our fermenter cultures are generally performed and completed during one working day, there is time for refilling of the storage tanks overnight. One working volume of water is required for the medium prior to sterilisation, with at least another working volume required for re-sterilisation and cleaning, although these are not needed on the same day in our case.

For the 100 L fermenter we have two 75 L tanks and an ELGA Purelab Option purifying unit capable of supplying 15 L/h. For the two 30-L fermenters we have a 75 L tank and a unit capable of producing 10 L/h.

3.4.11   Computer Monitoring and Control

With most fermenters it is the local control console and not the supervisory control and data acquisition (SCADA) software that is responsible for regulating the growth parameters, allowing its use in the absence of a computer. However, batch data are always useful, and the possibilities of automation to enhance batch reproduction may be of importance. These aspects are discussed in more detail in Chapters 2 and 10.

Computer monitoring and supervisory software can be purchased with practically all fermenter systems. The software must be capable of graphically logging all of the data and allowing the export of both data and graphs as usable files. However, the software varies drastically between companies and software versions. The software often differs in simplicity and functionality, i.e. the more complex the software, the more powerful it can be with respect to programming and data processing. Since supplied software does vary and is regularly updated, the most suitable software for use is best discussed with the fermenter supplier in detail. It is advisable to take the time to arrange a demonstration of the SCADA system so that its capabilities, and especially its 'user-friendliness', can be assessed, since the majority of use may be by nonspecialists.

3.5   Harvesting the Cells

Conventional centrifuges have the advantage of being commonplace in laboratories and have a relative ease of use. They also have the advantages, over some systems, of temperature control and variable speed. However, their main limitation is the capacity of the bucket, which will generally not exceed 1 litre, allowing processing of between 5 and 6 L per centrifugation cycle. Although this may be sufficient for small scale work, it is unsuitable for the harvesting of large quantities of cell broth since this becomes very time consuming. The ability to harvest the cells rapidly is desirable, since this allows them to be quickly stored under appropriate conditions, thus restricting the degradation of cellular components.

Continuous centrifuges require a continuous flow of broth into the rotor. This can be achieved using hydrostatic pressure, although this diminishes as the vessel empties, and is therefore not constant. It can also be achieved with overpressure, although this is not always controllable, and poses a hazard when using glass vessels. Ideally a separate pump, suitable for the desired flow rate, is used. With respect to harvesting of our cells sterility is not an issue. However, should sterile harvesting be necessary then some kind of containment of the equipment in a sterile environment will have to be arranged. This will be easier using closed centrifuge bottles or a cross-flow filtration device (see below) than continuous flow rotors.

The pump we use for large scale centrifugation is an Autoclude V8/55/24HR Duty variable speed peristaltic pump, capable of flow rates up to 8 L/min. For lower volume harvesting we use a Watson Marlow 501 peristaltic pump for flow rates up to approximately 350 mL/min.

Heraeus Continuous Flow Centrifuge

The Heraeus Sepatech Contifuge 17RS centrifuge has a relatively small capacity and centrifugation rate, only 10 L/h, but gives excellent clarity of supernatant. An external pump is required to feed the cell broth into the centrifuge, with the supernatant simultaneously removed to waste. Once full (capacity is approximately 300 g), the bowl is removed and emptied. Similar to a bucket centrifuge, this is time consuming, although does generally result in a dry, compact cell pellet. This centrifuge is more suitable for low density cell broths than are conventional bucket centrifuges as a consequence of the high removal efficiency. Due to the low flow rates that can be used, however, this may not be suitable for cultures above approximately 10 L volume.

Tubular Bowl Centrifuges

For harvesting of our 30 and 100 L cultures we use a Cepa Z41 tubular bowl centrifuge (Figure 3.5) and high speed peristaltic pump. The centrifuge rotor has a capacity of approximately 750 g of wet cells. This centrifuge provides good clarity and high throughput, although some cell loss is expected. The amount of cell loss is dependent on several factors such as flow rate, pellet rheology, and the volume of the cell pellet within the rotor. Using this machine, a 30 L batch of cells can be harvested in 20 minutes, and a 100 L batch in around 1 hour. Since the tubular bowl centrifuge has no temperature control, although the bowl can be chilled prior to use, it is the rapid throughput that restricts cell overgrowth and protein degradation. In large volume batches it is customary to chill the batch immediately upon starting the harvest to aid this.

Cross-Flow (Tangential Flow) Filtration

An alternative to centrifugation as a means of cell harvesting is cross-flow (also known as tangential flow) filtration. These devices use membranes to concentrate a cell suspension, with the flow of the broth across the membrane surface preventing adhesion of cells and blocking of the membrane (Figure 3.6). Prevention of membrane fouling enables continuous operation at a constant rate, unlike dead-end filtration. This is a useful method for cell separation, allowing acquisition of secreted soluble proteins prior to a concentration step, and can be performed aseptically if necessary. Cross-flow filtration can also be used for cell washing or buffer exchange prior to centrifugation. The size exclusion membranes ensure that no cells or cell debris are present in the permeate. However, since

(a)

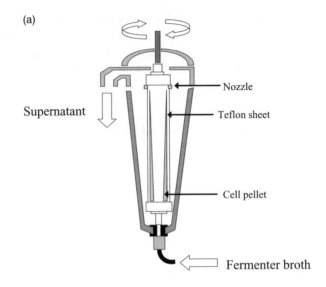

Nozzle

Supernatant

Teflon sheet

Cell pellet

Fermenter broth

(b)

Figure 3.5 *A tubular bowl centrifuge. (a) Schematic diagram of a tubular bowl centrifuge. Fermenter broth enters the rotor at the bottom via a pump. The cells are centrifuged onto an internal Teflon sheet, whilst the supernatant passes from the top of the system through a series of nozzles. (b) Photograph of cells sedimented onto the Teflon lining of the rotor following cell harvesting and subsequent removal from the centrifuge*

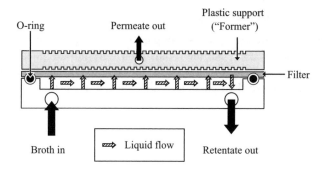

Figure 3.6 *Cross-flow filtration. Schematic of a cross-flow (tangential-flow) filter. The fermenter broth passes over the surface of the filter via a pump. Pressure applied by the pump forces a proportion of the liquid through the filter, with the remainder passing over the filter surface along with the retained cells. The retentate is then recycled around the system as a closed loop. Cell debris cannot 'blind' the filter since the tangential flow washes the filter clear of solids*

this system relies upon recycling of the cell suspension by a pump, a completely dry pellet is impossible to achieve, with a slurry being the final product.

Conclusion

In our work we prefer the use of centrifugation to allow rapid cell harvesting, followed by resuspension of the cell pellet in a 'freezing buffer'. In the case of cross-flow filtration this would be much more time consuming, which may be detrimental to the protein. However, cross-flow filtration may be suitable for volume reduction and cell washing in some cases, such as when sterile cell harvest is necessary.

3.5.1   Storage of Bacteria

Although not related to the fermentation equipment, storage is of relevance to any facility. Membrane protein production by IPTG induction in bacterial cell culture has tended to result in relatively small cell pellets, approximately 4 g wet cells/L broth. The recent advent of auto-induction cultures, in which induction is triggered by lactose once the cells reach late exponential or stationary phase after growth on glucose, has enabled much greater yields, often ten times higher, to be produced. Under these conditions it is likely that several kg of cells will be produced in the 100 L batches, and since processing of all of a batch cannot be carried out immediately, the storage of these cells should be considered.

At present our cells are frozen in suspension as pellets of about 80 g in about 100 mL Tris-HCl, EDTA, glycerol (20 mM, 0.5 mM, 10%) buffer, and stored at −80 °C. The buffer is used to prevent cell lysis, and also to chelate the metals and prevent protease action, but drastically increases the amount of freezer space required. If cell lysis is unimportant, it is also possible to freeze the cells as a pellet in a suitable container or freezer bag, which is useful should space be limited. The buffer can then be added to the pellet prior to defrosting and processing.

3.5.2   How are the Facilities Integrated?

In Figure 3.7 we show how the items of equipment are arranged. This is not an ideal disposition, for the simple reason that the rooms used were not designed for the purpose. Nevertheless, a disciplined force of three staff are able to accomplish two or three fermentations per week, more than enough to saturate the downstream processing stages (Figure 3.1).

3.6   Optimisation Trials for the Production of Proteins

Optimisation of culture conditions is probably the most important aspect of a fermentation process, allowing high cell and protein yields to be achieved and minimising the need for large scale culture. Typically, initial strains are tested for stability, reproducibility and yield in shaker flasks, a process that should be repeated when scaling up into fermenter culture.

The limitations of shaker flask cultures, i.e. the inability to control parameters other than agitation rate and temperature, the relatively modest oxygen transfer rates, and the accumulation of carbon dioxide within the headspace, usually means that the scale up into a fermenter produces a completely different environment for cell growth, as do the differences in the vessel geometry. This may prevent direct scale-up from flask to fermenter, and hence optimisation is important to ensure that the fermenter culture provides a comparable, if not improved, biomass and protein yield than the shaker flask culture. More detail on scale-up strategies of fermentation processes is given in Chapter 8.

Various problems may be observed during scale-up, which can be separated into those of biological, chemical and physical nature. Biological factors include mutation probability, and the increased likelihood of contamination. Chemical factors include the purity of medium components and water, pH control (the concentration of acid and base to use to ensure safe working practices, prevent equipment damage, and restrict dilution of the medium, whilst maintaining pH), and foam formation. Physical factors incorporate the differences between aeration, agitation and temperature control, which may all be influenced by mixing, among other things.

All parameters examined during shaker flask optimisation, such as medium composition, inducer concentration, induction OD and expression time should ideally be re-examined in a fermenter. In addition, dissolved oxygen and pH control can be performed throughout the fermenter culture, and should therefore also be studied. Once suitable conditions have been found and tested for reproducibility, further studies into the mode of operation (i.e., batch, fed-batch, continuous culture) can be examined as necessary.

3.6.1   Parallel Fermenters

The scale up from small scale to large scale fermenter is often much more predictable than from flask to fermenter. This is a consequence of similar geometries and the similarity of control parameters between fermenters. Even though many parameters can be scaled directly, e.g. air flow, agitator tip speed, inoculum volume, there remains a possibility

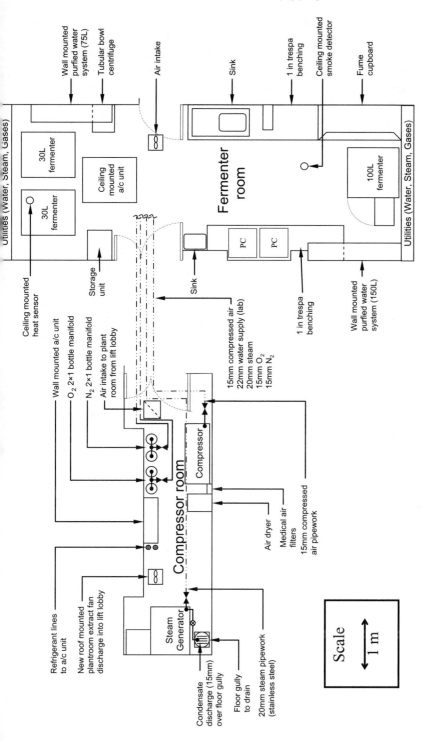

Figure 3.7 *Layout of the fermenter facilities. The left hand side shows the layout in a separate air-conditioned room containing the steam generators, air compressors, gas bottles and storage. This room is noisy and hot, and is separated from the main fermenter room, illustrated on the right of the diagram, by a corridor. Pipework through the ceiling space above the corridor connects the two rooms and a set of utilities is delivered to each of the fermenters through appropriate power outlets, valves, and taps. The air-conditioned room on the right contains the three fermenters, their purified water feeds, the centrifuge, a fume hood, a dry bench for the computers, and a wet bench with sink for cleaning and washing equipment. Most of the preparation of media and autoclaving of glassware and solutions is carried out in the main lab to the right of the fermenter room*

that conditions will not be identical. Microenvironments may form within the medium, preventing homogeneity of the solution and producing gradients of substrates, pH and temperature. Consequently it is necessary to ensure that the optimised conditions produce comparable yields and expression levels on a larger scale. Currently, our initial optimisation is accomplished in small volume shaker flasks. However, the conditions are very different from those in the large volume fermenter necessary for the high yields needed for structural studies. Manufacturers are now producing parallel series of fermenters for culture volumes of around 0.5–2 L, in which many growth parameters can be varied and controlled and the outcomes measured to seek the optimum conditions for scale up of protein production. A matrix of conditions can be designed and implemented to explore these.

Matrix Design and Implementation

A detailed study of the factors that affect the yield of membrane protein has previously been performed for yeast to show that carefully controlling two factors influencing an expressing culture (in this case temperature and pH) can result in a three-fold improvement in functional membrane protein yield. Moreover, it is clear that these two factors interact such that elevating the growth temperature can compensate for the reduction in yield caused by an elevated culture pH. This is an illustration of the complexity of fermentation systems in general, and of the interdependent nature of the process variables.

In order to extend and improve this study, the effects of additional factors can be examined simultaneously, and their corresponding effects on the host response measured with a microcalorimeter (see above). Two additional factors include stirring rate and oxygen flow rate as these contribute to the aerobicity of the culture, which is undoubtedly a critical parameter in dictating protein yield, and is as yet unexplored.

Design of a Matrix of Conditions to Optimise Production of Membrane Proteins

To maximise the output of the study, use of a design of experiments matrix (Table 3.2) is an appropriate way to quantify the effects of any chosen set of factors and all their interactions on yield for a panel of membrane proteins.

The methodology to be used centres on the collection of samples from high performance bioreactors, in which microbial cultures express a panel of membrane proteins while grown under tightly controlled conditions. Analysis is performed at the protein, transcript and metabolite level in order to obtain a complete picture of the host response. Online microcalorimetric measurements may also be recorded through independent channels to enable an accurate growth curve to be plotted, and give a full thermodynamic read-out of the cultures. pH, alkali/acid titration, aeration and gas analysis are also performed online. Samples are collected from the fermentations at defined points in the growth curves, allowing the analysis of membrane protein yield in total and membrane extracts. It is then possible to analyse mathematically the impact of each of the factors, and their interactions, on yield. This type of initial study can then be explored in more depth. Factors highlighted as being important can be more specifically studied by adding additional data points to the matrix, thereby allowing mathematical relationships to be devised and the importance of other parameters to be investigated.

Table 3.2 *Example of a coded matrix for a full factorial four-factor two-level design of experiments with centre points to examine the effect of varying four factors [temperature, pH, stirring rate (R) and aeration (F)] on yield for a given membrane protein. The matrix design includes blocking to account for fermenter-specific effects. + = high; − = low; c is the centre point. One run takes approximately 1 week*

Run	Fermenter 1 (T, pH, R, F)	Fermenter 2 (T, pH, R, F)	Fermenter 3 (T, pH, R, F)
1	c c c c	+ + − −	− − + +
2	c c c c	− + − −	+ − + +
3	c c c c	+ − − +	− + + −
4	+ + + −	c c c c	− − − +
5	+ + − +	− − + −	c c c c
6	− + − +	+ − + −	c c c c
7	c c c c	− − − +	+ − − −
8	+ + + +	− − − −	c c c c

3.6.2 The Production of Proteins Labelled with ^{15}N or ^{13}C for NMR Studies

Uniform labelling of bacterial proteins with a specific isotope is fairly easy to achieve. If the protein of interest is normally expressed well in minimal medium, ^{14}N→^{15}N or ^{12}C→ ^{13}C substitution can be achieved by growing the cells in the minimal medium with the isotopically-labelled nitrogen and/or carbon sources (e.g., ^{15}NH$_4$Cl and [U-^{13}C]glucose/[U-^{13}C]glycerol), all of which are readily available. Amino acid mixtures, as well as algal hydrolysates, labelled with any combination of isotopes, are also commercially available (e.g., from Cambridge Stable Isotope suppliers) for expression of proteins that require additions or rich media.

Although the costs of ^{15}N and ^{13}C labelled precursors have decreased substantially over the last few years, labelling enough material for several NMR experiments can still be quite expensive. However, efficient and cost-effective labelling with ^{15}N or ^{13}C isotopes is possible. Cultures are grown in limited amounts of natural abundance nutrients in a fermenter, while the oxygen levels are carefully monitored. Cellular metabolism is shut down upon nutrient depletion, which results in an abrupt drop in oxygen demand accompanied by a sudden rise in dissolved oxygen levels. The labelled compounds are added at this stage, allowing the oxygen demand to rise again. Protein production is initiated by induction when a second sharp rise in the dissolved oxygen signals the exhaustion of any remaining small amounts of unlabelled nutrients. The amount of protein per unit of labelled glucose increased threefold using this procedure. Another way of reducing labelling cost is to generate the bulk of the biomass in a rich medium before exchanging the cells to the labelling media prior to induction. The best results to date yielded a phenomenal 6 mg protein per mL reaction volume using an opportunistic cell-free production method. We have successfully produced fermenter scale, isotopically labelled, samples of the *E. coli* glucuronide and galactose transport proteins, GusB(His)$_6$

and GalP(His)$_6$. Both these proteins are used for solid state NMR studies of ligand binding.

In the case studies below, we discuss some of the challenges in these studies.

3.6.3   Optimisation of Growth and Expression Conditions for ^{15}N Labelling of Bacterial Membrane Transport Proteins

Many bacterial membrane transport proteins are homologues of those found in all higher organisms, including protozoan parasites, fungi, plants, animals and man. Currently, it is much easier and cheaper to produce these prokaryote homologues in bacteria or yeasts than to isolate the original protein from higher eukaryote cells. The results of studies on the structure–activity relationship of the prokaryote protein can then be legitimately extrapolated to understand, for example, sugar transport proteins in man. Many other membrane transport or receptor proteins are unique to bacteria, and these are of inherent interest for understanding features of prokaryote metabolism, and sometimes because they are potential targets for the generation of novel antibacterials, in a world where resistance of prokaryote pathogens to antibiotics is becoming a very serious problem. In our laboratory we study a membrane transport protein for glucuronides, GusB, of *E. coli* as an example of a prokaryote transporter, and one for galactose, GalP, also from *E. coli*, as an example of an homologue found in all other genera, including mammals. They are overexpressed in fusions with six histidine residues at their C-termini in order to facilitate purification.

We usually express GusB(His)$_6$ in M9 minimal medium supplemented with casamino acids (Table 3.3). To avoid the addition of expensive ^{15}N-labelled amino acids to the growth medium for [U-^{15}N] labelling, we investigated the growth and expression of GusB(His)$_6$ on M9 salts with NH$_4$Cl as sole nitrogen source. We grew *E. coli* strain NM554 (pWJL24H) in 50 mL M9 medium containing carbenicillin (100 μg/mL) in 250-mL baffled flasks, but without the addition of any casamino acids. NH$_4$Cl concentrations ranged from 2–20 mM and L-Leu (50 μg/mL) was added to permit growth of the leucine auxotrophic *E. coli* strain NM554. It is evident that an NH$_4$Cl concentration of

Table 3.3   *M9 medium and supplements. Autoclaving is performed at 121 °C and 15 psi for 20 minutes*

Nutrient	Final concentration
Autoclaved together:	
KH$_2$PO$_4$	3.0 g/L
Na$_2$HPO$_4$ (anhydrous salt)	6.78 g/L
NaCl	0.5 g/L
Glycerol	40 mM
Autoclaved separately to avoid precipitation of phosphate salts and degradation of amino acids:	
MgSO$_4$·7H$_2$O	2.0 mM
CaCl$_2$·2H$_2$O	0.2 mM
Casamino acids	0.2%
NH$_4$Cl	1.0 g/L

2 mM is not enough to sustain growth and a final A_{680} of only 0.638 was reached (Figure 3.8a). However, no significant difference could be detected for growth or expression of GusB(His)$_6$ (Figure 3.8b) on any of the other NH$_4$Cl concentrations used.

3.6.4   Growth and Expression of [U-^{15}N]GusB(His)$_6$ or GalP(His)$_6$ in a Fermenter

To prepare [U-^{15}N]GusB(His)$_6$, growth of NM554 (pWJL24H) is performed using M9 medium containing carbenicillin (100 μg/mL) in a 10 L fermenter, in the absence of casamino acids, as described in the previous section, and with a final ^{15}NH$_4$Cl concentration of 5 mM. As expected, the ^{15}N nitrogen isotope does not alter the growth of the cells (Figure 3.8c). Similarly to the small-scale trial experiments, GusB(His)$_6$ was expressed to 4–5% of the total membrane protein fraction. The inner membranes were obtained from these mixed membrane preparations with sucrose density ultracentrifugation. GusB(His)$_6$ constitutes 8% of the total inner membrane proteins, owing to the successful removal of a prominent outer membrane protein.

Uniform labelling of GalP(His)$_6$ was easily achieved, since the normal growth conditions for overexpression of this protein are on minimal medium (Table 3.4) with NH$_4$Cl as the only nitrogen source. *E. coli* strain JM1100 (pPER3H) was grown in minimal medium (Table 3.4) containing tetracycline (15 μg/mL) in a 10 L fermenter, but with 12.5 mM ^{15}NH$_4$Cl instead of the 42 mM unlabelled NH$_4$Cl. No addition of 2TY was made.

[U-^{15}N]GalP(His)$_6$ is expressed to the same high level as unlabelled GalP(His)$_6$ (not shown).

3.6.5   Determination of ^{15}N Labelling Efficiency

Amino acids and growth substrates, e.g. sugars, labelled with stable isotopes for incorporation into proteins for NMR studies are very expensive. Consequently, it is important to monitor the efficiency of their incorporation and optimise their concentrations, growth conditions, and culture volumes to achieve economy. Monitoring is best achieved by mass spectrometry, an example of which is given here for GalP(His)$_6$. [U-^{15}N]GalP(His)$_6$, produced by growth of an overexpressing *E. coli* strain on ^{15}NH$_3$, was purified with affinity chromatography, and further purified from residual lipids and detergents with size exclusion chromatography (Venter, 2001). Electrospray ionisation mass spectrometry (ESI-MS) on the purified intact protein reveals a molecular weight of 52 211.60. With 611 nitrogens in GalP(His)$_6$, that implies an overall incorporation of 86% labelled nitrogen. None of the three nitrogens in the 12 histidine residues is labelled, as unlabelled histidine was added to the growth medium of the histidine auxotrophic strain JM1100, so the resulting nitrogens in GalP(His)$_6$ are labelled to a highly satisfactory extent.

3.6.6   Optimisation of the Labelling Medium for the Production of [50%-^{13}C]GalP(His)$_6$

Owing to the cost of [U-^{13}C]glucose, a careful assessment of the minimum amount of glucose needed to support growth and expression of GalP(His)$_6$ is required. Some of the 28 mM glucose habitually used for growth of this strain, may be in excess, since the residual glucose concentration in the medium after growth has not been determined.

(a)

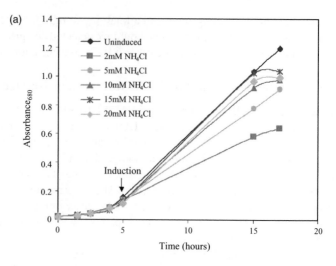

(b)

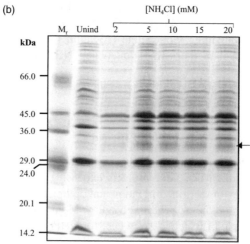

(c)

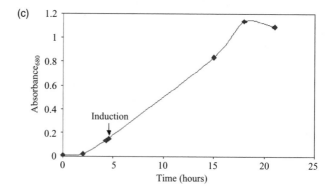

Figure 3.8 *(a) Growth of NM554 (pWJL24H) expressing GusB(His)₆ on M9 salts with NH₄Cl as the sole nitrogen source. Inocula were made from mid-log phase cultures grown on LB medium and diluted (1%) into the M9 medium. GusB(His)₆ expression was induced after 5 hours of growth by addition of IPTG to a final concentration of 0.6 mM. The uninduced flask was grown with 5 mM NH₄Cl. (b) SDS-PAGE of GusB(His)₆ containing cell membranes prepared by water lysis (Venter, 2001) of the cells described in Figure 9(a). The concentration of NH₄Cl in the growth medium is indicated on top of each lane. Uninduced NM554 (pWJL24H) membranes were included as GusB(His)₆ negative control. SDS-PAGE was performed and the gel was stained with Coomassie blue (as described in Venter, 2001). The position of GusB(His)₆ is indicated with an arrow. GusB(His)₆ expression is absent in the uninduced sample and in the cells grown on 2 mM NH₄Cl, while all the other membrane preparations have GusB(His)₆ expressed to 4–5% of the total protein. (c) Growth of E. coli NM554 (pWJL24H) expressing [U-¹⁵N]GusB(His)₆ on M9 salts with ¹⁵NH₄Cl as the nitrogen source. Inocula were made from mid-log phase cultures grown on LB medium and diluted (1%) into the indicated culture media. GusB(His)₆ expression was induced after 4.5 hours of growth by addition of 0.6 mM IPTG*

◄

Table 3.4   *Minimal growth medium for* E. coli *JM1100. Autoclaving is performed at 121 °C and 15 psi for 20 minutes*

Nutrient	Final concentration
Autoclave together	
NH_4Cl	2.25 g/L
KH_2PO_4	1.59 g/L
$Na_2HPO_4 \cdot 2H_2O$	4.26 g/L
$CaCl_2 \cdot 2H_2O$	0.11 g/L
Thiamine	0.083 mg/L
$MnCl_2 \cdot 4H_2O$	3.33 mg/L
$FeSO_4 \cdot 7H_2O$	3.33 mg/L
$MgSO_4$	66.67 mg/L
HCl	16.67 μM
Autoclave separately to avoid salt-catalysed caramelisation	
Glucose	3.6 g/L (20 mM)
Filter sterilise to avoid degradation by heat	
Thymine	20 mg/L
L-Histidine	80 mg/L
Set pH to 7.5	

E. coli strain JM1100 (pPER3H) was grown in 50 mL minimal medium (Table 3.4) containing tetracycline (15 μg/mL) in 250 mL flasks. The glucose concentration varied from 10 to 20 mM (Figure 3.9) and the inoculum volume was 1% of the final medium volume. Membrane preparations were then made (the procedure can be found in Ward *et al.* (2000)). The amounts of protein in each preparation were determined and the expression level of GalP(His)₆ in all samples assessed by SDS-PAGE (Venter, 2001).

The final density of cells declined progressively as the glucose concentration was lowered, so that an A_{680} of only 0.64 was reached for growth on 10 mM glucose

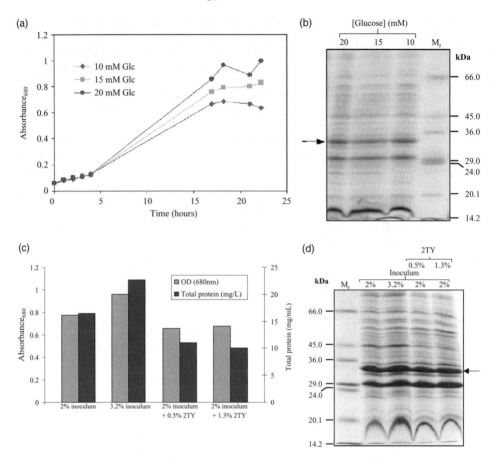

Figure 3.9 *(a) Growth of JM1100 (pPER3H) expressing GalP(His)*$_6$* in minimal medium with different quantities of glucose. Cells were cultured in minimal medium as described in the accompanying text and in Venter (2001). Inocula were made from mid-log phase cultures that were grown on 2TY and diluted (1%) into the final growth medium. (b) SDS-PAGE of GalP(His)*$_6$* expressing mixed membranes obtained by water lysis of JM1100 (pPER3H) grown on different glucose concentrations and 10 μg/mL L-histidine. SDS-PAGE was performed as described in Venter (2001) and the gel was stained with coomassie blue (Venter, 2001). The position of GalP(His)*$_6$* is indicated with an arrow. The amount of glucose present in the media is indicated on the top of each of the lanes. No difference in the expression levels of GalP(His)*$_6$* could be detected between the different preparations. (c) Growth and protein production of JM1100 (pPER3H) in minimal medium with various inoculum volumes and 2TY additions. (d) SDS-PAGE of GalP(His)*$_6$*-expressing cell membranes obtained by water lysis of JM1100 (pPER3H) grown from different inoculum volumes and 2TY additions. The position of GalP(His)*$_6$* is indicated with an arrow. The inoculum volume and amount of 2TY added are indicated on the top of each of the lanes. No difference could be observed between the expression levels of each preparation*

Table 3.5 *2TY medium. Autoclaving is performed at 121 °C and 15 psi for 20 minutes*

Nutrient	Final concentration
Tryptone	10 g/L
Yeast extract	10 g/L
NaCl	5 g/L
Set pH to 7.0 before making up to 1 L. Autoclave	

(Figure 3.9a). However, the expression level of GalP(His)$_6$ was not influenced by the reduced levels of carbon source (Figure 3.9b).

The level of growth obtained on 10 mM glucose was not enough to supply the quantity of membranes needed for NMR studies. However, increasing the amount of glucose from 10 to 15 mM would increase the cost disproportionately to the gains in cell mass. Standard growth of strain JM1100 (pPER3H) for GalP(His)$_6$ expression was carried out in a 25 L fermenter on minimal medium (Table 3.4) containing 28 mM glucose, with the addition of 1.3% 2TY (Table 3.5) and a 3.2% inoculum. Therefore, it was decided to investigate growth and expression on the lower glucose concentration but with either or both the addition of 2TY or larger inoculum, even though that would mean a lower labelling level. *E. coli* strain JM1100 (pPER3H) was grown in 800 mL minimal medium (Table 3.4) containing tetracycline (15 μg/mL) in 2 L baffled flasks with inoculum volumes and 2TY additions as indicated in Figure 3.9(c). The highest cell density and most protein were obtained from the culture that was started with a 3.2% inoculum. Even considering the inequalities in growth between medium with and without 2TY addition, it is evident that total cell mass is not so much benefited by the addition of 2TY as by a substantial inoculum volume. Again, the expression level of GalP(His)$_6$ was not influenced by the growth pattern (Figure 3.9d).

Sufficient quantities of [50%-^{13}C]GalP(His)$_6$ could be produced by growth in minimal medium with 10 mM glucose as sole carbon source and a 3.2% inoculum. *E. coli* does not utilise histidine as a carbon source, so the amount of histidine is of no concern, but should be higher than 10 mg/L.

3.6.7 Growth and Expression of [50%-^{13}C]GalP(His)$_6$

E. coli strain JM1100 (pPER3H) was grown on minimal medium (Table 3.4) containing tetracycline (15 μg/mL) in a 5 L fermenter. A mixture of equal amounts of unlabelled and [U-^{13}C]glucose (final concentration of 10 mM) was used as the carbon source with histidine supplementation (40 μg/mL). A 3.2% inoculum was made from a mid-log phase culture grown on 2TY. An inner membrane preparation containing 68 mg total protein was attained from growth in 5 L of this labelling medium.

When the regular strain for GalP(His)$_6$ expression was used for uniform labelling of the protein with both ^{15}N and with ^{13}C, the expression levels and transport activity of the labelled proteins were still consistent with those of unlabelled protein (results not shown).

3.7   Concluding Remarks

We are not an industrial-scale fermentation unit, and the advice in this article is limited to practical considerations of setting up and operating a relatively small-scale fermenter facility with the aim of producing membrane proteins in amounts sufficient for structural studies. At the present time, optimisation of the efficiency of the recombinant constructs for expression are a consideration that has not been fully explored. Also, we have not yet been able to undertake a systematic investigation of all the factors – pH, aeration, temperature, cell density, etc. – that might improve productivity, but we have illustrated a parallel fermenter, design-matrix strategy that will enable this to be implemented. Nevertheless, we are succeeding in the production of at least 17 different membrane proteins in amounts required for crystallisation trials that have yielded four cases where the crystals diffract X-rays for elucidation of 3-D structure of the proteins. In addition, we have successfully labelled three membrane proteins with ^{15}N and ^{13}C, both generally and in specific amino acids for NMR measurements.

Detailed protocols for the fermenter operations can be obtained from the authors.

Acknowledgements

Our fermenter equipment and room conversions have been funded by grants from Wellcome Trust, BBSRC, Advantage West Midlands, the EU and the Universities of Leeds and Aston. The ongoing operation is supported by BBSRC, especially the Membrane Protein Structure Initiative, MPSI, and by the EU European Membrane Protein Consortium, E-MeP.

Abbreviations

A_{680}	absorbance at 680 nm
CIP	clean-in-place
DO$_2$	dissolved O$_2$
ESI-MS	electrospray ionisation mass spectrometry
IPTG	isopropyl-β-D-thiogalactoside
MWV	maximum working volume
NMR	nuclear magnetic resonance
SIP	sterilisation-in-place

References and Additional Reading

This selection of material contains references that describe in much more detail some of the practical methods we introduce in this chapter, and some references that give some additional supporting or underpinning theory.

Battley, E.H. (1987) *Energetics of Microbial Growth*. John Wiley & Sons, Inc., USA.

Blomberg, A., Larsson, C. and Gustafsson, L. (1988) Microcalorimetric monitoring of growth of *Saccharomyces cerevisiae*. Osmotolerance in relation to physiological state. *J. Bacteriol.* **170**, 4562–4568.

Bonander, N., Hedfalk, K., Larsson, C., Mostad, P., Chang, C., Gustafsson, L. and Bill, R.M. (2005) Design of improved membrane protein production experiments. *Prot. Sci.* **14**, 1729–1740.

Cai, M., Hunag, Y., Sakaguchi, K., Clore, G.M., Gronenborn, A.M. and Craigie, R. (1998) An efficient and cost-effective isotope labelling protocol for proteins expressed in *Escherichia coli J. Biomol. NMR* **11**, 97–102.

Clough, J., Saidijam, M., Bettaney K.E., Szakonyi, G., Meuller, J., Suzuki, S., Bacon, M., Barksby, E., Ward, A., Gunn-Moore, F., O'Reilly, J., Rutherford, N.G., Bill, R.M. and Henderson, P.J.F. (2006) Prokaryote membrane transport proteins: amplified expression and purification. In *Structural Genomics of Membrane Proteins* (ed. Lundstrom, K.) pp. 21–42. CRC Press, USA.

Hewitt, C.J., Nebe-von Caron, G., Axelsson, B., McFarlane, C.M. and Niehow, A.W. (2000) Studies related to the scale-up of high cell density *E. coli* fed-batch fermentations using multiparameter flow cytometry: effect of a changing microenvironment with respect to glucose and dissolved oxygen concentration. *Biotech. Bioeng.* **70**, 381–390.

Larsson, C., Blomberg A. and Gustafsson L. (1991). The use of microcalorimetric monitoring in establishing continuous energy balances and in the continuous determinations of substrate and product of batch-grown *Saccharomyces cerevisiae*. *Biotechnol. Bioeng.* **38**, 447–458.

Liang, W-J., Wilson, K., Xie, H., Knol. J., Suzuki, S., Rutherford, N.G., Henderson, P.J.F. and Jefferson, R. (2005) The *gusBC* genes of *Escherichia coli* encode a glucuronide transport system. *J. Bacteriol.* **187**, 2377–2385.

Marley, J., Lu, M. and Bracken, C. (2001). A method for efficient isotopic labelling of recombinant proteins. *J. Biomol. NMR* **20**, 71–77.

Sinclair, C.G. and Kristiansen, B. (1987) *Fermentation Kinetics and Modelling* Open University Press, UK.

Studier, W.F. (2005). Protein production by auto-induction in high-density shaking cultures. *Prot. Expr. Purn.* **41**, 207–234.

Teusink, B., Larsson, C., Diderich, J., Richard, P., van Dam, K., Gustafsson, L. and Westerhoff, H.V. (1996) Synchronized heat flux oscillations in yeast cell populations, *J. Biol. Chem.* **271**, 24442–24448.

Venter, H. (2001) *Applications of Biophysical Techniques to the Analysis of Membrane Transport Proteins*. PhD Thesis, University of Leeds.

Venter, H., Ashcroft, A.E., Keen, J.N., Henderson, P.J.F. and Herbert, R.B. (2002) Molecular dissection of membrane transport proteins: mass spectrometry and sequence determination of the galactose-proton symport protein, GalP, of *Escherichia coli* and quantitative assay of the incorporation of [$ring$-2-^{13}C]histidine and ^{15}NH$_3$. *Biochem. J.* **363**, 243–252.

Von Stockar, U., Auberson, L. and Marison I.W. (1993). Calorimetry of technical microbial reactions. *Pure Appl. Chem.* **65**, 999–1002.

Walsh, M.C., Smits, H.P., Scholte, M. and van Dam, K. (1994). Affinity of glucose transport in *Sacchoromycis cerevisiae is* modulated during growth on glucose. *J. Bacteriol.* **176**, 953–958.

Ward, A., Sanderson, N.M., O'Reilly, J., Rutherford, N.G., Poolman, B. and Henderson, P.J.F. (2000) The amplified expression, identification, purification, assay and properties of histidine-tagged bacterial membrane transport proteins. Chapter 6 in *Membrane Transport – A Practical Approach*. pp. 141–166. Blackwell Press, Oxford.

Ward, A., Hoyle, C.J., Palmer, S.E., O'Reilly, J., Griffith, J.K., Pos, K.M., Morrison, S.J., Poolman, B., Gwynne, M. and Henderson, P.J.F. (2001) Prokaryote multidrug efflux proteins of the major facilitator superfamily: amplified expression, purification and characterisation. *Molec. Microbiol. Biotech.* **3**, 193–200.

Westerhoff, H. V. and van Dam, K. (1987) *Thermodynamics and Control of Biological Free-energy Transduction*, Elsevier, Amsterdam.

4

Modes of Fermenter Operation

Sue Macauley-Patrick and Beverley Finn

4.1   Introduction

The purpose of this chapter is to describe the various modes of operation available for the fermentation of large quantities of cells via submerged liquid culture in vessels usually referred to as fermenters or bioreactors. There are three main modes of fermentation technique: batch, fed-batch and continuous. In industry, batch and fed-batch fermentation techniques have been used for the production of alcoholic beverages and fermented food-stuffs since before 3000 BC in Egypt and Sumeria. At the beginning of the Twentieth century, other industrial applications became popular, such as the production of acetone, butyl alcohol and ethyl alcohol, later still (1940s onwards) came the production of anti-biotics by submerged culture of selected strains of filamentous bacteria and fungi. Each of these new classes of product led to technological changes within the fermentation industry, and, in particular, to novel methods of bioprocess operation. Most fermentation processes were batch for much of human history, with fed-batch becoming common only with production of baker's yeast and antibiotics. Most industrial fermentation processes still operate as simple batch or fed-batch.

There are few continuous fermentation examples in industry: vitamin C, propionic acid and Quorn production, for example. However, as a laboratory tool for studying the physiology of microorganisms, metabolomics, proteomics, etc., continuous fermentation techniques are well established and currently enjoying a resurgence of interest amongst researchers. The reader is referred to Chapter 12 for further details on the uses of continuous culture for research into microbial physiology.

In the pharmaceutical and biotechnology industries a great deal of development work is required from the concept to large-scale production of the desired product (see Chapter

Practical Fermentation Technology Edited by Brian McNeil and Linda M. Harvey
© 2008 John Wiley & Sons, Ltd

8). Bioreactors are used throughout this process, ranging from small-scale bench-top (~1–2 L), to pilot scale (20–100 L) for feasibility studies, to large-scale reactors that are capable of cultivating thousands of litres of microbes, and more.

In order to develop and optimise these processes, it is important to understand the three main modes of fermenter operation in order to determine which is most suited to the process. Here we discuss these modes of operation and provide information on their advantages and disadvantages, with emphasis on how, why and when to choose each.

4.2   Batch Culture

Batch fermentation is the simplest mode of operation, and is often used in the laboratory to obtain substantial quantities of cells or product for further analysis. A batch fermentation is a closed system, where all of the nutrients required for the organism's growth and product formation are contained within the vessel at the start of the fermentation process. The vessel can take the form of a shake flask, single use disposable system, or, for tighter control of parameters such as oxygen transfer, pH, agitation, etc., a bioreactor can be used (Figure 4.1). Historically, these processes would have involved nonsterile systems with self-selecting or natural inoculants. However, nowadays nearly all fermentation processes involve inoculation of a selected and specially bred strain of microbe, plant or animal cell into a sterile medium held within a sterile fermenter vessel. After medium sterilisation, the organism is inoculated into the vessel and allowed to grow. The fermentation is terminated when one or more of the following has been reached: (i) microbial growth has stopped due to the depletion of the nutrients or the build of toxic compounds; (ii) after a fixed predetermined period of time; (iii) the concentration of desired product has been achieved.

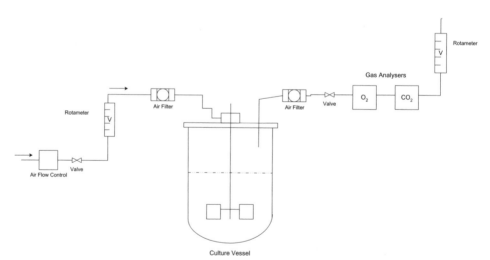

Figure 4.1   *A diagram of a simple batch fermentation. The system is 'closed', containing all the nutrients required by the organism prior to inoculation, except for the gaseous requirement, which is continuously added to, and removed from, the reactor via sterilising-grade hydrophobic filters*

4.2.1   The Batch Culture Growth Curve

When cells are grown in a batch culture, they will typically proceed through a number of distinct phases (Figure 4.2). The lag phase, which may or may not be present, is described as 'little or no growth at the beginning of the fermentation due to the physio-chemical equilibrium between the microorganism and the environment following inocula-tion'. The lag phase can be time consuming and costly and so it is highly desirable to minimise this phase. This can be achieved by growing the inoculum in comparable medium to the bioreactor and under similar growth conditions (pH, temperature, etc.). A minimum of 5% by volume inoculum of exponentially growing cells should also be used for inoculation. Once the cells have adapted to the new conditions of growth, they enter the exponential phase. Then, key to minimising the length of the lag phase lies in making sure that the culture being transferred undergoes the minimum levels of stress possible. In practical terms, this implies keeping the environments in the two fermentation systems as similar as possible. In reality, this is sometimes very difficult, e.g. a late exponential stage shake-flask culture will typically exist in an environment where substrate levels are reduced, oxygen levels low, and carbon dioxide levels elevated; pH may also be very different from the process start point. Clearly, the transfer of such a culture to a fully charged fermenter with fresh medium, highly aerated, low CO_2 environment. Inoculation of fermenters to minimise lag phase is something of a compromise and often involves empiricism and experience of that particular culture character.

Nutrient depletion and the formation of inhibitors (typically excreted products such as ethanol, lactic acid, acetic acid, methanol, and aromatic compounds) have the effect of decelerating cell growth, and the cells then enter the stationary phase where the rate of cell growth equals that of cell death. The kinetics of this are described more fully in Chapter 7, which deals extensively with mathematical modelling. Eventually, the cells

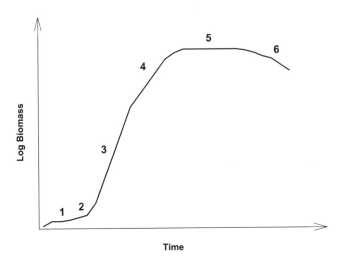

Figure 4.2   *Characteristic growth curve of a microorganism in batch culture. 1, Lag phase (not always present); 2, transient acceleration phase; 3, exponential phase; 4, deceleration phase; 5, stationary phase; 6, death phase*

enter death phase and this is characterised by a drop in optical density and biomass levels in most cultures.

The batch culture growth curve gives a good indication of when to stop the fermentation. Growth-associated products (primary metabolites) are produced during the exponential phase with their formation decreasing when growth ceases. Typically the rate of product formation directly relates to the rate of growth (Figure 4.3). The fermentation can be terminated at the end of the exponential growth phase before the cell enters stationary phase. This growth phase is sometimes referred to as the trophophase. Examples of primary metabolites are shown in Table 4.1. Nongrowth-associated products (e.g., classic secondary metabolites) have a negligible rate of formation during active cell growth. These secondary metabolites are produced as the cells enter stationary phase (Figure 4.4);

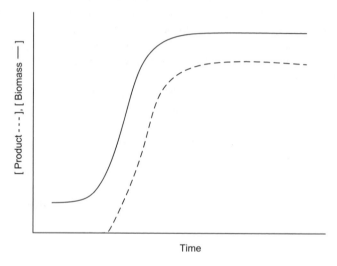

Figure 4.3   *Growth associated product formed during the period of active culture growth*

Table 4.1   *Examples of commercially produced primary metabolites*

Microorganism	Primary metabolite	Commercial use
Saccharomyces cerevisiae	Ethanol	Alcoholic beverage
Corynebacterium glutamacium	Amino acids – glutamic acid and lysine	Food industry – flavour enhancer
Ashbya gossipii and *Eremothecium ashbyi*	Riboflavin	Food industry – vitamin
Aspergillus niger	Citric acid	Food industry – flavour enhancer and preservative
Xanthonomonas campestris	Xanthan gum	Food industry – food additive and rheology modifier
Pseudomonas denitrificans and *Propionibacterium shermanii*	Vitamin B_{12}	Food industry – vitamin

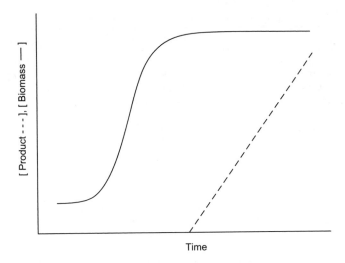

Figure 4.4 *Nongrowth associated product formed during the period of nonculture growth (stationary phase)*

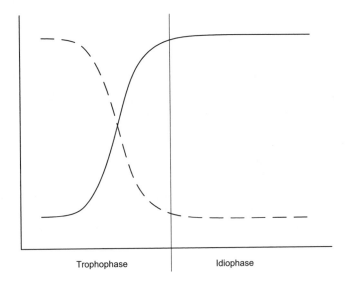

Figure 4.5 *Trophohase and idiophase production phases within a typical cell culture*

this can also be described as idiophase (Figure 4.5). Most antibiotics are produced as secondary metabolites (Table 4.2). The fermentation can then be stopped just before the cells enter the death phase.

Of course, it is as well to be aware that these textbook definitions do not always rigidly hold in actual practice, and the formation of fermentation products such as antibiotics and other secondary metabolites is often seen during growth phase.

Table 4.2 *Examples of commercially produced secondary metabolites*

Microorganism	Primary metabolite	Commercial use
Penicillium chrysogenum	Penicillin	Antibiotic
Streptomyces erythreus	Erythromycin	Antibiotic
Streptomyces griseus	Streptomycin	Antibiotic
Cephalosporium acrimonium	Cephalosporin	Antibiotic
Gibberella fujikuroi	Gibberellin	Antibiotic
Tolypocladium inflatum	Cyclosporin A	Antibiotic

Table 4.3 *Features of batch and fed-batch cultivations in the laboratory*

Small quantities of product can be obtained for laboratory studies
Products may not be able to be stored for long periods
High product concentrations may be required in order to optimise the downstream
  recovery of the product
Some metabolic products are produced only in the stationary phase of the organisms
  growth cycle
Instability of some production strains would make the use of continuous culture
  impossible
Batch and fed-batch processes can have fewer technical difficulties than continuous
  processes

4.2.2   Advantages of Batch Culture

• Simplicity of use. A batch culture can be easily readied, and, depending on the microorganism used, can be finished in less than 24 hours.
• Operability and reliability: less likely to have instrument failure on short batch runs;
• production of secondary metabolites that are not growth-related (i.e., produced when the organism enters stationary phase);
• fewer possibilities of contamination: all of the materials required for the bioprocess are present in the vessel and sterilised before the run starts. The only material added (with the exception of the inoculum at the beginning of the bioprocess) and removed during the course of a batch fermentation are the gas exchange, and if using a bioreactor, sterile antifoam and pH control solutions if required.
• It is easy to assign a unique batch number to each run, generating high confidence in the history of each batch of product. This is critically important in a highly regulated environment

Further advantages of batch cultivation, in common with fed-batch cultivation are listed in Table 4.3.

4.2.3   Disadvantages of Batch Culture

• Culture ageing, and more importantly differentiation, can be a specific problem, especially so with growth-related products;
• build up of toxic metabolites can restrict cell growth and product formation;

- initial substrate concentrations may have to be limited due to problems with inhibition and repression effects, therefore affecting the amount of product that can be obtained from such simple systems;
- batch-to-batch variability;
- the use of batch cultures in industrial systems can lead to an increased nonproductive period due to down time required for cleaning, resterilisation, filling and cooling of equipment;
- if using the organism from one bioprocess to seed another culture, degeneration or differentiation may occur, which could affect the bioprocess and product formation;
- cellular autolysis may occur during the decline and stationary phase, affecting the amount of product, its composition and potentially adding to downstream processing challenges due to release of autolytic breakdown products, activation of proteases;
- from a physiological viewpoint the use of batch cultures actually contributes greatly to the complexity of the experiments since the cell population is heterogeneous and constantly changing. This makes the use of such systems for clearly identifying cause and effect relationships in cell physiology rather unattractive (see Chapter 12 for a more complete discussion).

4.3 Fed-batch Culture

Fed-batch culture is essentially similar to batch culture, and most fed-batches begin life with a straightforward batch phase. However, unlike batch these cultures do not operate as closed systems. At a given point during the fed-batch process one or more substrates, nutrients, and/or inducers are introduced into the bioreactor. Fed-batch cultures can be run in different ways, e.g. at a fixed volume where at a certain time point, a portion of the fermenter contents (consisting of spent medium, cells, product, and unused nutrients) is drawn off and replaced with an equal volume of fresh medium and nutrients (withdraw and fill), or at a variable volume where nothing is removed from the bioreactor during the time course of the process, with the cells and product remaining within the vessel until the end of the fermentation period, and the addition of fresh medium and nutrients having the effect of increasing the culture volume. This feeding strategy allows the organism to grow at the desired specific growth rate, minimising the production of unwanted by-products, and allowing the achievement of high cell densities and product concentrations.

The addition of the feed can be over a short or long period, starting immediately after inoculation or at a predetermined point during the run. The feeding strategy can be continuous over a long period of time or incremental, with the addition of fixed volumes at given time points (Figure 4.6). All are determined either from past fermentation data allowing the process operator to permit the same predetermined feed, or from the organism's physiology and the concentrations of key metabolites within the fermentation broth.

It was during the production of bakers' yeast biomass in the early 1900s that fed-batch culture was originally developed. Yeast producers observed that when the concentration of the malt carbon source (primarily maltose) was too high within the culture medium, the yeast cells used this for the formation of the by-product ethanol, and not biomass. However, if the concentration of malt was too low, yeast growth would be restricted and

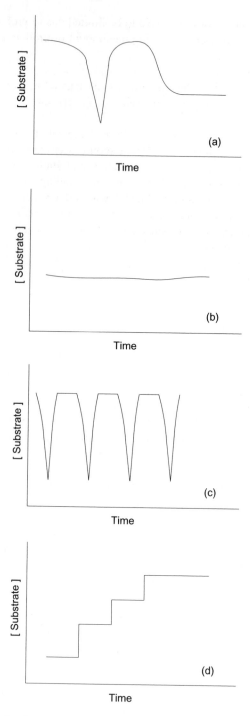

Figure 4.6 *Different feeding regimes in fed-batch processes, (a) Variable feeding regime; (b) continuous feeding regime; (c) intermittent feeding regime; (d) incremental feeding regime*

the final biomass yield would be reduced. To overcome this problem the feeding of the malt was carefully controlled during the exponential growth phase, and maintained at an optimal concentration, minimising ethanol production and maximising biomass production. Today, molasses is the main carbon source, and is added incrementally during the process to just balance the oxygen supply capability of the fermenter system. Typically, the RQ (respiratory quotient) is widely used in fed-batch yeast processes to control feeds. As well as the production of bakers' yeast, fed-batch in more recent years is routinely used for recombinant protein and antibiotic production.

Escherichia coli, like bakers' yeast, when grown aerobically with excess glucose present excretes copious amounts of acetate, as the ready availability of substrate exceeds the oxidative capacity of the bacterial cells. This can lead to culture inhibition, and slow fermentations. However, using fed-batch, the glucose feed can be kept minimal, thus allowing growth at the highest growth rate, but minimising the production of acetate. Accumulation of acetate during recombinant *E. coli* fermentations has been correlated with a reduced production of recombinant protein, demonstrating the importance of controlled feeding of glucose during the run. Other metabolic by-products that can accumulate during unrestricted aerobic batch growth with carbon in excess are lactate for *Lactococcus lactis*, and propionate for *Bacillus subtilis*. Therefore, fed-batch is superior to batch, as it allows both high cell density and a high production of product without inhibiting its production.

To summarise, fed-batch can be used to minimise or prevent flow of nutrients to waste products, and by this means to extend the productive phase of the process.

Pichia pastoris, another commercially important organism widely used in the production of recombinant proteins, also requires a careful feeding strategy. Initially the feed consists of a substrate to achieve high cell density quickly, followed by very careful feeding of methanol, the usual inducer of the recombinant gene expression. If the methanol is present in high concentrations it can be toxic for the cells and therefore its presence must be carefully controlled.

The production of the antibiotic penicillin by *Penicillium chrysogenum* is an example of the use of fed-batch to produce a secondary metabolite. The penicillin process has two stages, the first is the tightly regulated growth phase of the fungal culture, the second is the production phase or 'idiophase', when there is no net growth of the organism but the production of the antibiotic is as fast as possible and to the exclusion of other products. During stationary cell growth, phenyl acetic acid, a toxic precursor of penicillin, is fed continuously into the bioreactor. The feeding regime must be such as to maintain it at a low level, while still allowing it to be readily available for incorporation into the penicillin molecule. This is another example of the balancing act that is possible using fed-batch, namely, the carefully metered feeding of an inducer or toxic intermediate at rates that still permit active growth and synthesis to proceed. It is easy to see how difficult this kind of balance would be to achieve in batch. Figure 4.7 shows a diagrammatic representation of a simple fed-batch system.

4.3.1 Fixed and Variable Fed-batch Fermentations

Within a fixed volume fed-batch fermentation, the limiting substrate is fed without diluting the culture. The volume within the bioreactor is maintained at around the same level

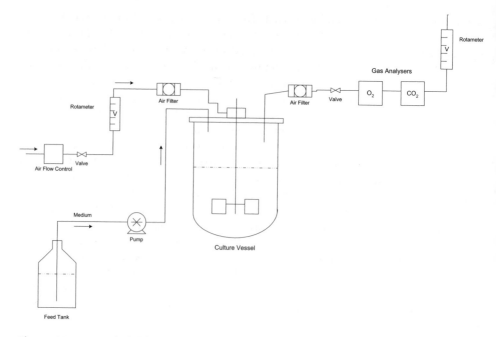

Figure 4.7 *A simple fed-batch fermentation set-up. Feeding begins at a predetermined point during the process, often at the end of the exponential growth phase. The feed consists of a super-concentrated version of the process medium, allowing control of the growth rate and carbon flux through the system*

by feeding either a super-concentrated medium containing the limiting substrate or the addition of a gas such as oxygen. Advantage can also be taken of the loss of volume via evaporation, since the volume can be kept constant by addition of an equal volume of liquid feed. A draw and re-fill strategy can also be adopted where a predetermined volume of the fermenter contents are drawn off at a selected time, and replaced with an equal volume of fresh medium. In this way, the reactor volume remains constant and the fresh nutrients have the effect of decreasing the concentrations of inhibitory species present, allowing extended processing.

A variable fed-batch fermentation is one where the addition of the feed alters the working volume in the bioreactor. The feed can be the same medium as that used in the vessel at the beginning of the process, or it can be a concentrated solution of the limiting substrate. This allows the organism to continue growing at its maximum specific growth rate resulting in a higher concentration of final biomass. All kinetic calculations of the process must take into account the continuing change in culture volume (see Chapter 7).

4.3.2   Control Techniques for Fed-batch Control

The various control techniques for fed-batch cultivation have the same ultimate aims including: to control the organism growth rate within a specified range; to control and to

Table 4.4   *Advantages and disadvantages of various feeding strategies*

Feeding strategy	Advantages	Disadvantages
Variable	By controlling the feed the substrate is converted into either biomass/substrate or both, minimising by-product formation	Controlled by feedback control, requires accurate monitoring and operator control
Continuous	Allows bioprocess to run without operator control	Must have historical process trend data Unexpected changes during the run will not be accounted for
Intermittent	Allows feed to be fully utilised before further additions, reducing catabolite repression	Controlled by feedback control, requires accurate monitoring and operator control
Incremental	Allows feed to increase with biomass formation, optimizing growth rate and product formation	Controlled by feedback control, requires accurate monitoring and operator control

direct the flux of carbon metabolism, and to reduce or eliminate the overflow of carbon to unwanted metabolic by-products.

The feed rate can be controlled by use of several control techniques (with or without feedback), each has inherent their advantages and limitations as outlined below, and an overview in Table 4.4.

No Feedback Control

Also known as an open loop strategy, this is a simple control mechanism that is not controlled on any specific parameter (Figure 4.8). The feeding regime has been predetermined by past observations or predicted data. The feed can be added continuously, incrementally, or exponentially with increasing biomass. Advantages are that this is a simple method of control to implement. During the process little or no operator intervention is required, nor removal of samples or use of *in-situ* probes, thereby minimising the risk of introducing contamination into the bioprocess. However, the major disadvantages are that any unforeseen or unexpected changes during the process cannot be accounted for within the realms of the predetermined control.

Feedback Control

This is a control mechanism that relates to the concentration of a key substrate or substrates in the fermentation culture, and can be either direct or indirect (Figure 4.9). The control strategy involves measurement of the levels of that key analyte, typically either by *in-situ* probe technology, or by grab samples with subsequent conventional analysis. Based on the outcome of that measurement, a control action may be required to keep the analyte concentration within set limits, e.g. for many cell culture systems glucose levels must be fairly closely controlled between too much, leading to lactate accumulation (and culture toxicity), and too little, leading to cell viabilities dropping.

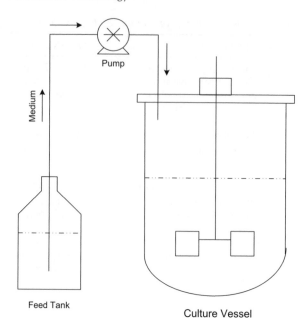

Figure 4.8 *Fed-batch without feedback control. The feeding regime has been predetermined from past data, requiring very little operator input, and reducing the chances of contaminating the process*

Direct feedback control. This is based on the concentration of the substrate within the culture medium, which is calculated by either online or off-line measurements. This can give accurate control and the information can be used to influence the feeding regimes. Online measurements are more desirable due to the reduced analysis time, limited operator assistance, and reduced risks of fermenter contamination. Advantages are that any unforeseen or unexpected changes within the bioprocess can be accounted for. However, the availability of accurate online systems is limited, and if using off-line measurements it should be noted that the time between the sample point and the result often means that real-time measurements of conditions within the bioreactor are not achievable. Examples of such measurable systems that could be or are routinely used online are calorimetry (temperature based control); capacitance or optical density probes (measures biomass), and redox probes (measures chemical reactions). More details are available in Chapter 9. Near-infrared spectroscopy can also be used online or off-line to determine the concentration of key analytes within the fermentation broth.

Indirect feedback control. This involves the measurement of observable fermentation parameters that are indirectly related to the substrate such as biomass, pH, dissolved oxygen, respiratory quotient (carbon evolution rate/oxygen uptake rate), etc. The main advantage is the speed at which the measurement can be made. The major disadvantage is that these parameters can be process or microorganism specific and must be predetermined if necessary.

The control of the feeding strategy is paramount to the bioprocess and to optimal yield of the desired product. Dual-measurements of more than one analyte can give better control over substrate concentration and help minimise the formation of by-products. The

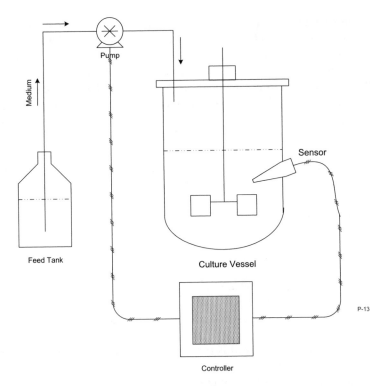

Pump

Medium

Feed Tank

Sensor

Culture Vessel

P-13

Controller

Figure 4.9 *Fed-batch with feedback control (direct). Process parameters are monitored via a sensor that is directly linked to a controller. The controller responds to the information received from the sensor by increasing or decreasing the rate at which feed is introduced into the reactor, thus controlling the growth rate of the organism*

choice of which analytes to measure is both process and system dependent, but should reflect the availability of equipment and the skill of the operator. However, both overfeeding and underfeeding the organism may be detrimental to the process, so careful control must be implemented. This is evident in the bakers' yeast and *E. coli* feeding strategy previously described, where overfeeding of the carbon source results in catabolite repression and the production of by-products that reduce cellular growth and product formation. Underfeeding reduces the final biomass that is achievable from the bioprocess, and may trigger autolytic events in some cultures.

4.3.3   Advantages of Fed-batch Control

- Controlling the concentration of the limiting substrate prevents the repressive effects of high substrate concentration and avoids catabolite repression.
- By careful feeding strategy the organism's growth rate and subsequent oxygen demand can be controlled. This was one of the original aims of developing a bakers' yeast fed-batch process: to balance the oxygen transfer rate of a given fermenter with the rate of nutrient feeding in order to minimise substrate flow to the ethanol.
- High cell density (up to ten times greater) can be achieved by use of fed-batch over batch culture. Batch culture limits the final cell growth due to no extra carbon being added

during the run, and the carbon that is present at the beginning of the bioprocess results in catabolite repression and inhibition of growth. Using fed-batch with a careful feeding strategy, *E. coli* and *Pichia* can achieve very high cell densities of over $100\,gL^{-1}$.

- Increased production of non-growth-related secondary metabolites. Many secondary metabolites are produced from intermediates and end products of primary metabolism. Others are formed after introduction of key precursors after the growth phase, as is seen in penicillin production with phenylacetic acid or phenoxyacetic acid, precursors of penicillins G and V respectively, being added prior to the stationary phase to allow the formation of penicillin.
- Reduction of broth viscosity. This is particularly important in filamentous fungal fermentations, or where the product is highly viscous, such as the polysaccharide products of *Sphingomonas elodea* – gellan gum; and *Xanthomonas campestris* – xanthan gum. The addition of fresh medium during the fermentation run, leads to the broth being 'diluted', and a brief viscosity drop, allowing better aeration and agitation within the system.

4.3.4 Disadvantages of Fed-batch Control

- Detailed knowledge of the organism's growth and product formation pattern is required, especially when no feedback loop is used.
- Deficiency of reliable online sensors for accurate substrate determination in near real time.
- Without feedback control, the feed is predetermined and therefore does not allow for any fluctuations within the bioprocess. This can potentially lead to mismatches between feed rates and culture metabolism, in turn leading substrate levels to become depleted or rise to undesirable levels.
- The process operator must be fully trained and highly skilled.

4.3.5 Ancillary Equipment

Holding Tanks

The medium reservoirs required for fed-batch processes are essentially the same as those required for batch culture. Vessels for acid and base, antifoam, inoculum and extra nutrient additions are essentially the same. At research lab scale, titrant vessels capable of holding 0.5–1 litre are usually sufficient. However if a holding vessel for feed is required, a 5, 10 or 20 litre aspirator can be used. To supply or withdraw any of these liquids to or from the bioreactor, sterile lines, usually consisting of surgical grade silicon tubing, are required. This tubing must be sterilisable and cleanable, ideally should be clear for visual inspection of the presence of blocks, air bubbles, etc., and can be connected to the bioreactor via a needle connection, which must also be sterilisable. When connecting the needle to the bioreactor, the headspace or back pressure in the vessel must be reduced slighty by reducing inflowing air rate. The needle is then removed from its sterile housing and pierced through the septum usually located in the lid of the reactor.

Pumps

When choosing a pump suitable for the aseptic pumping of liquids into the bioreactor, a number of parameters must be considered:

- flow rate range;
- ability to be self-priming;
- steady continuous flow rather than pulsating flow;
- ability to handle abrasive/viscous media and shear-sensitive cells;
- ability to run dry;
- cost.

Various types of pump are available; however, pumps such as centrifugal and piston pumps are generally not used due to their inability to pump small volumes. An example of some pump speed, duration and reservoir volumes for different bioreactors can be found in Table 4.5.

Peristaltic pumps. These consist of a rotating unit of three or six rollers, a drive motor and electrics contained within a main body. Liquid in the tubing advances along the length as the roller turns, squeezing the tubing and drawing in the liquid that is then trapped by the next roller. Liquid is expelled from the pump as the roller unit moves round (Figure 4.10).

The pump can run dry and provides a constant flow as long as the correct diameter of tubing for the roller is used. Flow rate can be varied by either the speed setting on the pump or by changing the bore diameter of the tubing.

Table 4.5 *Examples of required pump flow rates and duration for various working and reservoir volumes for three different fermenters*

Fermenter type	Initial working volume (litre)	Final working volume (litre)	Total reactor volume (litre)	Reservoir volume (litre)	Duration (hours)	Pump rate (mL min^{-1})
Glass Autoclavable	0.5	1.0	1.5	0.5	2.0	4.2
Autoclavable/SIP	5.0	10.0	15.0	5.0	10.0	8.4
SIP	10.0	20.0	25.0	10.0	25.0	6.6

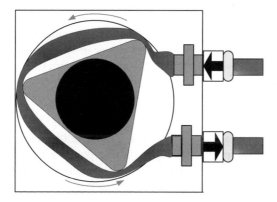

Figure 4.10 *Schematic diagram of a peristaltic pump. As the roller turns, liquid in the tubing is pushed forward by a squeezing motion. Liquid is expelled from the pump as the rollers move round*

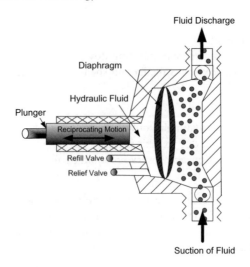

Figure 4.11   *Schematic diagram of a diaphragm pump. Fluid is sucked in via the suction inlet tube to the pump head with the pressure discharge tube being closed. The suction inlet tube then closes and the pressure discharge tube opens, forcing liquid out. The suction and pressure forces within the pump head are generated by the action of both the diaphragm plunger and the return spring*

The Watson-Marlow 520S/R is capable of giving flow rates from $0.004\,\mathrm{mL\,min^{-1}}$ (using 0.5-mm bore size) up to $2400\,\mathrm{mL\,min^{-1}}$ (using 8.0-mm bore size), and is suitable for most fed-batch operations in lab scale fermenters.

Diaphragm pumps.   These consist of a main body and a detachable heat-sterilizable head. Fluid is sucked in via the suction inlet tube to the pump head with the pressure discharge tube being closed. The suction inlet tube then closes and the pressure discharge tube opens forcing liquid out. The suction and pressure forces within the pump head are generated by the action of both the diaphragm plunger and the return spring (Figure 4.11).

The advantage of this type of pump is that it does not require frequent calibration and can maintain low flow rates, but it can stall if run against too great a head pressure. A major disadvantage of this type of pump is its requirement for clean, dry, compressed air, and its inability to run dry.

4.3.6   Feed Flow Rate

In order to maintain accurate control of feed rate, frequent flow rate checks are required. This is particularly important when using peristaltic pumps since inaccurate flow rates can be observed when the tubing in the pump head becomes worn over long periods of operation. The following calculation can be used to determine the feed flow rate for a substrate-limited fed-batch process:

$$FS_{\mathrm{o}} = \frac{\mu X}{Y} \tag{4.1}$$

Where F is the feed flow rate (Lh^{-1}), S_o is the concentration of limiting substrate in the feed (gL^{-1}), μ is the specific growth rate (h^{-1}), X is the biomass concentration (gL^{-1}) and Y is the yield of cells on the substrate (g biomass per g substrate).

In-line flow meters are useful for determining flow rates and can be constructed simply by using a 10-ml graduated pipette situated between the feed reservoir and the pump below the liquid level in the feed reservoir. The tubing to the flow meter would normally be closed off with a gate clip that can be opened to allow the pipette to fill. The gate clip is then closed off, isolating the feed reservoir, and the flow rate determined by measuring the level drop in the pipette over a given time interval.

The reservoir can also be kept on a balance that is linked to a supervisory control and data acquisition (SCADA) unit (see Chapter 10 on features of SCADA systems). The feed that is removed can be monitored by the change in weight and reported to the operator.

Graduated or calibrated medium aspirators may also be used as the feed reservoir and flow rates can be determined by measuring the level drop within the aspirator over a given time interval, although this method is generally lengthy, usually taking hours to perform.

4.4 Continuous Culture

Historically. continuous culture techniques have not been widely used in laboratory scale, but are more common in industry where these techniques are used for such processes as vinegar production, waste water treatment, ethanol production and single cell protein production. In the laboratory, these techniques have been used increasingly to study the growth and physiology of microorganisms (Chapter 12). In particular, continuous systems have been used to study proteomics, transcriptomics and flux analysis, showing increased reproducibility and accuracy of data when compared with similar studies in batch cultures. Continuous cultivation is a method of prolonging the exponential phase of an organism in batch culture, whilst maintaining an environment that has less fluctuation in nutrients, cell number or biomass. This is known as *steady state*. The organisms are fed with fresh nutrients, and spent medium and cells are removed from the system at the same rate. This ensures that several factors remain constant throughout the fermentation, such as, culture volume, biomass or cell number, product and substrate concentrations, as well as the physical parameters of the system such as pH, temperature and dissolved oxygen.

4.4.1 Control Techniques for Continuous Culture

Several control techniques can be used with continuous culture. The most commonly used continuous culture technique, the chemostat, operates on the basis of growth being restricted by the availability of a limiting substrate, whilst the turbidostat is operated under no limitations. Other continuous cultivation techniques include the auxostat, and will be discussed briefly here.

The Chemostat

Figure 4.12 shows a schematic representation of a typical chemostat. The feed medium contains an excess of all but one of the nutrients required for growth of the culture. The

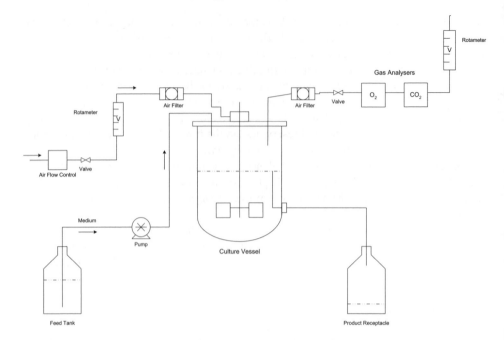

Figure 4.12 *A continuous fermentation set-up that can easily be used in the laboratory. Feed is pumped into the reactor, increasing the volume within the reactor. At the same time, the overflow weir allows the excess volume to flow into the product tank. This ensures that the volume within the reactor remains constant*

supply of the nutrient that is not in excess therefore determines growth rate of the microorganism.

In steady state, the flow of medium into the bioreactor equals the flow of spent medium and cells out. In fact, if there are no cells in the system at all, this is still the case. There is therefore a mass balance across the system. The biomass, normally calculated as dry cell weight and denoted as gL^{-1}, remains constant because the addition of fresh medium allows the formation of new biomass, and this is balanced by the loss of biomass in the spent medium outflow.

When running a chemostat culture, it is imperative that the condition of steady state is calculated and monitored; this means that it is necessary to calculate dilution rate, specific growth rate, yield of product on substrate, etc., and a number of equations have been derived in order to do this; see Chapter 7 for more detailed descriptions. Here we simply deal with the most basic of these relationships that anyone seeking to run chemostats should have an awareness of.

The dilution rate (D) describes the relationship between the flow of medium into the bioreactor (F) and culture volume within the bioreactor (V):

$$D = \frac{F}{V}$$

(4.2)

Flow rate is expressed in Lh^{-1}, and volume is expressed in L, therefore dilution rate is expressed in units per hour (h^{-1}). Residence time (t) is the inverse of dilution rate and is also related to the reactor volume and flow rate:

$$t = \frac{V}{F} \tag{4.3}$$

Residence time is measured in hours. In general, the continuous culture must go through four or five residence times before it can be considered to be in a steady state. When a reactor is perfectly mixed, all nutrients and cells in the bioreactor are equally distributed throughout the vessel. A small scale stirred tank bioreactor with moderate concentrations of cells present is assumed to be perfectly mixed and this assumption is used in the derivation of the continuous culture equations. Therefore the following statements are assumed to be correct throughout a chemostat when at steady state:

$$\frac{dX}{dt} = 0 \tag{4.4}$$

and

$$\frac{dS}{dt} = 0 \tag{4.5}$$

i.e. change in biomass (X) over time (t) is zero, and change in substrate concentration (S) over time (t) is zero, that is, no net accumulation of cell mass or substrate.

Some systems never actually reach a real steady state, but a pseudo-steady state. This is typically where the concentrations of key analytes oscillate around a single value.

The specific growth rate of a culture at steady state is set by the dilution rate (i.e., $\mu = D$), which is in turn determined by the rate of flow of nutrient solution to the culture. However, when the dilution rate is greater than the maximum specific growth rate (μ_{max}) of the microorganism, the result is *washout*. This can be calculated using the biomass equation for the continuous reactor:

$$\frac{dX}{dt} = \mu X - DX \tag{4.6}$$

So when DX is greater than μX, dX/dt becomes negative, i.e. the dilution rate, or the rate at which fresh medium is added to and spent medium is removed from the system, exceeds the maximum specific growth rate of the organism, resulting in a decrease in the number of cells in the bioreactor over time. The number of cells in the bioreactor will therefore eventually become zero (washout).

Most chemostat cultures become progressively more unstable as the dilution rate approaches the critical dilution rate (above which washout occurs). Hence, physiological and other studies involving this type of continuous culture system should not be operated near this region, as the results may be highly variable. Likewise, production systems using chemostats rarely run near the critical dilution rate because of instability problems and the risk of inadvertent washout occurring. This obviously limits biomass productivity since it imposes a ceiling on the rate of culture growth.

The Turbidostat

In a turbidostat, the feed medium contains all of the required nutrients in excess. Growth is therefore not substrate limited as it is in the chemostat, and the microorganism can grow at its maximum specific growth rate (μ_{max}). The system can be controlled at a desired cell density by monitoring the turbidity, and therefore the biomass, continuously. This can be achieved by measuring optical density using a spectrophotometer. If the monitor detects a deviation from the cell density set-point value, a signal is relayed to the controller, and so the rate at which the feed medium is added to the bioreactor can be adjusted. When the turbidity increases above a set point, the feed rate is increased in order to dilute the culture and bring the turbidity back to its set point. If the density of the reactor population falls, the feed rate is decreased, allowing the population to grow until the turbidity set point is reached, thus avoiding washout of the organism.

The turbidostat is commonly used for the selection of antibiotic resistant mutants and the degradation of toxic wastes, where nutrient limitation is not desirable. It is also routinely used to avoid the washout effects that are more common in chemostat systems, and to produce cells of approximately uniform morphology and composition over prolonged periods. However, this system has two major disadvantages: first, fouling of the optical surfaces of the probe used to measure turbidity caused by unwanted cell growth, and second, gas bubbles trapped in the circulating medium result in inaccurate optical density measurements and therefore the system is often difficult to control.

The Auxostat

Auxostats use feed rate to control a state variable, such as pH or dissolved oxygen, in continuous culture. The microorganisms themselves establish their own dilution rate, and auxostats tend to be much more stable than chemostats at higher dilution rates, especially at dilution rates approaching the maximum specific growth rate of the microorganism. The high dilution rates exert a selection pressure upon the microbial population, which leads to more rapidly growing cultures. This method of continuous culture control is therefore ideal for such applications as high-rate propagation and detoxification of waste materials at maximum rate concentrations. A typical auxostat system is represented in Figure 4.13.

A simple feedback control based on nutrient concentration is used, and there is a wide variety of feedback variables that can be used. The pH-auxostat is easily set up using commercially available bioreactors, making it accessible to most researchers. The pH-auxostat couples the addition of fresh medium to pH control. As the pH drifts from a desired set point, fresh medium is added to bring the pH back to the desired set point. The rate of medium addition is not directly determined by the pH set point, but rather by the buffering capacity and the feed concentration of the limiting nutrient. For example, cultures of filamentous fungi have been grown in pH-auxostats where it was shown that biomass production could be controlled by the addition of alkali.

Other feedback variables that have been used to develop auxostat systems include dissolved oxygen (DO), specific gravity, nutrients such as glucose and ammonium ion, inhibitory substrates, carbon dioxide evolution rate (CER) and oxygen uptake rate (OUR). For further reading on these systems, see Gostomski *et al.* (1994).

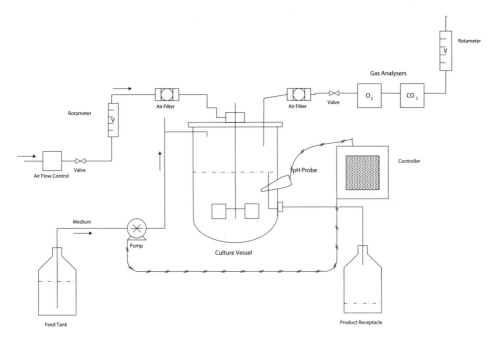

Figure 4.13 *Typical layout of an auxostat using pH control coupled to the addition of fresh medium*

4.4.2 Advantages of Continuous Culture

- Productivity and growth rate can be optimised by changing the feed flow rate during production;
- Longer periods of productivity with less down time. In theory, a continuous process can be operated indefinitely; however, because long periods of operation can result in mechanical failure, the process must be stopped occasionally to allow for system maintenance;
- It can take advantage of cell immobilization, which allows the maintenance of high cell concentrations in the bioreactor at low substrate concentrations. In these systems, cells are perfused via a membrane with a steady and continuous flow of fresh medium. The cells are supplied with oxygen and nutrients, while waste and desired products are continuously removed. The cells are retained by means of the membrane barrier and are recycled into the bioreactor (Figure 4.14) rather similar effects are achievable using acoustic perfusion systems (see Chapter 2);
- The effects of environmental or physical factors are more easily analysed in a continuous system, where any changes in the constant steady state are observed and can be attributed solely to the change in those factors.
- Evolution in these cultures can be readily studied. Basically, continuous culture of a given strain allows us to 'direct the evolution' of the strain. As the cells reproduce in an invariant environment, the operator can select a particular evolutionary pressure,

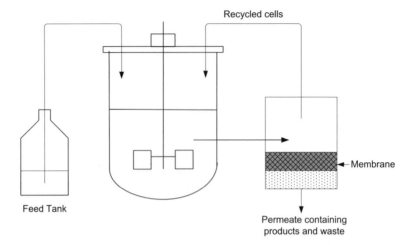

Figure 4.14   *A continuous cell recycle bioreactor, consisting of a stirred tank bioreactor and a membrane separation unit. Fresh nutrients are added to the system, whilst products and waste are removed via a membrane system that retains the cells and the cells are recycled into the reactor*

e.g. temperature, nutrient limitation or pH, and study how the culture evolves in response to that selective pressure in terms of physiological, metabolic or genetic change.

4.4.3   Disadvantages of Continuous Culture

- The US Food and Drug Administration (FDA) does not accept continuous culture in the production of therapeutic products as a Current Good Manufacturing Practice (cGMP). This is because they require the manufacturer to segregate such products into batches for traceability purposes, precluding continuous culture as a means of production.
- Not all products are produced optimally in continuous processes, e.g. some fermented foods and beverages require cellular products released from different phases of batch culture growth for full flavour development. Nongrowth-associated products such as antibiotics, monoclonal antibodies and toxins are also not produced well in continuous culture. This is because antibiotics and toxins can exert a selective pressure in the chemostat, and because there are risks of plasmid loss in genetically modified cultures over a number of generations.
- Contamination can be a major problem in continuous cultivation, and can result in the wash out of the desired organism and therefore a loss of product.
- Culture mutation can easily occur in continuous processes, often resulting in the microorganism 'shedding' genes required for product formation, or in the case of genetically engineered organisms, a 'back mutation' can occur, causing the slower-growing recombinant population to wash out of the vessel. For example, *Clostridium acetobutylicum* loses its ability to make acetone butanol in contenous culture and instead makes acetate and butyrate.

4.4.4 Additional Accessories and Ancillary Equipment

Holding Tanks for Feed Medium and Product

The amount of feed medium required depends upon the volume of the bioreactor. When using a reactor working volume of 2 L, glass aspirator bottles of 10 or 20 litre capacities should suffice. Table 4.6 can be used as a guideline for volumes of feed required at varying dilution rates for a 2-L working volume. Glass and Nalgene aspirator bottles can be obtained from commercial laboratory suppliers. The Nalgene aspirators are usually used for waste collection, and unlike the holding tanks for feed medium, do not need to be vented via a hydrophobic vent filter when using cotton wool bungs. It is good practice to prepare several sterile feed tanks containing the feed medium and several sterile waste/product tanks before initiating the continuous process.

Feed Pumps

Two separate pumps may be required for continuous culture, one for introducing fresh nutrients to the system, and one to remove product and waste for the system if using a pumping out method of level control. The type of pump required depends on the size of the vessel, the flow rate, the nature of the feed (viscous or corrosive), and the speed at which the feed should be delivered. The most commonly used pumps for continuous culture are peristaltic, as mentioned in Section 4.3.5. To set up the continuous culture and operate the overflow into the product receptacle, a pump capable of producing a faster flow rate than the feed pump is necessary. Bioreactor vessels that come supplied with peristaltic pump modules are often not capable of producing the speeds necessary in the outflow pump. This can cause difficulties with level control and therefore with the maintenance of a steady state.

Level Control Mechanisms

Controlling the culture volume is essential to obtain and maintain a steady state within the bioreactor. Constant agitation and aeration of the culture medium leads to false level detection, which can have detrimental effects on the steady state of the culture. The choice of level control mechanism depends on several factors:

Table 4.6 *The feed requirements of a laboratory chemostat with a 2 l working volume*

Dilution rate (h^{-1})	Feed rate ($mL min^{-1}$)	Litres/day	Residence time (t) (h)
0.025	0.84	1.2	40.0
0.050	1.67	2.4	20.0
0.075	2.50	3.6	13.3
0.100	3.33	4.8	10.0
0.250	8.33	12.0	4.0
0.500	16.67	24.0	2.0
0.750	25.00	36.0	1.3
1.000	33.33	48.0	1.0

- microorganism (e.g., bacteria, fungi, yeast);
- anaerobic/aerobic fermentation;
- medium constituents (e.g., solubility and miscibility can be important factors);
- culture volume and size of bioreactor;
- degree of accuracy of the chosen method (5% variance is generally acceptable for most microbial continuous cultures);
- cost;
- degree of operator skill.

There are five main types of level control device, as given below.

Overflow weirs.   These are extremely cost effective and are simple and easy to install. They are sterilised along with the bioreactor and can be adjusted to the correct height within the bioreactor to correspond with the culture volume required (Figure 4.15 a). Foaming can result in a disproportionate number of cells being removed from the system, but can be overcome by adjusting the take-off point. This method of level control is not suitable for filamentous fungi as the mycelium tends to aggregate around the top of the weir, resulting in blockage or a straining effect.

Pumping-out mechanisms.   These regulate the culture volume to the depth at which the harvest tube is placed (Figure 4.15 b). This method is particularly useful for small bioreactor sizes, and reactors that have no base plate, and is more expensive to achieve than with overflow weirs since two pumps are required. Level control in this way can be difficult to achieve in vigorously agitated fermenters since the rate at which the spent culture is pumped out must equal the rate at which the feed is pumped in.

Manometric systems.   These were designed specifically for use with filamentous fungal cultures. They consist of a level monitor, an air-actuated solenoid, a pinch valve, and a pressure transducer with a flexible membrane. Feed flowing into the bioreactor causes an

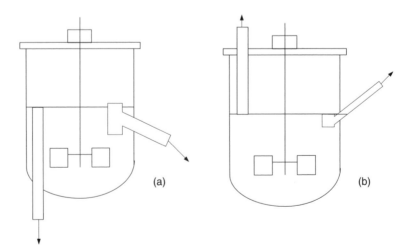

Figure 4.15   *Examples of (a), overflow weirs and (b), pumping out mechanisms for level control in continuous culture vessels*

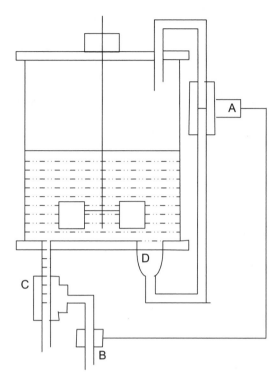

Figure 4.16 *Level control mechanisms for continuous culture, an example of a manometric system. A, Level monitor; B, air-actuated solenoid; C, pinch valve; D, pressure transducer*

increase in biomass that displaces the flexible membrane of the pressure transducer (Figure 4.16). This in turn causes an increase in fluid level that is detected by the level monitor, triggering an air-operated pinch valve that allows the culture to flow out until the original volume within the bioreactor is achieved. The flexible membrane and tubing required for operation of this system are subject to wear and have to be replaced frequently, meaning that this can be a costly method of level control.

Level probes. These can be used in conjunction with manometric and pumping-out mechanisms, or a solenoid operated drain valve (Figure 4.17). They are available as either conductivity or capacitance probes, are expensive and are not suitable for volumes of less than 5 L. Level probes are subject to false level detection when foaming occurs within the bioreactor; however, they have been reliably used for filamentous fungi.

Load cells. These are essentially an elastic body, such as a steel cylinder, which distorts when compressed. The distortion is measured by electrical resistance strain gauges, and the changes of resistance in response to the amount of distortion are proportional to the load on top of the body or cylinder. When this load (usually measured in kg) exceeds a set point value, a signal is sent either to open a pinch valve or start a pump, and discharge a given volume of culture from the bioreactor. This method is not affected by foaming; however, it is not very accurate for small culture volumes (less than 30 L) and can be costly.

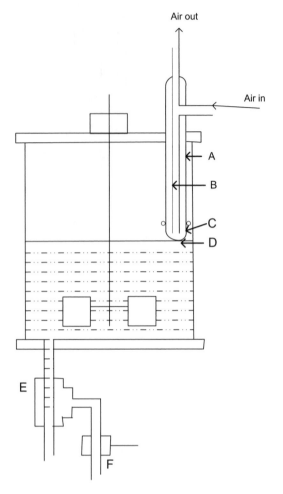

Figure 4.17 *Level control systems for continuous culture, an example of a level probe assembly. A, Outer case of probe; B, inner tube of probe; C, membrane; D, disc; E, pinch valve; F, solenoid valve*

Whichever level control system is used, thought needs to be given to the physical characteristics of the culture in designing the culture exit line. Cultures of filamentous organisms, fungi and Streptomycete bacteria, will typically tend to clump or aggregate, and may be difficult to get out of the fermenter if the exit line is of narrow bore. For such systems, the exit line should be 12-mm internal diameter where possible. Diameters of less than 5 mm will lead to potential blockages of the exit line by culture clumps.

4.4.5   Running a Continuous Process

The reactor is set up to run a batch culture at a volume at least 20% greater than the volume at which the continuous process is to be run. This allows the 'excess' volume to be drawn off to a specified level before the continuous process begins. The batch culture medium should have a nutrient concentration around half that of the feed medium to reduce any detrimental effects seen in the transition phase between the batch and continuous processes. The transition phase is a period of adaptation for the organism, similar to the lag phase that is sometimes seen in batch culture. Feeding begins when the organism enters the exponential growth phase. The dilution rate selected should not exceed the maximum specific growth rate of the organism. This value can be obtained from literature or can be estimated from data obtained from previous batch culture experiments with the organism.

Once the organism has adapted to the new growing conditions, a steady state is reached. Typically, steady state is reached when the organism has undergone between three and six residence times (*t*). When changing a process parameter (such as dilution rate, pH or temperature) during continuous cultivation, at least three residence times must be allowed to elapse for the culture to adapt to the new conditions. Frequent measurement of process parameters and culture analytes during the continuous process is advised for monitoring the steady state. During steady state, two or three sample measurements per day will suffice. If, however, the researcher is particularly interested in the functional genomics aspect of continuous cultivation, more frequent measurements can be taken.

Dynamic Phase Studies

Most of the preceding discussion assumes the adoption of steady state cultures since these give the clearest link between cause and effect, e.g. a single change in culture pH can be unequivocally linked to subsequent changes in culture physiology. This is not possible in batch.

However, chemostats offer the possibility of using the steady state as a defined reference point, in effect a defined physiological state, and then designing experiments that examine the culture response to the deliberate planned disturbance of that state during the dynamic phase (the culture phase between steady states). The power of such techniques to generate high quality data is generally not fully recognised or exploited sufficiently. This is an area of research that clearly needs much more attention than at present.

References and Further Reading

Cinar, A., Parulekar, S. J., Undey, C., Birol, G. (2003) *Batch Fermentation: Modelling, Monitoring and Control.* Chemical Industries Series, Vol. 93. Marcel Dekker, Inc., USA.

Gostomski, P., Muhlemann, M., Lin, Y-H., Mormino, R., Bungay, H. (1994) Auxostats for continuous culture research. *J. Biotechnol.* **37:** 167–177.

Hoskisson, P. A., Hobbs, G. (2005) Continuous culture – making a comeback? *Microbiology,* **151:** 3153–3159.

Simpson, D. R., Wiebe, M. G., Robson, G. D., Trinci, A. P. J. (1995) Use of pH auxostats to grow filamentous fungi in continuous flow culture at maximum specific growth rate. *FEMS Microbiol Letts.,* **126:** 151–158.

Tempest, D. W., Wouters, J. T. M. (1981) Properties and performance of microorganisms in chemostat culture. *Enz. Microb. Technol.,* **3:** 283–290.

Waites, M. J., Morgan, N. L., Rockey, J. S., Higton, G. (2001) In *Industrial Microbiology: An Introduction.* Blackwell Science, Oxford, p. 103.

5

The Design and Preparation of Media for Bioprocesses

Linda M. Harvey and Brian McNeil

5.1   Introduction

It is worth keeping in mind before we begin to discuss how we grow selected strains of cells in bioreactors, and get them to make the products we want, that most microorganisms are not culturable by standard techniques. From molecular genetic evidence we know that there is an immense diversity of these nonculturable cells, so those we culture in the laboratory represent a tiny subsection of the population, adapted to rapid growth, usually in nutrient rich media, at temperatures around 37 °C (+/− 20 °). In other words, fermentation technology focuses on the exceptions. The normal means of operating a fermenter/ bioreactor involves inoculation with a pure culture, and the maintenance of this monoseptic condition throughout the process. Again, this is not representative of the natural condition of most microbiota.

Microorganisms, plant, insect and animal cells, when cultivated in a laboratory or industrial environment, grow on nutrient solutions containing all they require in order to grow and produce any primary or secondary products the cell has been designed to generate. Such nutrient solutions are called media (singular: medium).

It is imperative that the correct medium is used, for both laboratory investigations and industrial bioprocessing activities, if one is to make the best of the expensive, often laborious, investment that has been made in developing the strain of microorganism or cell line being utilized. The choice of correct medium can make or break a research programme or an industrial process.

Practical Fermentation Technology Edited by Brian McNeil and Linda M. Harvey
© 2008 John Wiley & Sons, Ltd

Medium design should be a relatively simple process, but is often complicated by factors such as cost, availability of substrates, reliability of substrate supply, handling, storage, ease of preparation and storage, transportation of components, not to mention health and safety considerations, which are becoming progressively more important in today's climate. It is often easy to forget that the primary consideration is really quite simple: to supply the microorganism or cell line with a source of necessary nutrients in a *readily utilisable* form. In an ideal world all medium components should readily dissolve in water in order to be available to the cell. Many media, especially at industrial scale, depart from this ideal, and contain suspended solids and oil phases that add to the complexity of sterilisation and fermenter operation.

5.2   Where Do We Start?

This chapter aims to answer the above question and help the reader to decide how best to approach their own particular medium design problem. The obvious start point is to consider what has been published in the scientific press and, if industrially based, to consider the historical records available. In the latter case, this information tends to be very patchy, particularly with the loss of older staff who know processes well and who have had years of experience with particular cell lines, and with replacement of computers holding archived information.

Once media have been identified in the literature it is important that a critical analysis is carried out on the different media identified. It is worth remembering that just because material has been published it doesn't necessarily follow that it is correct. In any literature search for a suitable medium for culturing cells, one will inevitably find a medium that is used simply because it is what everyone else uses (L- broth for cultivation of *Escherichia coli* comes to mind here – used everywhere, but for no logical reason), the excessively formulated medium containing many more substrates than is absolutely necessary, the medium that has been used with a complete lack of thought and the, frankly, completely outrageous (the authors have come across papers indicating that it is essential for a specific type of baby food to be included – balanced it may be, but necessary for cell culture – undoubtedly not!). It is much rarer to come across the perfect medium for cultivating specific cell lines – there are many that are highly suitable but the reader requires to be able to identify them. Consequently a range of questions needs to be asked, for example:

(i)   What is the elemental composition of the cell line to be used?
(ii)   What energy source does the cell line require?
(iii) What is the purpose of the bioprocess under consideration? Biomass/product/ both?
(iv) Will the formulation of the medium impact on the product? For instance, pH can have a significant effect on products of a proteinaceous nature and substrates that can affect pH must be considered carefully, even where pH control is utilised, as localised pH hotspots can potentially denature the protein even at short exposure times.

(v)   Will the process be batch, fed-batch or continuous (see Chapter 4), and will different media be required at different stages of the process?

(vi) Will the type of medium chosen impact on any other stages in the process, e.g. downstream processing activities such as filtration, centrifugation, purification, etc.?

(vii) Does the laboratory have the correct sterilisation equipment for, say, heat labile substrates or substrates that are particularly viscous or heavy with particulate matter?

This is not an exhaustive list. The use of different types of cell line may mean that a different set of questions needs to be posed. So before commencing the search, think about what the requirements are, and what is to be achieved from the search. Draw up a list of salient points and questions and then start the search, source the information and carry out the critical evaluation. This can be a very time consuming activity but it will be time well spent as it will save expensive and time consuming mistakes at a later stage.

Many scientists have applied the technique of response surface methodology in order to optimise medium. Although undoubtedly a useful technique, especially in the very earliest stages of process definition, the same improvements in yields can be obtained simply by applying sound logic to the answers to the above questions. Select the correct carbon and nitrogen sources, add the required macro- and micronutrients and growth factors, and the yield of cells should be very good. Applying Plackett–Burman design to identify significant factors (i.e., identify key nutrients – which can be carried out by finding out the elemental composition of the cell line chosen), followed by central composite design (CCD) to determine the optimal level of medium constituents and their interaction is unnecessary if the physiology of the organism is understood. A simple calculation of how much biomass is wanted and how much product is required, and applying simple microbial kinetics (Chapter 7) can determine the optimum levels of the substrates added. The key point here is that ideally, though, understanding of cell physiology, and quantitative approaches to the latter, are probably all that is needed to formulate a suitable medium for most fermentations. In the very early stages of process development, little may be known in detail about culture physiology, and about interactions between nutrients, but this state of affairs should not be allowed to continue. Nowadays, it is essential to understand the process physiology, and this is equally applicable in industry, where you may wish to demonstrate the depth of this understanding to regulators in order to assure them of product reproducibility and quality, or in the research laboratory itself, where lack of process understanding can backfire in terms of experimental variability or variability in the protein being made by fermentation. So, RSM has its place, but should not be a substitute for process physiology knowledge and thinking!

5.3   Media Types

Media can be solid, semi-solid or liquid. For the purposes of this chapter solid and semi-solid media will not be considered, only liquid media used in bioprocessing will be covered. Most liquid media can, of course, readily be converted to solid media by addition of a suitable gelling agent, such as agar or gellan. Just make sure the gelling agent is fully dissolved and dispersed before attempting to sterilise the medium.

5.3.1   Microbiological Culture Media

Media used in bioprocessing can be broadly divided into three categories based on their composition: synthetic, semi-synthetic and complex.

Synthetic Media

Synthetic media are fully chemically defined. All the components are known, as is the specific concentration of each of the components. Such media are usually quite simple, containing a carbon/energy source, a nitrogen source, and a range of salts (usually inorganic); however, if the cell line chosen is particularly fastidious and/or has specific growth requirements, the complexity of the synthetic medium may increase significantly, e.g. in the drive to replace foetal calf serum for cultivation of animal cells the synthetic media derived have between 70 and 80 different components, depending on the cell type grown.

Synthetic media are useful in research and laboratory situations where experimental accuracy is paramount and data interpretation needs to be clear. In general, such media are much more costly than other media types, especially if specific components such as vitamins and growth factors are required, as these ingredients tend to be very expensive when supplied in the pure form. Yields of cells tend to be much lower than those obtained when the same cell line is grown in either semi-synthetic or complex media, as the synthetic media often do not match an organism's or cell line's exact requirements. It is, of course, easier to investigate the impact of single nutrients upon cell growth and physiology using these media than the more complex ones described below. Likewise, since medium composition can be defined and delivered in a reproducible fashion, in principle, the fermentations carried out using such formulations should exhibit less variability (see Chapter 11 on the sources and consequences of process variations).

Such media are generally fully soluble and provide the fewest downstream processing challenges in an industry context or purification/recovery challenges in a small laboratory context (see Chapter 3).

Semi-synthetic Media

Semi-synthetic media are largely chemically defined, as above, but include one or more poorly specified component(s) of variable but controlled composition, for example, yeast extract, which is particularly useful if a cell line requires a range of B vitamins. Such media are useful in research and laboratory situations where a particular organism has a requirement for a substrate that would be too expensive to supply in the pure form on a routine basis. Such components are also valuable where the specific nutritional needs are not well known, and a rich source of many nutrients such as yeast extract is used in a 'shotgun' approach to provision of minor nutrients.

Plant, animal, fish and microbial extracts have routinely been used in the past to supply vitamins and essential growth factors for specific organisms. The current trend in medium design is to avoid extracts of animal origin, especially bovine sources, as there are potential health risks associated with these products. There is thus a void in the market that is waiting to be filled by a substrate that is cheap, reliable and contains a range of vitamins and growth factors.

It is wise to remember that, when using media of this nature, batch-to-batch variability of the poorly specified component is a very real possibility. Where accuracy is paramount it may be necessary to analyse the substrate to define the specific component that the material is supplying. This obviously has a cost implication, as routine assays have both consumable and labour costs associated and, perhaps more importantly, time needs to be factored in to carry out these assays. Such control of the quality of raw materials is standard practice at industrial scale, but has time and equipment resource implications. Given that variation between batches of these raw materials exists, and can affect the fermentation progress, how can the impact of this be minimised in a small laboratory without the time and resources to monitor 'quality' of raw materials? Simple batch mixing is one route to keep this to a minimum.

Use of semi-synthetic media can also impact on the routine analyses required in monitoring the process. For example, using a substrate such as yeast extract to supply vitamins to the cells would mean that a routine, relatively simple, assay for quantifying the nitrogen in the medium (e.g., NH_3 content using an ammonia electrode) becomes a lot more complex as the extract will also contain complexed nitrogen sources. The samples would therefore require additional assays to be carried out (in this case a Kjeldahl digestion assay) to determine the total nitrogen available to the cells. Yeast extract also adds some fermentable carbohydrates to the medium, again complicating the clear identification of C substrate usage in these media.

Critical evaluation of the medium to be used should therefore balance the gain achieved in using undefined components against the disadvantages associated with the more complex analyses. It may, after all, be more cost effective to use the expensive pure compound!

Complex Media

Complex media are largely composed of substances that are usually of plant or animal origin, and that have ill defined and variable composition. These materials vary from batch to batch, and composition is influenced by time of year, location of origin, and small changes in production methods. Seasonal availability may change appreciably. They are usually relatively cheap, can be supplied in bulk and the source should be reliable. Substrates such as beet and cane molasses, corn steep liquor, soya bean meals and extracts, whey powders, hydrolysed starches and a range of many different oils all fall into this category. Complex media are used for many biotechnological processes, especially those that produce high volume, low value products, e.g. single cell protein (bakers' yeast, Quorn). These complex substrates will normally meet more than one requirement of the cell, perhaps supplying carbon and nitrogen, e.g. rapeseed (canola) oil will supply both carbon and nitrogen to the process, and many of the other complex substrates will supply vitamins, growth factors, and many elemental needs of the cell in addition to being a source of carbon. Many industrial antibiotic processes still employ varying amounts of oils such as rapeseed, as in addition to the nutrients added, the oil may specifically enhance antibiotic synthesis, and also help to suppress foam formation. Foam formation

Table 5.1   *Typical sugar content of molasses*

Type of molasses	Sucrose (% wv)	Invert sugar (% wv)
Beet	50	1
Cane	35	20

tends to be far worse in complex and semi-synthetic media than in fully synthetic, due to the presence of amino acids, and proteinaceous material in the former. This greater need for antifoam must be anticipated with these medium types.

At research level it is rare to use such substrates, unless an industrial process is to be mimicked exactly in the research laboratory (see Chapter 8 on scale up and scale down), as each one may contain a wide range of impurities and the composition of the substrate can be highly variable depending on what part of the world the substrate is grown in, soil type, climate, etc. If the use of such materials is unavoidable at small scale, it implies high in-process variability, and even with careful planning of experimentation (e.g., using design of experiment software packages), you should expect that more replicate fermentations will be required to obtain clear results.

For example, the composition of molasses varies according to whether the molasses is obtained from beet or cane sources. Values of typical sugar content are listed in Table 5.1, however, these values are subject to considerable variation between samples from different batches and should not be taken as absolute limits. Routine use of such a substrate implies routine analyses of every batch used. Molasses may also contain a high concentration of iron, as well as other elements, and iron can be toxic to many cells. The substrate may therefore be subject to ion-exchange chromatography before it can be used, and thus incurs capital and labour costs. Again the critical evaluation needs to balance the role and advantages of the complex medium against the disadvantages.

When dealing with complex media it is useful to consider what is scheduled to happen downstream of the bioreactor, as the physical characteristics of the substrate can impact greatly on factors such as filtration, centrifugation and purification stages.

5.4   Medium Components

The starting point, when considering which medium components to supply to a cell, is a consideration of the elemental balance of the cell. This may not necessarily be readily available in the literature, often being considered too routine to be worthy of publication. A well washed, dried sample of the cells can be analysed by any well found chemistry laboratory. Cells grown for this purpose should not be grown on a minimal synthetic medium as the cells may not grow to their full potential on this type of medium. Instead cells should be grown on a complex medium with the *potential* to supply most of the cells' requirements. The cells must be thoroughly washed in an osmotically favourable solution to ensure that any residual sugars, etc., are removed from the cell surface and dried under vacuum before sending for analysis. Table 5.2 shows the typical average composition of bacteria yeast and fungi.

Table 5.2   Average composition of bacteria yeast and fungi

Component	Bacteria (%)	Yeast (%)	Fungi (%)
Carbon	48	48	48
Nitrogen	12.5	7.5	6

Table 5.3   Average macromolecular composition of bacteria, yeasts and fungi

Macromolecule	Bacteria (%)	Yeast (%)	Fungi (%)
Protein	55	40	32
Carbohydrate	9	38	49
Lipid	7	8	8
Nucleic acid	23	8	5
Ash	6	6	6

It can be seen that while the carbon content of the different groups is very similar, the nitrogen content varies considerably. This is a reflection of the way in which the different organisms grow. Fungi, having large areas of the hyphae that are empty and no longer involved in biochemical activities, and thus contain no nucleic acids or proteins, have considerably less nitrogen than yeasts or bacteria. Keep in mind also that different stages of culture growth, and different specific growth rates may produce a cell composition that varies significantly. A good example is seen in chemostat (see Chapter 4) cultures of the yeast (*Saccharomyces cerevisiae*) where cells grown at low dilution rates may typically have 50% more protein content than those produced at high dilution rates.

Table 5.3 shows the composition of the three groups of organism based on macromolecular composition. Again the differences observed reflect the differences in growth of the different groups of organism.

There are obviously variations in the composition of specific organisms but the above can be a useful guide point from which to start. The major component of any cell is, of course, water.

5.4.1   Carbon Source

A carbon source is necessary to provide the cell with energy as well as the material with which to grow and synthesise arange of primary and secondary metabolites. The best energy source depends on the type of organism utilised, e.g. autotroph, chemotroph, etc.

There is obviously a wide range of carbon sources and the one chosen should be appropriate to the organism but also to the economics of the process. At research scale the latter tends to be less important, but it should be borne in mind if the objective of the programme is to develop an industrially relevant process. This section will look at several carbon sources and indicate the advantages and disadvantages of each. A typical microorganism is approximately 50% carbon (Table 5.2), making carbon the most significant substrate and care should be taken to ensure that the carbon concentration does not become limiting. The yield of biomass on carbon is approximately 0.5 (Chapter 7), which means that if a biomass concentration of $50\,\text{gL}^{-1}$ is required, $100\,\text{gL}^{-1}$ of the carbon source must be supplied, albeit not necessarily all at once if fed-batch culture is being utilised.

The great majority of laboratory and industrial fermentations tend to use a very limited range of easily utilisable substrates to supply the energy and C requirements of cultures. This does not imply that other C and energy sources could not be used, just that this limited range is usually available and methods for preparing and analysing the consumption rates of such substrates are well understood. The production of pharmaceuticals, especially biopharmaceuticals by fermentation, takes place in the context of an industry that is, with the possible exception of the nuclear industry, perhaps the most regulated in the world. This fact accounts for a degree of 'conservatism' in terms of nutrient supply.

Glucose

Glucose is universally acceptable for growth of most cell lines, be they animal, plant or microbial. Supplied as a powder in the pure form, the substrate is readily available, reliable, easily stored, easily handled and has no significant implications for health and safety. These qualities make glucose a popular choice of carbon source. There some drawbacks to using glucose, notably:

(i) the possibility of the organism suffering from the 'Crabtree effect' if glucose is over supplied in the initial stages of growth;
(ii) the loss of available substrate to the Maillard reaction, which can occur if glucose is sterilised with a nitrogen source present.

Both of these situations can be overcome readily, by careful monitoring and control of glucose feed in the first case and by either sterilising the substrates separately or by using filtration sterilisation methods in the second.

A popular method of supplying glucose on a larger scale is to use glucose syrups, manufactured by the hydrolysis of starch (a substrate which itself tends to be insoluble and consequently unavailable to the cells, difficult to handle, but is readily hydrolysed to glucose). The drawbacks remain the same, with the additional complications of transportation, storage and handling of a liquid. Formulation of the medium and delivery to the bioreactor may also be considerations that have to be taken into account. The main advantage to using glucose syrups is cost, as it is significantly cheaper than powdered glucose.

Sucrose

Often the sugar of choice in research laboratories, sucrose, is utilised by many but *not all* cell lines. Commercially available sucrose comes in many forms and grades, from pure granulated forms to complex molasses solutions (above). Sucrose does not tend to suffer from the same drawbacks as glucose and, although it can be subject to caramelisation when over sterilised, generally it can be autoclaved/sterilised with nitrogen compounds without the same problems as glucose. If using molasses, sufficient cold storage must be available to handle the needs of the laboratory as the range of different substrate available means that the molasses can be readily contaminated, and although sterilisation can destroy any contaminating organisms, the amount of available sucrose to the organism of choice will decrease. Again mixing of molasses from several batches can help iron out some variability contributed by this source.

Lactose

Lactose can only be utlilised by a few cell types, e.g. *Escherichia coli*, and is usually only metabolised very slowly. The sugar is only really useful for some commercial processes that use the complex substrate whey, a by-product of the dairy industry, which contains both lactose (approximately 50 g lactose per litre of whey and 4% w/w nitrogen). Available as a liquid or as whey powder there are choices available regarding storage and handling. Some fermentation media in the past typically batched in both glucose, to achieve rapid culture growth, and lactose, which was only utilised after glucose exhaustion, and the slow rate of usage fuelled, e.g. antibiotic synthesis. Controlled feeding of glucose usually obviates the need for this nowadays.

Other Sugars

Other sugars can be used as substrates but tend to be very specific to the cell line chosen, e.g. melobiose is sometimes used for the cultivation of yeast cells. Sugars falling into this category tend to be too expensive to use on a routine basis. If choosing an unusual sugar the questions to be posed are 'Why is this sugar required?' and 'What advantage will the organism gain?'

Oils

A number of oils are now routinely used as carbon sources in the bioprocess industries. Rapeseed oil (canola oil in the USA/Canada) and methyl oleate are two popular choices. Representing both a carbon and an energy source, oils are economical and readily available. The plants grown to produce the oils will grow in many climatic zones and are readily cultivated, so supplies will be reliable. Variations do exist in the composition/quality of the oils and they must be regularly assayed to check for variation in component content between batches. In a laboratory situation care must be taken to ensure that suitable storage facilities (4 °C) are available.

On a small scale delivery of the oil can be an issue. If operating a batch bioreactor, the oil can be added at the start, before sterilisation and appropriate mixing within the reactor will disperse the oil throughout the vessel. However many small scale reactors are poorly mixed and there may be difficulties that ensue as a consequence of this. Before selecting oil as a suitable substrate, the mixing in the bioreactor should be checked to make sure it is suitable (particularly at 1–2 litre scale). If using fed batch culture at laboratory scale, calculation of the amount of oil required per unit volume per minute must be very accurate and should take into account all substrates in the oil.

Flow rates need to be checked to ensure that sufficient oil is delivered – sometimes it can be very difficult to achieve if the correct pumps for substrate addition are not chosen. Low flow rates often involve working at the lowest speed of the pump and this is usually where pumps are most inaccurate. Choosing the correct type of low speed pump is essential – it is not necessarily going to be the pump supplied by the bioreactor manufacturer, who tends to consider only nonviscous fluids such as routine feeds and pH titrants. To complicate the issue even further the viscous nature of the oil can make additions difficult and often inaccurate. To maximise accuracy, the following should be observed:

(i)   know the exact substrate concentrations in the oil;
(ii) try to use a combination of tubing bore (internal diameter), which lets the pump be operated in the middle of its range/speed – this may involve purchase of a special pump fit for purpose – the internal and external diameter of the tubing round the pump head can always be altered to help improve control;
(iii) regularly check the substrate reservoir and calculate the delivery rate. With slow speed pumps delivering a small volume per hour this can often be achieved more accurately by taking a reading every 4 hours rather than every hour.

An additional benefit of using oil is that it can be used as an antifoaming agent, thus avoiding the use of nonmetabolisable antifoams that can potentially interfere with downstream processing activities. The oil is broken down as it is utilised by the cells and thus does not cause problems later. A few drops of oil can be sufficient to prevent foaming, particularly when dealing with a stirred, aerated proteinaceous broth. If using oil for this purpose it is advisable to know exactly how much of the oil is used, so that the carbon and nitrogen contribution to the medium is known. Sometimes this knowledge can only be gained from experience of the process.

Oil is often filtered/sterilised at the large scale for the sake of simplicity, and increasingly this is a convenient method for small scale processes too.

5.4.2   Nitrogen Sources

The amount of nitrogen in any medium really determines the amount of biomass that will be achieved for the particular cell line, given that plenty of carbon is available and all required nutrients are present in the initial medium. Nitrogen is required for growth and synthesis of, for instance, proteins and nucleic acids. The type and sources are variable. We have already seen that some nitrogen can be delivered as a side benefit to the addition of complex nutrient sources.

Points to consider before selecting the nitrogen source are as follows:

(i)   What do we want the process to do – manufacture cells (i.e., biomass), intracellular or extracellular product, or both (see Chapter 7)? In many fermentation processes growth alone is not the target, the synthesis of a specific product is required.
(ii) Do we want the nitrogen to run out at any point in the process? For example, if dealing with a true secondary metabolite, such as an antibiotic, it is necessary to design the medium so that one substrate becomes limiting in order that growth ceases or slows down considerably to allow production of the product. This is often achieved by allowing nitrogen to become limiting once sufficient biomass has been achieved – this cannot be the case if a protein is the required product for obvious reasons.
(iii) How will pH control be achieved? In a well controlled bioreactor this is not an issue, but if a buffered system is being employed it is worthwhile considering what contributions salts in the medium may make to the overall pH of the broth. In some instances, the culture may be fed N source in the form of NH_3 or NH_4 OH at a rate just matching its metabolic activity. This also achieves pH control in acid-forming fermentations.

A traditional rule-of-thumb guide to a balanced growth medium suggests an approximate ratio of nutrients required of carbon:nitrogen:phosphorus of 100:10:1. So for every $10\,gL^{-1}$ carbon source, $1\,gL^{-1}$ nitrogen is required. A common error here is to assume that $1\,gL^{-1}$ $(NH_4)_2SO_4$ contains $1\,gL^{-1}$ N, which it obviously does not, so remember to take into account the other elements that may be in the chosen nitrogen source. Likewise, this is important when assessing the effects of different N providing salts: make sure the absolute amounts of N being provided are equal otherwise what is observed will be a function of N availability not necessarily the identity of the N source.

Typical nitrogen sources are discussed in the following sections.

Ammonia

Often used in industrial processes, ammonia is not a common source of nitrogen used in research laboratories because of the volatility of the liquid and the associated handling problems. Health and safety issues are a real risk with this nitrogen source, and considerable thought is needed if this is to be the substrate of choice. Special handling equipment, storage reservoirs and breathing equipment are all required, as are additional safety precautions should there be a leak.

The advantages of using ammonia are that it readily dissolves in the medium, it is immediately available to the cell, and it is cheap (although storage and handling equipment may not be).

Nitrogen Based Salts

Ammonium sulfate, $(NH_4)_2SO_4$, and ammonium chloride, NH_4Cl, are common nitrogen sources found in the literature. Ammonium is very utilisable by many organisms and use of a salt is a cheap and convenient method of supplying the cell with what it requires. In addition, it is easy to calculate the exact amount of nitrogen being supplied to the cells. Storage of the substrate is easy, requires no special storage areas or equipment, the salts just needing to be stored in a cool dry place. Analysis of the nitrogen content of the resultant fermentation broth is usually very simple and can be achieved using a simple assay and an ammonia probe.

It is useful to consider other requirements the organism may have, for example if the cell has a requirement for sulfur, then ammonium sulfate would be the salt to choose, so that one salt satisfies more than one requirement. This is not only economical it also makes formulation of the medium much simpler. As a rule, try to keep the medium formula as simple as possible to minimise the chance of errors and to reduce medium generated variability.

A word of warning when dealing with any salts required in bioprocess media is to watch the combination of salts used, and the order in which they are added. An incorrect combination can lead to a batching vessel full of salt crystals. Be systematic and make sure the order of addition of ingredients is consistently adhered to.

Complex Nitrogen Sources

There is a number of excellent complex nitrogen sources available, e.g. yeast extract, soya bean meal (8% w/w nitrogen), and corn steep liquor (4% w/w). All supply a significant amount of nitrogen, in addition to other essential nutrients such as carbon, vitamins, and minerals.

As with the complex carbon sources, there is significant batch-to-batch variability and every batch of the substrate has to be analysed for component composition before use. Analysis of the fermentation broth is also more complicated and often involves acid hydrolysis of the broth in order to get the nitrogen into a readily assayable form. This involves considerable time, labour and extra cost.

5.4.3   Other Substrates

Elements

There are several elements that are essential fro the growth of microorganisms. Each must be provided, generally in the form of an inorganic salt, in order for the cell to grow and function properly.

Calcium.   This is required to stabilise cell walls, and is particularly important if the cells form endospores (which contain a high concentration of calcium dipicolinate).

Magnesium.   This often acts as a cofactor for the activity of enzymes, can play a significant role in membrane structure and function.

Phosphorus.   This is required for the production of phospholipids, nucleic acids and the generation of energy (ATP, ADP).

Potassium.   This is involved in many reactions within cells and is required by all cells, animal or microbial.

Sodiums.   A few species have an absolute requirement for sodium, e.g. marine species and halophiles. Many cells however, have no obligate requirement for this element.

Sulfur.   This required for amino acid synthesis (cysteine, cystine and methionine), for the production of some vitamins, e.g. biotin, and for coenzyme function in many, but not all, cells.

Again, if complex media are being utilised, there may be a reduced or even no need to supply these elements separately. In defined media it is often necessary to add the elements if the cell line is to grow and function normally.

Trace Elements

Trace elements are regarded as micronutrients, being required only in tiny amounts, usually μg quantities per litre. The functions of trace elements within the cell are many and varied but are usually associated with enzyme activity, where the trace element forms part of the enzyme or functions as a catalyst for the biochemical reaction the enzyme is involved with. Again, where complex or semi-synthetic media are utilised, it may not be necessary to supply these elements separately. Local water supplies may even contain adequate amounts of these elements. If, however, a very pure synthetic medium utilised in the laboratory, or if a cell has an absolute requirement for a particular element or elements, they *must* be added to the medium. Very high cell density fed-batch cultures (e.g., *Pichia pastoris* for synthesis of recombinant proteins) can run out of trace elements unless care is taken to feed these at rates matching feeds of macronutrients. Think ahead, and do not simply assume that there will be such surplus of these elements batched in at the start that it obviates the need to add these during the process.

Typical trace elements include iron (Fe, probably one of the more important elements considering its role in cellular respiration), copper (Cu), cobalt (Co), manganese (Mn), molybdenum (Mo), chromium (Cr, animal cells only) to name only a few. Analysis of the literature should reveal which trace element or elements are required by the organism or cell line chosen.

It is usual to make up a trace element solution containing all the elements that are required by the cell, and to add a few millilitres of the resultant solution to the medium. This can readily be carried out before sterilisation of the bulk medium. Stored carefully, the trace element solution can last many weeks but should evidence of precipitation be seen, e.g. one can often see the iron salts starting to coat the container, a new solution should be made up. The caveats above regarding order of addition in batching are also applicable here.

Growth Factors

Growth factors, like trace elements, are required only in tiny quantities by the cell. Often these can be manufactured by the cell itself; however, some cells do not have the ability to manufacture certain key growth factors and they must be supplied to the cell in the medium. They are invariably organic compounds, such as vitamins, purines, pyrimidines, and amino acids. Such compounds are usually very expensive to add in the pure form and it is more normal in bioprocessing to use a cheaper version, which may not be as pure but that provides all the cell needs. If added in pre-form, they will usually not be sterilised using pressurised steam in an autoclave or in situ in an SIP fermenter, as they may well be heat labile.

A very good example of such a substrate is yeast extract. Synthesised from yeast cells (many microorganisms have to ability to manufacture many growth factors) yeast extract contains a number of growth factors, the foremost and most important of which are the B vitamins. Vitamins are the most commonly used growth factors in cell culture media and supplying them in the form of yeast extract, or other plant and animal extracts, is cheap and convenient.

5.4.4   Inhibitors

Obviously any medium formulation should be free of compounds that are known to inhibit the cell to be cultivated. However, most cell types, if overfed, will exhibit undesirable activities and may well excrete toxic by-products or waste products that can lead to slow growth or product synthesis, or termination of the fermentation process itself. In general, with synthetic or semi-synthetic media, known inhibitors are easily avoided. With complex media, however, there can be inhibitors present that have to be removed prior to use. For example, molasses contains a high concentration of iron, which can be toxic to the cell and this has to be removed (or the concentration significantly reduced) if molasses is to be used as the substrate.

Care also has to be taken where a substrate can adversely affect the cell line if present at too high a concentration. Glucose is a particular culprit here as desirable metabolic activities in many organisms are subject to repression by glucose, or excess glucose may simply cause 'overflow' metabolism to occur, e.g. acetic acid excretion by *E.coli*. If this is the case then glucose will usually have to be delivered in a fed-batch mode such that the concentration of residual glucose in the vessel is maintained below levels likely to cause repression or inhibition.

The best view of fermentation processes (especially fed-batch) is that of a balancing act, with the aim being to supply a medium of the right composition, at the right rate, in line with the oxygen transfer rate of the vessel (see chapters 2 and 8) to encourage the desired culture activity and to avoid or minimise side or waste product accumulation. This isn't always easy.

5.5   Medium Formulation

Typical production media contain complex mixtures of dry solids (carbon and nitrogen sources, elements and perhaps oils). Correct storage and handling is a must, for instance, powders must be stored in cool dry places in such a fashion that they cannot mix or get damp and form insoluble aggregates. Liquid materials such as oils, corn steep liquor and molasses need to be kept at sufficiently low temperatures to prevent microbial growth and spoilage (typically refrigeration temperatures).

Laboratory usage of such compounds is little different, apart from the scale of operation. If using complex substrates, it is useful for research purposes that, in order to reduce batch-to-batch variability, one batch is sufficient to cover a complete range of experiments. This may involve storage of a fair amount of liquid and a cold room facility is very useful under these circumstances. Substrates such as corn steep liquor and molasses should be well mixed before the substrate is used as it is common for settling out of particulate matter to occur, which can potentially affect results.

If the batch of substrate is to be kept for any length of time it can be useful to take a viscosity measurement of substances like molasses and corn steep liquor, as this can indicate when the substrates are starting to break down. As the media age, even under refrigerated conditions, they are subject to enzymic and microbial degradation. As degradation occurs the substrates become much less viscous, i.e. the substrates thin considerably and should be discarded at this point.

If the intention is to use a semi-synthetic or complex medium, order and store carefully all the components needed for a at least one full experimental block (including replicates), in order to minimise variability. Nowadays, many commercial companies supply presterilised liquid nutrient media ready for use. These are very convenient, but come at a premium price. Although the range of media is expanding not all are covered, and, of course, it is difficult to carry out research into effects of medium composition upon cell physiology.

5.5.1   Batching Up

Batching instructions should be carefully followed in order to prevent the formation of aggregates (which limits the availability of the substrate to the cells). This can be a particular problem when a number of salts is being used together. Phosphate salts in particular can be difficult to handle and it can be beneficial to sterilise phosphates separately in order to prevent salt formation in the medium. The phosphate solution can be aseptically added to the medium after sterilisation.

While tempting, it is not advisable to use a bioreactor as a batching up vessel for the medium, especially if it is a new formulation that is being utilised. If the formulation is incorrect and salt crystals precipitate out within the bioreactor, they can be extremely difficult to remove and may cause damage to the mechanical seals in a bottom driven bioreactor.

The appropriate concentration of medium components should be dissolved in a larger mixing vessel before being added to the bioreactor, where water can then be added to reach the correct volume for the bioprocess. If necessary the mixture can be gently heated to help dissolve any materials that tend to be difficult to get into solution.

5.6   Preparing the Bench Top Bioreactor for Autoclaving

Once all medium constituents that can be sterilised by autoclaving have been dissolved in water, they can be poured into the bioreactor. This can either be carried out through a port or directly into the reactor before the top plate is put into position. Care should be take to ensure that the following apply.

(i)   The o-ring between the glass and steel section is in place.
(ii)   Baffles, if mobile ones, are not directly below any probe ports.
(iii)   When the vessel has been filled, either prior to the top plate going on or through a port if the top plate is already in place, the liquid level is marked with an indelible marker if the vessel itself does not have volume markings.
(iv)   When tightening the bolts that hold the top plate in position, bolts that are directly opposite are tightened in order to maintain an even pressure over the entire glass edge of the vessel. If this is not carried out there is a real chance of cracking the glass due to an uneven pressure being exerted. This should be carried out in stages, tightening opposites until resistance is met. Once resistance is met a quarter turn on each opposite bolt is sufficient. The bolts should not be over tightened as this can crack the vessel and/or strip the thread of the bolt.
(v)   All electrical connections, e.g. tops of pH, DO_2, CO_2 probes, heating elements, and mechanical seal, have the connection covers in place before autoclaving. Damp connections can mean that probes, etc., may respond erratically or not at all. If the connection covers have been lost, the connection can be covered with aluminium foil.
(vi)   All lines from the vessel are clamped off using Hoffman or appropriate clamps. Care should be taken to ensure that the tubing is clamped directly under the central post of a Hoffman clip to ensure a perfect seal.
(vii)   One line, usually the air exit line with in-line filter, is left open to allow pressure equilibration on both sides of the vessel. If this is not carried out the vessel may implode. The open line and filter should be positioned such that it does not hang down below the level of the vessel, if this occurs, the medium will siphon out.

(viii) Tubing lines are fitted with an appropriate connector to connect to the tubing on inoculation, feed and titrant reservoirs. The connector should be of an appropriate size (bore and wall) to fit the tubing on the reservoir. Tubing ends and connectors should be covered with nonabsorbent cotton wool and aluminium foil in order to make a sterile cover once autoclaved.

(ix) The correct sterilisation time is achieved. The temperature sensor in the chamber of the autoclave should be placed in a cold water reservoir that is of the same volume and dimensions as the vessel. Once the cycle has been completed, the water reservoir must be replenished with fresh cold water in order to ensure that the next cycle is of an appropriate length. It may also be necessary at this point to change the reservoir to a suitable volume to match the items to be sterilised next.

(x)   Once the cycle is complete and the chamber door opened, the air line is clamped until the vessel is on stand and ready for use.

(xi) Any significant water loss that has occurred during the sterilisation process is replaced. Top up with sterile water, to the previously marked level, before aerating or agitating the vessel.

A bioreactor, when it is off stand for sterilisation and cleaning is often top heavy, due to the presence of probes, agitator shaft and condensers. It is thus very vulnerable to being knocked over and consequently should not be left in an area of the laboratory where this could happen.

The quality of the medium is greatly affected by sterilisation. Too little and contamination will occur, too much and some medium constituents will be lost to the organism. Following the steps indicated in point (ix) should result in a satisfactory medium being delivered to the organisms.

5.7   Water Quality

Depending on the bioprocess, water quality can be significant. There is a wide range of water purification systems available and the one selected should match the bioprocess: should the water be distilled, double distilled, produced by reverse osmosis, etc., etc.?

Where complex medium components are in use, simple tap water is perfectly acceptable – why use expensive reverse osmosis water when molasses or corn steep liquor are present that contain all the elements present in tap water only at much higher concentrations?

For animal cell culture the quality of the water has to be extremely good and it may be worth investing in a suitable water purification system. As well as capital cost investment, a consumable budget is required as the filtration systems need to be maintained and, at the very least, filters changed on a regular basis.

5.8   Cell Culture Media

Animal cell culture media can be classified according to their composition into (i) low serum media; (ii) serum free media, and (iii) animal component free media. In all cases it is usual to add antibiotics to the media in order to prevent bacterial growth. The usual antibiotics added are penicillin and streptomycin. Such media are usually filter sterilised and there is a number of systems available for doing this, see below.

Some cell lines show a limited life span when grown in media containing 10% foetal calf serum. Limiting the contribution of foetal calf serum to 1% and adding additional growth factors has improved cell proliferation. Thus, in addition to improved proliferation, improved downstream processing is an added benefit.

There are several advantages to selecting serum free media. The performance of the medium is more consistent, regulatory documentation is simplified, the requirement to pre-screen serum lots is eliminated and downstream processing problems, particularly purification become a lot simpler.

For low-serum or serum-free media a range of supplements may have to be added, specifically growth factors, hormones, animal or plant proteins and trace elements. These supplements are required in order that the cells attain their optimal growth. The concentration of each component is specific for each cell type, e.g. different cell types may have different receptors on their membrane and thus have different requirements for optimal growth.

There is a growing number of readily purchased, ready-to-use media for cell culture, mammalian or otherwise. Common media types are:

(i)   Dulbecco's Modified Eagle Media – used for a broad range of cell types. Variations on the medium include high and low glucose concentrations, +/− L-glutamine, +/− sodium pyruvate, +/− HEPES.
(ii) Minimum Essential Medium – used for a broad range of mammalian cells.

The latter can be modified by supplementing with Earle's or Hank's salts depending on the culture conditions wanted. For example, addition of of Earle's salts will result in rapid rise in pH of the culture medium, whereas addition of Hank's salts will result in a rapid drop in pH of the culture medium. There is a large number of MEM formulations and the selection of the most appropriate formulation for the required application requires an understanding of the way these formulations are constructed and offered. For example, the medium can be formulated in powder or liquid form, and can have a range of different salts, growth factors, etc., added. There is a vast choice of different medium formulations on the market, each specific for a given cell line. There are several companies who supply these media and these can readily be found in your local area by searching on the Internet. Companies such as Sigma Aldrich, Invitrogen, ScienCell and Lonza have a broad range of media to choose from, each with its own specialist function. In today's market, where good manufacturing practice dominates, media should be ISO compliant, and cell culture medium components can be validated by a range of specialist companies; often the company supplying the medium can supply the necessary validation certificates.

One final comment about media and fermentation: remember that as a fermentation process evolves over time it does so in parallel with medium development. An acquired or specifically engineered trait in the cell line only imparts a potential; the correct medium is needed to realise that promise.

5.9   Sterilisation of Media

5.9.1   Preparation of the Culture Vessel or Medium Reservoir

The first step in the sterilisation of any medium is the selection of an appropriate vessel in which to carry out the sterilisation process. This is obviously volume dependent and can range from something as simple as a screw cap bottle, through shake flasks for inoculum development, to the complexity of the bioreactor itself.

For autoclavable or sterilise-in-place (SIP) bioreactors (see below) the manufacturer's instructions should be followed carefully. It can be very dangerous to try to cut corners for the sake of saving a few minutes. Remember bioreactors are pressure vessels and during sterilisation the temperature of the medium will reach 121 °C or above. The manufacturers have developed the perfect sterilisation cycle for the design of bioreactor and this must be adhered to.

It is useful with glass-walled vessels to mark the liquid volume level in the bioreactor, without aeration or stirring, with an indelible marker as there is frequently a loss of water due to evaporation during the process of sterilisation by heat. This water loss should be corrected by aseptically adding sterile water of an appropriate quality to the level marked, before agitation and aeration commence. This ensures that the substrate concentrations are accurate, as liquid loss during sterilisation can be significant.

This can also be carried out for medium reservoirs used in fed batch and continuous culture. Medium reservoirs should be carefully calibrated before use, some can be bought with scales already marked on the side, but more often than not the scale is too large to be of use. Calibration of the reservoirs can easily be achieved by adding known volumes of water to a vessel and marking the volume with an indelible marker. Although this process is tedious, it can ease the calculation of flow rates and dilution rates as the bioprocess progresses and can ultimately save a lot of time. The placing of reservoirs on balances is also an accurate, if more costly, means of monitoring nutrient supply.

For inoculum development the shake flask is often utilised. Mistakes are often made here as the volume of liquid medium added to the flask is often too much to allow for good oxygen transfer. Oxygen transfer in a shake flask is poor anyway, but compound this with overfilling of the flask and the resultant inoculum will probably be sub-standard. For a typical shake flask the volume of liquid medium should be no more than 40%, i.e. 200 ml in a 500-ml flask. For highly aerobic organisms or those that tend to foam, this may have to drop to 20% (100 ml in a 500-ml flask). Care should also be taken to ensure that the bung placed in the neck of the flask is not too loose or too tight. Too loose and contamination may result, too tight and air cannot move through the bung and the cells will suffer a lack of oxygen and a build up of carbon dioxide. Foam bungs can be purchased, but the fit of these is often questionable. It is simple to make a bung from non-absorbent cotton wool, which fits the flask perfectly (it should be possible to pick the flask up by the bung but it should not be so tight that air cannot pass through. Again, the liquid level can be marked and any water losses replaced aseptically with sterile water prior to use.

Medium reservoirs can be filled and capped with silicone rubber bungs, screw caps, or absorbent cotton wool bungs with appropriate entry and exit holes. Each reservoir requires an exit route for the medium (usually stainless steel and silicone rubber tubing) and an air vent (usually silicone rubber tubing attached to an air filter). Any tubing must be clipped to avoid the medium siphoning off when not in use. Any tubing ends should have tubing connectors with an appropriate cover (aluminium foil is very good for this purpose) to allow an aseptic connection to be made to the vessel.

5.9.2   Sterilisation

Once the medium has been formulated and placed in the appropriate receptacle, it is then necessary to sterilise the medium in an appropriate fashion in order to maintain the quality of the substrate components. This may mean that all components are sterilised together, or that some components are sterilised separately in order to prevent losses through (i) denaturation of heat labile compounds, and (ii) reactions between compounds such as glucose and nitrogen (Maillard reaction), both of which limit the availability of the substrates to the cells.

Sterilisation is normally carried out by heat or filtration. Although other sterilisation methods exist, such as irradiation and chemical sterilisation, they are unsuitable for the purpose of media sterilisation. Filtration is much more common for fermentation media than previously, due to rapid advances in filtration technology, and the advantages of the on-thermal sterilisation route. Historically, most microbiology labs had an autoclave, so most medium sterilisation for fermentation tended to involve the use of this convenient method.

5.9.3   Heat Sterilisation

At laboratory scale heat sterilisation is usually achieved by (i) bench top autoclave; (ii) autoclave, or (iii) sterilise in place (SIP) reactors.

Bench Autoclaves

Useful for small volumes and for sterilising small numbers of bottles or flasks containing medium or for items that require careful monitoring of sterilisation times (standard autoclaves have long cycle times), the bench autoclave is an invaluable tool in any bioprocessing laboratory, especially when items are required urgently, as cycle times are relatively short. The bench autoclave has several advantages:

(i)   it is robust;
(ii) it is relatively cheap to buy and operate;
(iii) it has relatively short cycle times;
(iv) it is useful in emergencies.

and some disadvantages:

(i)   The control of cycle time is usually operator dependent. Lack of automation means that the introduction of human error is a possibility.
(ii) They have small volumes – only one or two shake flasks can be sterilised at one time.
(iii) They require monitoring – no monitoring devices available so no records can be obtained. This is problematical in operating under GMP conditions.
(iv) They can dry out if insufficient care is taken.

When using any steam sterilising equipment it is essential to make sure that there is sufficient space for free penetration of steam in and around the items to be sterilised. If an autoclave is overfilled, the steam may not adequately reach some items, resulting in contamination. It also makes sense to check the functioning of small bench autoclaves regularly via thermocouples or biological indicators such as spore strips. This will avoid potential failure of sterilisation-type contamination incidents. Be careful about the maintenance of seals on such small vessels also.

Autoclaves

Autoclaves are ideal for the sterilisation of (usually glass) bioreactors that have a less than 10 L working volume (see Chapter 2). Care is required to ensure that the operational chamber is sufficiently large. If small bioreactor vessels are to be sterilised it is important to consider the height of ancillary parts, such as drive shaft, condenser, probes, etc., when positioned in the bioreactor. Such items can add half the height again to the vessel (see Figure 5.1). It is useful to have a little headroom above the total height of the reactor to allow for ease of loading and unloading.

Purchase of a horizontally loading autoclave is essential. Although bioreactors can fit into some vertically loading autoclaves, there is a greater manual handling risk when unloading and loading a reactor that is full of medium or spent broth. It is also necessary to consider the total throughput of the system, as sterilisation of a bioreactor normally implies a long cycle time because of the volume of the reactor and or medium reservoirs. This can impact considerably on others using the system.

Figure 5.1 *Typical bench top cell culture bioreactors. Note the height of condenser, which significantly increases overall height and thus determines type of autoclave required. (Reproduced by permission of Sartorius Biotechnology Ltd)*

Advantages of autoclaves:

(i)   They are useful where larger volumes or throughputs are required.
(ii) Large autoclaves are very useful when operating continuous culture systems, where throughput of medium reservoirs is high.
(iii) They have automated cycle times.
(iv) Monitoring devices are available on newer machines that allow records to be generated for GMP purposes.

Disadvantages:

(i)   They have high capital and installation costs.
(ii) A regular, dependable steam supply is required.
(iii) They have high maintenance costs.
(iv) They require variable cycle times for fermentation laboratories as volumes to be sterilised can be up to 20 L (where 15 mins at 121 °C does not work).
(v)   They generate exhaust steam that needs to be ventilated from the laboratory – the laboratory design/location must take this into account.
(vi) Cycle times for large volumes can be long – damage to medium components, and subsequent loss of quality, can therefore be a problem, especially with caramelisation/Maillard reaction.

One of the biggest problems with sterilising medium components using an autoclave is the long cycle times involved in sterilising significant volumes of media. It can take as long as half a day, depending on the volume to be sterilised, to sterilise the bioreactor/reservoir properly by the time heating and cooling cycles are added to the sterilisation time itself, which can significantly affect the quality of the medium. At the end of the cycle the medium can still be very hot and it may not be practical immediately to move the reactor or the reservoir.

Particular care must be taken with larger combination glass–steel reactors (>5 L) and glass reservoirs. The heating and cooling cycle can leave the glass in a fragile condition, particularly if the vessel has been bumped during preparation – the risk of shattering is very real and appropriate risk assessments are required to ensure safe removal from the autoclave and subsequent transfer to the laboratory. Large glass medium reservoirs can be replaced with polycarbonate ones; however, it should be noted that the plastic becomes brittle with repeated autoclaving and the life span of such vessels is consequently much shorter.

It is sensible to check that the temperature and pressure records match closely. If there is any obvious discrepancy between these it implies that the autoclave is not functioning properly. A routine should be in place for checking function of such vital equipment even in the smallest laboratory.

Continuous Sterilisers

A continuous steriliser can be used to deliver high quality medium to the bioreactor. Useful where large volumes of media require to be generated in a short period of time, these are not often found in fermentation laboratories but are commonly used on the industrial scale for the production of media for large volume, low value products such as antibiotics. Basically a simple heat exchange system comprising concentric stainless steel tubes carrying high pressure steam in one direction and a stream of medium in the other, continuous sterilisers have the advantage of a very high temperature and a short time exposure, thus providing media in which the substrates have not been compromised.

As with SIP bioreactors (below), a source of steam generation is required and capital costs are high initially.

Sterilise in Place Bioreactors

In the perfect fermentation laboratory SIP bioreactors (Figure 5.2) should dominate. The bioreactors, usually 10 L working volume and above, are usually jacketed and thus require an external steam supply to the vessel jacket. This may imply a considerable extra cost should steam not be available to the building. In addition extra floor space is needed in order to incorporate additional pipe work and frame.

State of the art reactors have excellent control and monitoring of the sterilisation cycle and relatively few operator intervention steps are required. Sterilisation protocols, supplied by the manufacturer, must be strictly adhered to in order to obtain the best performance and to work within safety regulations. The main advantage here is that the temperature of the medium is controlled at all times. Heating times are usually fast, as

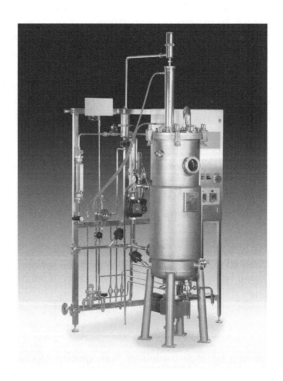

Figure 5.2 *Biostat D 100-L SIP bioreactor. Note the additional framework required for the pipe work (Reproduced by permission of Sartorius Biotechnology Ltd)*

are cooling times, and this means that the medium is not 'overcooked', as it often is when sterilised in a standard autoclave. The medium is normally stirred slowly throughout the sterilisation process, thus ensuring good heat distribution and avoidance of local hot spots. Thus medium quality is assured.

SIPs often have the additional benefit that, with the correct independently sterilisable port installed, medium addition points can be independently sterilised, a distinct advantage when operating in fed-batch or continuous mode.

SIP sterilisation has the additional benefits that the air supply can be turned on to the vessel during cooling, so generating a slight positive pressure inside the vessel and its associated lines. Hence, any potential weakness in the sterile envelope may be overcome. Of course, any significant compromise of the sterile envelope implies that fermenter contents will be aerosoled into the laboratory. It is therefore essential to check physically the condition of all lines/addition points and tubing leading to and from the vessel pre- and post-sterilisation.

5.9.4 Filtration

Filtration is extremely useful in the production of media for many bioprocesses, and although it can be used for all processes, filtration dominates the cell culture industry, where components often cannot be heat sterilised. In microbial systems the filtration step may involve the simple filtration of one medium component, e.g. heat labile vitamin or growth factor, which can be added after the bulk medium has been sterilised in the autoclave or in an SIP bioreactor, or indeed, it may be necessary to filter sterilise the medium in its entirety. Again the filtration system chosen should match the process. Companies involved in filtration, e.g. Millipore, Sartorius and many others, are usually very happy to discuss the needs of the individual process and will work with the client to engineer and manufacture suitable liquid filtration equipment for that process. The level of sterility required in important. If a system requires bacterial removal then a filter pore size of $0.22\,\mu$m is required; if viruses are to be removed then a pore size of $20\,$nm is needed.

Filtration systems can be disposable or not, as the operation requires. Use of disposable filter units eliminates the need for cleaning validation, which is both timely and costly. The companies supplying the filters will validate the filter system, again reducing the need for this to be carried out in house.

Typical applications of filtration include medium, additive and buffer sterilisation, removal of cell debris, endotoxin removal, cell culture clarification, serum clarification and plasma clarification.

Housings for the filters again are varied and include vacuum cups, syringe filters, bench top filters, and a range of housings that vary in size and complexity, depending on the application.

Bottle Top Filters

Useful for sterilising small volumes (10–1000 mL), bottle top filters (Figure 5.3) have a range of membranes and pore sizes and are suitable for most general purposes. They are ideal for sterilising vitamins, growth factors, trace elements and buffers.

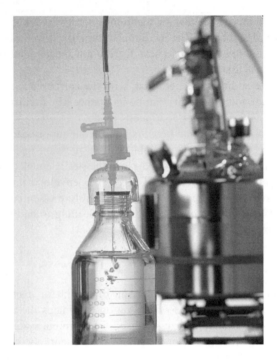

Figure 5.3 *Bottle top filter used for the sterilisation of buffers (Reproduced by permission of Sartorius Biotechnology Ltd)*

Cellulose Nitrate and Cellulose Acetate Membrane Filters

Cellulose nitrate and cellulose acetate filters of $0.22\,\mu$m have been used for the removal of bacteria for many years. In bioprocessing the application of such filters includes sterilisation of buffers and solutions, and harvesting sample volumes of culture broth for the calculation of dry weight, e.g. analytical filters with cellulose acetate membranes are available for the recovery of microorganisms at laboratory and industrial scale. Available as discs or as filter units, such filters have general purpose functions, are easy to use and are disposable. The main drawbacks to such filters are (i) low flow rates; (ii) low throughput, and (iii) high protein binding.

High protein binding makes such filters unsuitable for cell or microbial culture material containing proteinaceous material. Protein binding reduces the available filtration area and thus reducing flow and ultimately causing the filter to clog. In practical terms this usually means that, if using these filters to carry out a dry weight, only a very small volume (5–10 mL) can usually be filtered before problems are encountered.

PES Membrane Filters and Bottle-top Filters

Originally developed by Millipore but now used across the industry, polyethersulfone (PES) membranes are the filters of choice for the sterilisation of cell and tissue culture media, and are ideal for biological and pharmaceutical applications. Importantly these filters have high throughputs, fast flow, low protein binding, and low extractables. Available for sterilising small or large volumes these filters are ideally suited to laboratory use. The low protein binding, achieved by complex surface chemistry applied during the manufacture of the membrane, makes these filters ideal for any cell culture application. Figure 5.4 shows a range of filters that are disposable, can deal with a wide range of applications and volumes, and are reasonably priced. Using PEF membranes, such filters can be used for development and clinical production (Figure 5.5), from small volumes up to pilot scale.

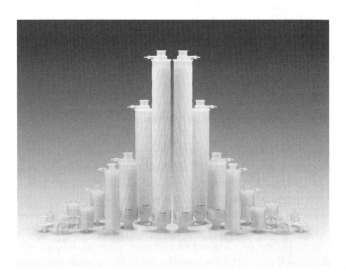

Figure 5.4   *Midicap filters (Reproduced by permission of Sartorius Biotechnology Ltd)*

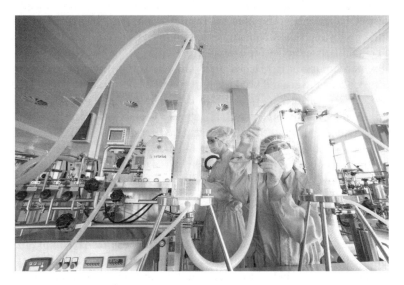

Figure 5.5   *Example of Midicap filters being used in production (Reproduced by permission of Sartorius Biotechnology Ltd)*

Care should be taken to select the correct filter type and membrane for the application required, i.e. fit for purpose. Volume, pore size and effective membrane diameter are all process dependent. The main objectives, apart from sterility, are fast sample processing and low protein binding. It is often good practice to use a combination of pore sizes (e.g., $0.45\,\mu m + 0.2\,\mu m$) in order to enhance the retention ratings and to improve throughput. Pre-filters, often depth filters, can be put in line ahead of the downstream sterilisation filter in order to protect the sterilisation filters from clogging. Depth filters are useful for removing a range of material from particulates to lipids, and their use prior to the sterilisation filter can save time and money.

All materials used in the production of the filters are validated by the suppliers, and appropriate certificates can be supplied if necessary should validation be a requirement of the process.

5.10   Designing Media for Specific Functions

In many bioprocesses, growth alone is not the target. Synthesis of a metabolite, whether primary or secondary, is also required. The medium selected must be designed such that sufficient cells are produced in order to provide a high yield of product.

Where growth is the primary concern, as in the production of single cell protein, bakers' yeast, etc., the medium must be designed so that the maximum biomass that the bioreactor can support from a mass transfer viewpoint (see Chapters 2 and 8) is achieved. The cells must therefore have plenty of carbon and nitrogen sourced as well as other essential nutrients. Should the cell run out of nitrogen, for example, growth will cease. If the desired target is a primary metabolite, and is thus growth associated, the same rule applies.

If, on the other hand, a secondary metabolite is the required target, the medium must be designed such that sufficient biomass is achieved to give maximum productivity, but when the desired biomass concentration has been reached, one substrate becomes limiting. The limiting substrates are frequently either nitrogen (unless a protein or amino acid or nitrogen containing antibiotic, for example, penicillin, is the target) or phosphate. At the point at which limitation is reached there must be sufficient substrate (s) available for the cells to produce the desired product. It may well be necessary to adopt a fed-batch culture approach (see Chapter 4) in order to deliver the correct amount of a specific substrate. For example, many organisms are subject to catabolite repression by glucose. It may not be possible to add all the glucose required for growth and product formation at the start of the fermentation. Instead sufficient glucose is supplied at the beginning of the fermentation to produce the biomass that is required to produce the product, and once the growth phase is over and the organism is ready to produce the product glucose can be fed to the cells at a controlled rate, so that product is produced but cells are not inhibited.

Some secondary metabolites require a precursor to be added to the medium at the point at which growth ceases and product formation is expected, e.g. for penicillin production from *Penicillium chrysogenum*, phenyl acetic acid is required for penicillin G synthesis to occur. The phenyl acetic acid is not added until the growth phase of the fermentation has ceased and, as it can be quite toxic to the cells, it must be added in fed-batch mode. It is just pure serendipity that the original industrial production medium for penicillin synthesis used corn steep liquor, which contained reasonable quantities of phenyl acetic acid, and hence led to good antibiotic synthesis.

Another example of secondary metabolite formation is the production of polysaccharides by a range of microorganisms. In order for polysaccharide production to be induced, the medium must be designed such that nitrogen fuels some cell growth then becomes limiting, leading to diversion of the remaining excess carbon source (usually glucose except in xanthan production where sucrose is used) to biopolymer synthesis.

5.11 Summary

Medium design is a key factor in a successful bioprocess. Obtaining the correct substrates, in the correct concentrations, is not easy and considerable research has to be put into obtaining all the information needed to optimise the medium. Searching the literature, carrying out a critical evaluation and working out what is required of the process are all key steps to be taken before any work is carried out in the laboratory. Proper handling of the medium constituents (storage, sterilisation, formulation) will ensure that the constituents are available to the cells and maximum productivity, providing the physical conditions are correct, will be achieved.

Additional Reading

Atlas, R.M. (2004). *Handbook of Microbiological Media*. Third edition. CRC Press, Boca Raton, USA.

Davis, J.M. (2006). *Basic Cell Culture*. Practical Approach series, Oxford Univeristy Press, Oxford.

Freshney, R.I. (2005). *Culture of Animal cells: A Manual of Basic Techniques*. Wiley-Liss, New York.

Isaacs, S. and Jennings, D. (1995). *Microbial Culture: Introduction to Biotechniques*. BIOS Scientific Publishers Ltd (Garland Science), New York and Oxford.

Stacey, G. and Davis, J. (2007). *Medicines from Animal Cell Culture*. John Wiley & Sons, Ltd, Chichester.

Vinci, V.A. and Parekh, S.R. (2002). *Handbook of Industrial Cell Culture: Mammalian, Microbial and Plant Cells*. Humana Press, Totowa, New Jersey.

Zhang, J. and Gao, N.-F. (2007). Application of response surface methodology in medium optimisation for pyruvic acid production of *Torulopsis glabrata* TP19 in batch fermentation. *J. Zhejiasang University Science*, **8** (2), 98–104.

For Cell Culture Media and Filtration Options

Contact companies for the most up to date systems and information. Many have information on line.

6

Preservation of Cultures for Fermentation Processes

James R. Moldenhauer

6.1 Introduction

Cells are the building blocks of fermentation. Proper preservation of cells is paramount to the successful use of fermentation technology for production of biologically active proteins or peptides. In fact, cell cultures may be considered to be the most critical raw material used in fermentation. As such, they must be dutifully characterized and stringently controlled to ensure consistent output from the fermentation process. It is essential that cells be preserved in a manner ensuring that their genetic and cellular properties remain fixed in time and space. In other words, cells must be induced to achieve a metabolic state of so-called 'suspended animation' in which they remain unchanged and essentially timeless. The most common way to induce this metabolic state is by freezing cells to very low temperatures. Even when frozen, cells must retain their intrinsic biological potential for growth and be ready to jump into action whenever called upon to do so. In this way, they are much like an army of 'cellular soldiers' waiting to be called into action to sacrifice their lives vigorously for a greater purpose – fermentation!

In this chapter, I will attempt to present current understanding on issues and methods that affect the preservation and storage of cell cultures utilized for industrial bioprocesses. In order for the reader to place the practical techniques described in context, some of the underlying principles (with background references) are discussed, and some exciting future development trends in this area are pointed out. I hope that this information will provide an opportunity for the reader to gain a better understanding of laboratory

Practical Fermentation Technology Edited by Brian McNeil and Linda M. Harvey
© 2008 John Wiley & Sons, Ltd

techniques that impact the successful use of cells as tools in biotechnology today. Although other methods will be mentioned briefly (e.g., freeze-drying), the focus will be on the use of cryogenic temperatures to preserve and store cells. Temperatures in the range of about −196 degrees centigrade (°C) to −135 °C are considered cryogenic and will serve to preserve cells for many years. Liquid nitrogen (LN$_2$) is most commonly used to achieve these cryogenic temperatures. Typically, cells are stored in the vapor phase above the liquid nitrogen. Although, in some applications storage directly in LN$_2$ may be done, this technique can be used only when cells have been preserved in heat-sealed glass ampoules or heat-sealed plastic straws.

Three major categories of cells will be discussed – microbial, plant, and animal. Plant cells will be discussed based strictly on a cursory review of the literature, not on the direct experience of the author. Again, the focus of this treatise is the cryopreservation of cell cultures, that is, the maintenance of cell viability and biological properties through the application of freezing methodologies. Properly preserved cell cultures are usually taken for granted and may only gain attention when the fermentation process goes awry, and the foregone conclusion is that the cells 'don't work'. Oftentimes, rightly or wrongly, failures in fermentation will be assigned to the frozen cell cultures or cell banks. When correctly identified as the root cause of fermentation failure, frozen cell cultures will likely have been improperly characterized, preserved, or stored. Certainly, when mistreated, cell cultures can become the 'Achilles heel' of the fermentation process. Hopefully, I can provide some insights to help the reader learn how to treat their cells with the respect they so richly deserve!

6.2   Water, Ice, and Preservation of Life

6.2.1   Frozen Cells

Water is the universal solvent and is essential for life. Chemical reactions required for life depend on the presence of water molecules. Ironically, it is these same life-sustaining molecules that can become so deadly when transformed into the hydrogen-bonded crystalline lattice called ice. Fuller and Paynter [1] concisely summarized this problem as follows: 'Whilst low temperatures themselves have defined effects on cell structure and function, it is the phase transition of water to ice that is the most profound challenge for survival'. It is the mission of cryobiologists to meet the challenge of such physical laws and find ways to preserve life in the frozen state.

Cell cultures are preserved cryogenically in order to establish banks of frozen cells for use in a variety of laboratory applications ranging from animal breeding and human infertility, to research and development, and manufacture of novel, life-saving, medicines. Cryopreservation induces cellular metabolic stasis, which prevents or suppresses any genetic drift during storage. Exposure to low temperatures induces a strong inhibition of chemical reaction rates catalyzed by enzymes. As a matter of fact, any decrease in temperature will cause an exponential decrease in reaction rates as described by the Arrhenius equation [2]. At cryogenic temperatures in the range of about −196 °C to −135 °C cells could, in theory, be stored indefinitely if storage conditions remained stable. Historically, cells were aliquoted into glass ampoules, heat sealed, and stored directly in LN$_2$, i.e. in liquid phase. In most cases today, only the vapor phase of LN$_2$ is used for cryopreservation and glass ampoules are not in common use. The temperature of LN$_2$ is about −196 °C, varying only slightly with atmospheric pressure.

Cell banks are simply collections of cryopreserved cells. Cells are grown under defined conditions in liquid nutrient medium, harvested after an appropriate incubation time, aliquoted into sterile containers, and frozen. Cell banks may consist of glass ampoules, plastic vials, or thin plastic tubes, commonly referred to as 'straws'. These straws are used around the world in the animal breeding industry for storing semen specimens in the liquid phase of LN$_2$. Straws are also used at Eli Lilly and Company to prepare microbial cell banks but all straws are stored in the vapor phase of LN$_2$. All cell banks are prepared and filled under near-sterile or aseptic conditions using a laminar airflow biological safety cabinet.

All cryogenic containers can be referred to as cryovials. Cryovials must be robust enough to withstand sterilization by steam, irradiation, or gas, e.g. ethylene oxide. The most commonly used cryovial is the presterilized screw-capped polypropylene vial designed for storage only in LN$_2$ vapor phase and purchased from manufacturers such as Nalgene® or Corning® (Figure 6.1). Screw-capped vials are typically filled by hand to a volume of about 1 mL. Alternatively, these cryovials may be filled using sterilized tubing in a peristaltic pumping system. Straws are filled to a volume of about 1/2 mL by a special machine that fills by vacuum and seals by melting the straw tip with an ultrasonic probe. In the end, an individual cell bank may consist of only a few dozen screw-capped vials to thousands of straws.

6.2.2 Biological Unity

Preservation of cells by freezing seems like a rather simple task, and in many respects it is so. However, cells are complex biological machines that exhibit different structures

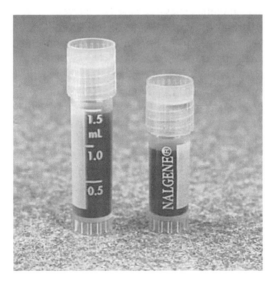

Figure 6.1 *A photograph of polypropylene screw-cap cryovials. Notice the external threads that facilitate sterile technique and the thin internal gasket for sealing cap. Remember, these can be used in LN$_2$ vapor phase only*

and functions depending on which branch of the phylogenetic tree of life they are found. Fortunately for those of us in the business of exploiting their biological properties, these very diverse groups of cells share the same basic components of life such as DNA and RNA for encoding life-sustaining proteins, bilayered lipid membranes (and cell walls in some cases) for protection of the cytoplasm and selective diffusion of molecules, and organelles for translation of proteins and production of energy currency in the form of ATP. It is the presence of these shared cellular characteristics that have allowed the use of remarkably similar laboratory methods for cryopreservation of cells that are very dissimilar from a phylogenetic or an evolutionary perspective. All cells, regardless of evolutionary status, are adversely affected (albeit to differing degrees) by the same metabolic and physical stresses induced by freezing and thawing. This has led to the realization that there is a remarkable commonality among these three diverse biological systems – animal, plant, and microbial – with respect to the methods used successfully to preserve the integrity of cellular function during and after exposure to freezing conditions. As a result, the basic principles learned from laboratory studies of one type of cell have facilitated the application of comparable methods to other cell types. Of course, freezing and thawing is normally a lethal process for most organisms, tissues, and cells. Consequently, it is fascinating to observe how nature has facilitated the evolution of some ingenious strategies used by certain organisms to overcome this deadly environmental insult.

6.2.3   Mother Nature

Cryobiology is the study of the effects of freezing temperatures on living organisms, tissues, and individual cells. Some animals in nature have evolved physiologic mechanisms, e.g. production of 'anti-freeze agents', to survive the lethal effects of freezing. One such example from nature is the wood frog, *Rana sylvatica* [3]. These wood frogs overwinter in leaf litter on the forest floor where they may experience multiple freeze – thaw cycles during the winter months. They are able to endure freezing temperatures as low as −6 °C to −8 °C for periods of two weeks or more. This adaptive mechanism involves the breakdown of liver glycogen to blood glucose in response to the onset of freezing temperatures. Once circulated to all organs and tissues of the frog's body, glucose functions as a cryoprotective agent to dehydrate cells by inducing osmotically driven diffusion of water across plasma membranes. This intracellular dehydration greatly limits (or prevents) intracellular ice crystal formation, which is uniformly lethal.

Another intriguing example of the use of a natural cryoprotective agent is observed in the larvae of the parasitic wasp, *Bracon cephi*. Triggered by the seasonal changes, the larvae accumulate 25% glycerol in their hemolymph (i.e., blood) during the autumn months. This clever adaptation allows these insects to tolerate temperatures as low as −50 °C by placing them into a state of so-called 'suspended animation' whereby molecular activity is minimized [4]. Again, the circulation of glycerol throughout the insect's tissues and subsequent diffusion into cells achieves the intracellular dehydration needed for cell survival during freezing.

6.2.4   Protective Agents

In their infinite wisdom, cryobiologists looked to nature to find solutions to the challenges of preserving cells by freezing. Early investigators in the field of cryobiology mimicked nature by introducing the use of glycerol as an antifreeze agent to maintain cell viability during freezing and thawing; we're not so smart, insects figured this out millions of years ago! Ever since Polge [5] reported in 1949 that glycerol could protect spermatozoa during freezing and thawing, it has become common practice to include one or more such cryo-protective agent (CPA) in freezing medium formulations. Using appropriate CPAs at proper concentrations, laboratory practitioners today can preserve cellular integrity and functions for long periods of time (i.e., years or decades) at cryogenic temperatures. The effectiveness of glycerol, or other similar low molecular weight CPAs, is a direct result of its high solubility in water and high permeability through cell membranes. In this way, CPAs are able significantly to reduce or depress the freezing point of water and biological solutes associated with it (i.e., freezing point depression). Essentially, CPAs have the capacity to interact with water molecules through hydrogen bonding and change the properties of water at low temperatures. Depression of the freezing temperature allows time for CPAs to displace water molecules and inhibit ice crystal formation. It is particularly critical to prevent or minimize the formation of intracellular ice crystals that disrupt subcellular structures and biological functions. Also, effective CPAs must exhibit a low biological toxicity when used at concentrations required to maintain cellular integrity during freezing and thawing. Suffice it to say that cryobiologists have only a limited understanding of the molecular mechanisms underlying the action of CPAs. Consequently, trial and error has been the mother of invention in cryopreservation.

Dimethylsulfoxide (DMSO) is the most widely and successfully used low molecular weight or permeating CPA. The use of DMSO as a CPA was first reported in the literature in 1959 by Lovelock and Bishop [6] as an alternative to glycerol for the cryopreservation of human and bovine red cells. It rapidly penetrates both cell membranes and cell walls and has found universal application in cryopreservation of viruses, spermatozoa, bacteria, fungi, yeasts, protozoa, and cells derived from both mammals and insects. Only high purity – tissue culture or spectroscopy grade – DMSO is used [7–9]. The optimal concentration of DMSO is determined empirically based on cell type but is typically in the range of 5–10% v/v for most cryopreserved cells. The ability of DMSO to scavenge oxygen free radicals, particularly the hydroxyl radical, may also contribute to its effectiveness as a CPA [10]. Another CPA success factor attributed to DMSO is its ability to maintain fluidity of plasma and mitochondrial membranes at temperatures below 5 °C [11]. Presumably, this allows diffusion of both solvents and solutes through cell membranes and counteracts cell shrinking and swelling due to osmotic pressure changes during freezing.

Similarly, glycerol penetrates both cell walls and cell membranes, but does so at a slower rate than DMSO. The permeability of glycerol is markedly affected by temperature. Therefore, both temperature and time of exposure to cells prior to freezing are important considerations when using glycerol. Like DMSO, glycerol at concentrations from 5–10% v/v has been used in cryopreservation of bacteria, viruses, mycoplasmas, fungi, protozoa, and yeasts. Unlike DMSO, glycerol has not been used as a CPA for more complex eukaryotic cells such as those derived from insects and mammals.

The other category of cryoprotectants include those nonpermeating compounds (i.e., higher molecular weight) such as sucrose, lactose, trehalose, mannitol, sorbitol, hydroxyethyl starch, and polyvinyl pyrrolidone, which cannot penetrate cell membranes. These solutes are often used in combination with the permeating cryoprotectants (e.g., DMSO) as adjuncts in freezing media to increase the recovery of viable cells. These extracellular CPAs may exert their protective effect through a combination of inhibiting ice crystal formation outside cells and mechanical stabilization of cell membranes and cell walls. Higher molecular weight polymers may protect cells by binding water molecules, increasing viscosity, and possibly inhibiting ice nucleation sites [10]. They may also exert their effects through alteration of membrane permeability induced by changes in extracellular osmotic forces [12].

It is important for laboratory practitioners to appreciate fully the reality that optimal concentrations and combinations of CPAs must be determined empirically for a given cell type. However, if in doubt, start with 5–10% DMSO unless either experience or literature advises otherwise. For many applications glycerol has been supplanted by DMSO. Nonpermeating CPAs and peptones or proteins (e.g, serum albumin) are useful in combination with glycerol or DMSO, for optimal cryopreservation of certain bacteria and yeast [13]. Some examples include:

- serum albumins for viruses and rickettsia;
- casein hydryolysate for certain cyanobacteria;
- bovine serum albumin for *Leptospira* spp.;
- peptones for yeast;
- yeast extract for protozoa.

Whereas CPAs are essential for cryopreservation of biological systems, they do not ensure 100% survival of cells after freezing and thawing. In fact, the utility of CPAs may be limited by their inherent toxicity at higher concentrations. This observation has been described by Fahy [14] as 'cryoprotectant-associated freezing injury'. Sufficiently high concentrations of CPAs (e.g., 40–80% w/w DMSO) can suppress all ice crystal formation in a biological system by a process called vitrification [10]. Vitrification involves complete cell dehydration with CPAs prior to freezing and induction of a vitreous or 'glassy' state (i.e., highly viscous semi-solid) during subsequent cooling at very rapid rates [4]. The critical step is the dehydration of cells to remove water molecules, which precludes any ice crystal formation. All commonly used CPAs will vitrify (i.e., form an amorphous solid) if used at sufficiently high concentrations. Unfortunately, the concentrations required for vitrification of cells, tissues, or organs can be toxic. Fahy [14] suggested that if there were no biological constraints (i.e., toxicity) on the use of CPAs at high concentrations, then 100% survival of cells, as well as more complex biological structures like tissues and organs would, theoretically, be possible. Fortunately for those of us propagating essentially single cells (or small cell aggregates) in suspension culture systems, vitrification is not necessary in order to achieve acceptable rates of cell survival. However, vitrification techniques are being actively developed for cryopreservation of economically important plant tissues [15].

6.2.5   Freezing and Thawing

Whether in nature or in the laboratory, the rates at which cells or organisms are frozen and thawed are paramount to survival. After choosing a single CPA, or combination of CPAs, the perennial challenge to the laboratory practitioner is finding the optimal cooling and warming rates for a given cell type. Optimizing the freezing and thawing process will help to mitigate the two main chemical and physical stressors on cells during freezing: intracellular ice formation and osmotic stress (i.e., hypertonicity) induced by high concentrations of solutes left behind as ice forms in the extracellular milieu. During cooling, ice crystallization occurs first in the extracellular solution. As ice crystals form, the solution left behind (i.e., unfrozen fraction), becomes more and more hypertonic as solutes concentrate in the residual water. If cells are cooled too slowly, they will be exposed to the hypertonic extracellular environment and shrink due to the osmotic gradient forcing water out of the cytoplasm. This cell shrinkage is caused by dehydration of the cytoplasm. If exposed long enough to this hyperosmotic stress, cell dehydration and shrinkage can cause irreversible breakdown of the plasma membrane and destabilization of membrane-associated proteins. Movement of water results in increasing invagination of the cell membrane due to reduction in membrane surface tension (i.e., decreasing cell volume and 'turgor pressure') with changes to the structure and function of phospholipid bilayers.

Also, the impingement of extracellular ice may exert additional mechanical stresses on the plasma membrane and cause irreversible damage. One explanation for this cellular damage is that the growth and expansion of ice crystals exert physical stresses on cells during freezing and cause deformation and disruption of cytoplasmic membranes [16]. If cells are cooled too fast, there is not enough time for sufficient water to diffuse out of the cytoplasm to prevent intracellular ice formation and cell death. The key to successful cryopreservation is to find a balance between these two forces (Figure 6.2). Cells must

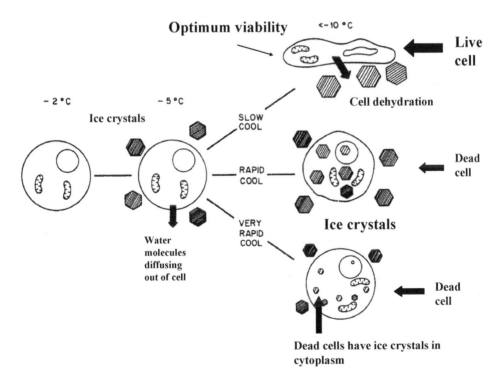

Figure 6.2   *Illustration showing the effect of different freezing rates on viability of mammalian cells*

be allowed to dehydrate sufficiently to avoid intracellular ice formation, but not so much that irreversible damage is caused by excessive cell shrinkage.

It is a fact of life that the same physical and chemical forces act on all cell types. Unfortunately, since all cells are not created equal, there is no universal freezing protocol to ensure cell survival. Again, we are back to the reality of experimental trial and error. However, since all cells share certain fundamental biological structures and functions, there does exist a number of standard protocols to follow. It helps that much of the trial and error has already been done by those cryobiologists who have already 'blazed the trail' of cryopreservation. The optimal freezing rate for a given cell type is achieved when a rate is found that is slow enough to prevent generation of intracellular ice (Figure 6.2) and, at the same time, fast enough to minimize exposure to the damaging hypertonic effects of the surrounding medium. For example, in *Saccharomyces cerevisiae* cells, the incidence of intracellular ice formation increased dramatically when the cooling rates were greater than 10 °C/min [16].

The rate of warming is the key factor in prevention of recrystallization of residual frozen water molecules, which can form from ice nucleation foci in cells as the temperature increases beyond the equilibrium freezing point. Even though dehydrated if frozen at an appropriate cooling rate with CPAs, cells still retain of the order of 10% residual water that has been referred to as unfreezable or 'bound water' [17]. Evidence has indicated that this residual water is chemically bound within the cytoplasm and cannot freeze at any temperature. As reviewed by Mazur [16], there is a correlation between recrystallization of intracellular ice and cell death as observed in yeast, higher plant cells, ascites tumor cells, and hamster tissue cells. The goal is to achieve a state of dehydration during freezing that minimizes the residual water molecules in the cytoplasm. The limit of dehydration is affected by the complex biological mixture of solutes (including a permeating CPA) and initial osmotic pressure, or osmolarity, of the intracellular solution. However, excessive dehydration can cause irreparable harm and a balance must be found between removal of water and time of exposure to the high intracellular solute concentrations and induction of the 'glassy' state where the intracellular solution solidifies into a highly viscous amorphous matrix. Once this 'glassy' state is achieved, residual water molecules are thermodynamically unable to form ice crystals and the cell enters its final resting state of suspended animation. Needless to say, these molecular events are complex and not well understood.

For the laboratory practitioner, it is most important to understand that thawing must be done as quickly as possible in order to avoid formation of ice crystals as residual water (ice nucleation foci) transitions from one thermodynamic state to another as the temperature is increased. Therefore, by controlling the rates of freezing and thawing, the laboratory practitioner can successfully cryopreserve cells by reducing or eliminating the osmotic and physical stresses caused by the chemical transformation of water into ice, ironically, during both freezing and thawing.

6.2.6   Water, Ice, and Heat?

As strange as it may seem, heat is generated during freezing. The release of heat during freezing of water is referred to as the latent heat of fusion. The latent heat of fusion or latent heat of crystallization, can be described as the release of energy in the form of heat (i.e., exothermic reaction) as ice crystals are formed from water molecules that are cooled below 0 °C. In chemical terms, as water molecules are cooled to the freezing point and form an orderly crystalline structure (i.e., ice) via hydrogen bonding, the associated decrease in entropy results in the release of heat energy as molecules move from a higher to lower energy level. When ice forms, each water molecule becomes hydrogen bonded to four surrounding water molecules, which is repeated to form a lattice. It is this high degree of hydrogen bonding that causes the relatively high level of heat generation.

An increase in the cooling rate just prior to the initiation of the latent heat of crystallization is required to compensate for the transient increase in the temperature of the cells and suspending medium. Without control over the cooling rate, the release of this heat can cause a temperature spike (increase) of the supercooled medium away from its freezing point and expose cells to a transitory freeze/thaw cycle, which can affect viability. The release of latent heat can significantly affect the cooling curve during freezing. In fact, temperature increases of up to 10 °C have been observed in cryovials during this early stage of freezing [18].

6.2.7   Storage Temperature

Once cells are frozen and 'resting' in their glassy matrix at about −40 °C or below (depending on CPA), temperature is the most critical factor in maintaining viability for future use. At temperatures at or below the glass transition temperature of pure water (about −139 °C), no recrystallization of ice will occur and rates of chemical reactions and biophysical processes are too slow to affect cell survival [19]. Essentially, liquid water does not exist at temperatures below about −135 °C. The only physical state that exists at those temperatures is crystalline or glassy with an associated viscosity so high as to prevent molecular diffusion. Cryogenic storage of frozen cell cultures below about −135 °C is universally accepted by cryobiologists. As discussed above, the vapor phase of liquid nitrogen is most commonly used to achieve these temperatures. At liquid nitrogen temperatures (i.e., −196 °C) essentially all chemical reactions cease. Therefore, if this temperature is maintained, then cell viability will be independent of the time held in storage. Theoretically, as temperatures are increased above this level some molecular motion will be initiated and chemical reactions will begin to take place. However, the rates of chemical reaction should be negligible below the glass transition temperature of water.

Even when stored at liquid nitrogen temperatures, cellular DNA is still susceptible to photochemical damage from background levels of ionizing radiation or cosmic rays. At cryogenic temperatures, the normal cellular DNA repair mechanisms (i.e., enzymes) are not functional and any genetic damage induced will be accumulated throughout the period of storage and be expressed upon thawing and subsequent growth of cells. It has been calculated that with the background irradiation at existing levels, it would take approximately 30 000 years to accumulate the median lethal dose (~63% cell death) for mammalian tissue culture cells cryopreserved in liquid nitrogen at −196 °C [20].

Conversely, many cells stored in mechanical freezers at temperatures of −70 °C to −80 °C are not stable and lose viability at rates ranging from several percent per hour to several percent per year depending on the type of cell and conditions of cryopreservation [16, 21]. Decreases in viability of cells stored at these higher temperatures is most likely due to damage caused by slow, but progressive, crystallization of residual intracellular water [21]. Also, as the temperature increases, the more thermodynamically favorable are chemical reactions that can result in changes to cellular macromolecules and lipid membranes. On a more practical level, storage at higher temperatures, for example in a −70 °C to −80 °C mechanical freezer, will expose frozen cells to detrimental temperature fluctuations as the freezer is opened and closed during normal laboratory activities. Also, it is important to realize that the equilibrium freezing point of DMSO mixed with water is in the temperature range of −70 to −80 °C (see above for CPA-induced freezing point depression). Cycling of temperatures around this freezing point can cause 'melting' or recrystallization of ice [11].

6.3   Specialized Cell Banks for Industry

Aside from careful preservation and storage, cell cultures employed for industrial use must be thoroughly characterized to confirm identity and freedom from any adventitious microorganisms. Regulatory agencies such as the Food and Drug Administration (FDA) recognize and expect the establishment of the Master Cell Bank (MCB) to be the first step in ensuring the purity, safety, and efficacy of biopharmaceutical products. In fact, the MCB is required for licensing a new product in any domestic or international marketplace. The MCB ensures that sufficient quantities of well-characterized, homogeneous, and stable cells are available to manufacture a licensed product throughout its commercial lifespan. The MCB is produced by filling vials with a homogenous cell suspension derived from a single pool of cells, propagated under controlled laboratory conditions, using a well-defined liquid medium.

Typically, a three-tiered cell banking system is used for biotechnology products (Figure 6.3). The first tier cell bank is a well-studied research cell bank or 'pre-Master Cell Bank' generated in research and development laboratories by any one of a number genetic technologies, such as transfection, transformation, or mutagenesis, followed by selection of desirable cell clones. The MCB (the second tier) is generally prepared from such a research cell bank that has been tested to confirm purity, phenotype, genotype, and protein expression.

The MCB serves as the cornerstone of the global regulatory process and is the foundation on which the company builds a product registration package for submission to regulatory agencies for review. The clinical phase of the product development process starts with creation of the MCB and ends with a final purified biologically active product. Therefore, the product license and continued commercial livelihood is dependent on the proper preparation, preservation, and storage of the MCB. For these reasons, MCB cryovials may be considered by many in industry as the 'crown jewels' of their company. Enough MCB vials, typically about 200 cryovials, are produced to last for the entire commercial life of the product. The MCB must undergo extensive testing for purity, absence of adventitious agents (e.g., bacteriophage, viruses, mycoplasma), phenotypic properties, and genotypic stability in order to become qualified for manufacturing use.

The third tier cell bank is the Working Cell Bank (WCB), which is derived directly from a single cryovial or, if necessary, multiple (i.e., pooled) cryovials of the MCB. If cell banks are managed properly, an almost limitless supply of WCB can be generated

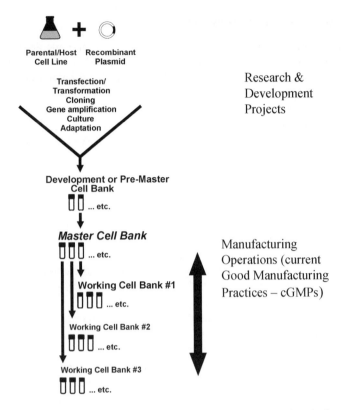

Figure 6.3 *Diagram of a typical three-tiered cell banking system used in the biopharmaceutical industry*

from a single MCB (Figure 6.3). The number of cryovials produced for a given WCB will depend on the manufacturing demands for a particular product. At Eli Lilly and Company, the number of WCB cryovials ranges from a few hundred for mammalian cell banks to a few thousand for some microbial cell banks.

The WCB is the first true building block in the manufacturing process. The WCB is used directly in the manufacturing scale-up process to achieve final production volume in the bioreactor or fermenter. Quite simply, success in manufacturing depends on proper preparation, preservation, and storage of the WCB. In this way, the WCB may constitute the most critical raw material to enter the flow of manufacturing operations. Consequently, the quality of this raw material must be rigorously controlled to the highest standards.

6.4   Laboratory Guide to Successful Cryopreservation

The following is a general guide to laboratory methods and practices based the author's 25 years of experience in the field of cell culture and cryopreservation. These techniques and practices are applicable for the wide range of commercially important cell cultures discussed in this chapter.

6.4.1 Cryogenic Containers (Cryovials)

- Select a suitable cryogenic container or cryovial (i.e., vial or ampoule) that is designed to withstand the rigors of freezing to temperatures produced by the vapor phase of liquid nitrogen (approximately −110 °C to −190 °C). The most commonly used plastic cryovials are supplied by Nalgene® and Corning®.
- Cryovials can either be presterilized by the manufacturer or be sterilizable by steam, gamma irradiation, or ethylene oxide gas. Cryovials for use with mammalian or insect cells must be pyrogen free, that is, free from Gram-negative bacterial endotoxin (i.e., lipopolysaccaride).
- To identify each cryovial properly, use a suitable label or indelible marking pen that can withstand the rigors of freezing and thawing at liquid nitrogen temperatures. The loss or obliteration of labels is a serious problem.
- Use some type of screw-cap plastic cryovial (e.g., high density polyethylene or polypropylene) to avoid the explosion hazards associated with improperly sealed or leaky glass ampoules that have been stored directly in liquid nitrogen. Pin-hole leaks or improper seals will allow seepage of liquid nitrogen into the ampoule during storage. Upon thawing, the rapid expansion of the liquid nitrogen to gaseous nitrogen can cause ampoules to explode.
- If glass ampoules are used they must be checked for leaks prior to freezing by immersion in 0.05% aqueous methylene blue at 4 °C for about 30 minutes [22]. Any ampoules containing traces of the blue dye are discarded.
- **Do not store plastic cryovials directly in liquid nitrogen as most screw-cap seals are not designed to prevent penetration of liquid nitrogen**.
- Due to the explosion hazard with glass ampoules, cell banks today should be stored in polypropylene cryovials with high density polyethylene closures.
  - For example, the American Type Culture Collection (ATCC) converted from glass to plastic ampoules for the liquid nitrogen storage of microbial cultures [23]. Interestingly, the ATCC reported that the contents of plastic ampoules took four times longer to thaw in a 35 °C water bath than in glass ampoules but the effect on viability of the thawed microorganisms was not studied.
  - Additionally, the ATCC initiated an active program in the mid-1990s of substituting plastic for glass in its animal cell culture repository laboratories.
  - Another consideration is the use of newly manufactured plastic cryovials that are used within their expiry date. It is possible that some plastics can deteriorate during storage at room temperatures and may leak after exposure to liquid nitrogen temperatures [24].

6.4.2 Cryoprotectants and Freezing Media

- All cryoprotectants (e.g., DMSO, glycerol) must be sterilized by filtration with a 0.2-μm pore size membrane prior to use, or else supplied sterile from the manufacturer.
- If sterilizing DMSO, ensure that the filter membrane is resistant to organic solvents. DMSO is available as a highly pure (>99%), presterilized, endotoxin and cell culture tested product from vendors such as Sigma®. It can be purchased in sealed 5- or 10-mL

glass ampoules having an extended shelf-life. DMSO should be stored at room temperature as it will gel or solidify at refrigerator temperatures.

- Cryoprotective freezing media containing glycerol or DMSO should be stored at refrigerator temperatures and used within a few days of preparation.
- Permeating CPAs such as glycerol and DMSO must be given time to equilibrate across cell membranes prior to beginning the freezing process. In actual practice this is not a problem as the time it takes to fill cryovials normally exceeds the minimum CPA exposure time, e.g. at least 15 minutes at 4 °C is required for DMSO [14]. On the other hand, glycerol is less permeable than DMSO and more time (e.g., 1–2 h) may be required for sufficient diffusion across cellular membranes.
- Since permeating CPAs have some intrinsic toxicity, even at lower concentrations, cell exposure time before freezing must be limited. Exposure of cells to CPAs should be done at lower temperatures to mitigate potential adverse effects, e.g. 4–8 °C.

6.4.3 Cell Harvest and Filling Cryovials

- Harvesting, transfer, and filling of cryovials with cell cultures must be performed in a properly certified laminar airflow biological safety cabinet. Biological safety cabinets are designed for the dual purpose of protecting the operator from the culture and the culture from exposure to extraneous microorganisms.
- Generally, most cell types should be harvested for preservation during middle or late exponential phase growth, when cell viability is maximal. However, some plant cells must be harvested in early exponential phase growth.
- After addition of cryoprotectants and during filling of cryovials, cell suspensions should be kept cold, e.g. 4–8 °C (or as cool as possible) with the use of flexible frozen cold packs that can be wiped clean and disinfected (e.g., 70% isopropyl alcohol). These cold packs can be wrapped or placed against the sides of the vessel holding the cell suspension. To avoid risk of contamination, do not use wet ice in the laminar airflow biological safety cabinet.
- The cell suspension must be mixed or agitated during the filling process to help ensure uniform distribution of cells into cryovials. This will help to minimize vial-to-vial variations within a given cell bank. This is done manually by simply swirling the culture dispensing vessel, using a tissue culture stir flask designed for gentle mixing of mammalian cells, or by more vigorous agitation with a stir bar for use with more robust cell types (e.g., bacterial cells).
- Cells can be transferred into sterile cryovials either manually using sterile pipettes with some type of pipetting aid, or semi-automatically using a peristaltic pump and sterilized tubing apparatus. In either case, the filling must be done under strict aseptic conditions in a certified biological safety cabinet. One type of semi-automatic filling device is the Cozzoli® Machine Company Model F400X (www.cozzoli.com) which has a footprint of only about 19 × 35 cm and is compact enough to be operated inside a biological safety cabinet. It is constructed of high quality 316 stainless steel and all cell culture contact parts can be sterilized in an autoclave. Even with this equipment each individual screw-cap cryovial must be opened and closed manually. As expected, successful cryovial filling always requires competent technicians with steady hands!

6.4.4   Freezing of Cryovials

- Most cell types have an optimal or limited range of cooling rates associated with maximum survivability. The use of a programmable instrument for controlled step-wise freezing of cells (i.e., controlled-rate freezer) is recommended to ensure optimal freezing conditions for any cell bank used in manufacturing processes. It is important to control cooling rates in order to minimize the formation of intracellular ice (see Section 6.2.4).
- Controlled-rate freezers achieve reproducible linear cooling and heating by controlled injection of atomized liquid nitrogen into a highly insulated chamber, and the pulsing of a heater. A fan (or fans) and chamber baffles are designed for uniform circulation of vaporized liquid nitrogen and temperature control across the cryovial racks (Figure 6.4).
- These freezers are equipped with thermocouples (e.g., platinum resistance temperature probes) to provide precise monitoring of chamber and sample temperatures during freezing. The equipment is programmable, and allows the laboratory practitioner to optimize the conditions for a given cell type and particular load or configuration of cryovials. Use of this equipment helps to ensure process reproducibility and lot-to-lot consistency of cell banks. The freezing program must be developed through a series of test runs using calibrated thermocouples interfaced with a temperature recording instrument (e.g., Validator® 2000, Kaye Instruments, Inc.). Thermocouples are placed inside cryovials and temperature data are gathered from representative positions within the freezing racks. It is advisable to test both minimum and maximum cryovial loads.
- The step-wise freezing program must be optimized to ensure that the cooling rate is controlled during the critical fluctuation of temperature induced by release of latent heat energy (i.e., latent heat of fusion) as water crystallizes. This capability is the advantage

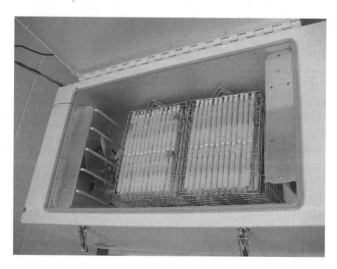

Figure 6.4 *A photograph of a controlled-rate freezer containing racks of straws ready for freezing. Notice the large capacity cooling fan on the left side – there are two fans, one on each end to facilitate uniform cooling*

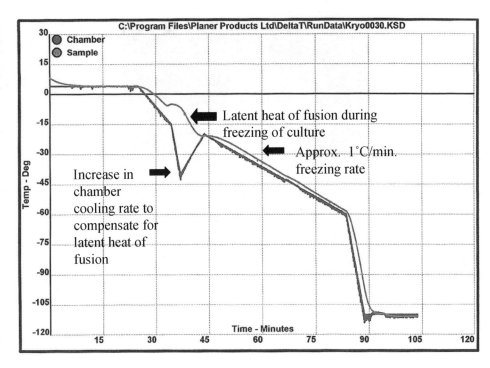

Figure 6.5 Graph showing cryovial freezing profile generated in a programmable, controlled-rated, freezer

of a programmable freezer over a manual method where cryovials are frozen in a static freezer. Figure 6.5 is a graph showing the temperature profile during cooling and freezing of a mammalian cell bank at Eli Lilly and Company. The cryovials are allowed time to equilibrate at 4 °C for 15–20 min before starting the cooling program. This hold step allows time for cells to reach osmotic equilibrium with the surrounding medium containing CPAs (i.e., DMSO). The freezing program was validated to achieve a freezing rate of about 1.0 °C to 1.5 °C per min in cryovials throughout the load. After traversing the latent heat of fusion at approximately −8 °C, and reaching a cryovial temperature of approximately −60 °C, the cooling rate is increased to about 10 °C/min. until a final holding temperature of about −110 °C is reached. At this point, cryovials are transferred to the vapor phase of LN_2.

- If a controlled-rate freezer is not available, then cryovials may be frozen manually by placing them in an insulated container within an ultra-cold mechanical freezer at −70 °C or lower. After several hours, or overnight in most cases, cryovials are transferred to the vapor phase of liquid nitrogen. It is important to understand that a static freezer cannot compensate for the release of latent heat energy during crystallization of water and, therefore, cannot generate a constant rate of cooling. The transient spike in temperature caused by this heat generation could affect cell viability. Although not optimal, this method can provide acceptable survival rates for many cells, even mammalian cell lines.

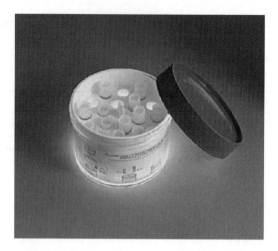

Figure 6.6   *A photograph of the Nalgene® Mr. Frosty. This device should approximate a freezing rate of 1 °C/minute when placed in a minus 70 °C freezer. Notice the cryovials placed in the insulated container*

   – One example of a commercially available insulated freezing container is the Nalgene® 'Mr. Frosty', an insulated container to which 70% isopropyl alcohol is added before placing in a −70 °C freezer (Figure 6.6). Nalgene® claims that this device will control the freezing rate at close to 1 °C/min, which is optimal for most mammalian cells. This is an inexpensive and simple device to use when a controlled-rate freezer is not available.
   – For those laboratories with only portable liquid nitrogen storage vessels (dewars), many cells after pre-chilling to 4–8 °C, even some mammalian cell lines, can tolerate direct immersion in the vapor phase at the top of the vessel. This rather harsh treatment must be used with caution and tested on individual mammalian cell lines or microbial strains to determine the effects on post-thaw cell viability and functionality.

6.4.5   Cryovial Storage and Shipping

• Once frozen, cryovials must always be stored where temperatures below about −135 °C (i.e., glass transition temperature of water) can be maintained, preferably in the vapor phase of liquid nitrogen. However, mechanical freezers are available (e.g., Sanyo) that are capable of achieving temperatures as low as −152 °C. For example, this type of freezer has been used successfully to store canine semen [25]. However, mechanical freezers must be equipped with a battery for back-up power supply. Also, if possible, they should have the option of installing a back-up LN_2 supply in the event of compressor failure.
• Regardless of whether a liquid nitrogen storage vessel (i.e., cryogen) or mechanical freezer is used, the equipment should be continuously monitored and alarmed to alert laboratory personnel to temperature excursions above −135 °C in locations where cryovials are stored.

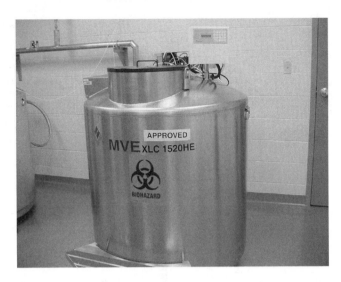

Figure 6.7 *Photograph of a high efficiency, automatic-fill cryogen. Notice the vacuum-jacketed piping for delivery of LN$_2$*

- Opening of cryogens (or freezers) should be minimized to prevent any repeated temperature fluctuations that could reduce cell viability upon thawing. It is preferable to use high efficiency cryogens designed for a low static rate of liquid nitrogen evaporation (Figure 6.7). This type of cryogen has a relatively small diameter neck opening (e.g., 12 in or 300 mm), which reduces the rate of liquid nitrogen evaporation to approximately 4 L per day (Figure 6.8). Other cryogens with larger diameter neck openings (e.g., 21 in or 525 mm) may have static evaporation rates of up to 6 L per day. However, the static evaporation rates can vary significantly with ambient temperature and barometric pressure. There is very limited temperature fluctuation in the vapor phase within the high efficiency cryogen, even when essentially empty during validation studies. Thermocouple mapping studies conducted at Eli Lilly and Company have indicated that temperatures measured at the top and bottom of storage racks differed by only about 14 °C (e.g., 178 °C vs 164 °C). Even opening the lid for 2 min during testing did not cause an increase in vapor phase temperatures of more than about 2 °C at any given point of measurement.
- Cryogens should be capable of filling automatically in response to liquid nitrogen level in the bottom of the vessel. In most applications, using a high efficiency cryogen, only 3–8 inches of LN$_2$ is needed to maintain a vapor phase temperature below −135 °C. Typically, this level of liquid nitrogen will maintain temperatures as low as −160 °C in the vapor phase storage compartment.
- If portable, manually filled LN$_2$ vessels or Dewar Flasks are used, then the level must be carefully monitored several times per week with a plastic (cleanable) dipstick and attempts made to maintain a fairly consistent range of liquid level at all times. The danger in using this type of storage vessel is the propensity to overfill and submerge cryovials. This increases the risk of cryovial leakage since the gaskets in the screw caps

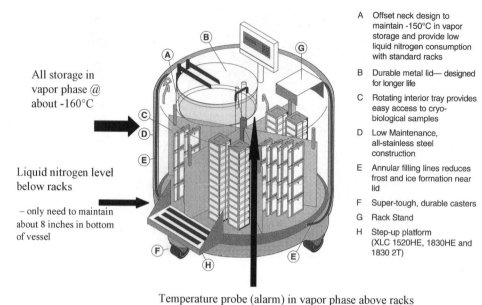

A   Offset neck design to maintain -150°C in vapor storage and provide low liquid nitrogen consumption with standard racks

B   Durable metal lid— designed for longer life

C   Rotating interior tray provides easy access to cryo-biological samples

D   Low Maintenance, all-stainless steel construction

E   Annular filling lines reduces frost and ice formation near lid

F   Super-tough, durable casters

G   Rack Stand

H   Step-up platform (XLC 1520HE, 1830HE and 1830 2T)

All storage in vapor phase @ about -160°C

Liquid nitrogen level below racks

– only need to maintain about 8 inches in bottom of vessel

Temperature probe (alarm) in vapor phase above racks

Figure 6.8 *Schematic drawing showing the inner components of a LN₂ storage vessel or cryogen (Reproduced by permission of MVE/Chart Industries, Inc)*

are typically not designed to withstand exposure to liquid nitrogen. Also, the ever-changing levels of LN_2 may produce substantial fluctuations in storage temperatures thereby increasing the likelihood of a drop in cell viability over time. However, these LN_2 (i.e., liquid) storage devices are commonly used, relatively inexpensive, and can provide a reliable means of cryopreservation. Two time-honored suppliers are Taylor-Wharton and MVE, a division of Chart Industries, Inc. It is advisable to purchase the liquid level alarm, which is sold as an accessory.

• Cryogens, portable laboratory Dewar Flasks, and ultra-low freezers should remain under lock and key and access limited to authorized laboratory personnel only. Once again, as the 'crown jewels' of the company, the MCB and WCB must be secured and protected. Even many research cell banks have a high intrinsic value and cannot afford to be lost or destroyed through accidents or carelessness.

• Shipping and transport of cell banks should always be done in a vapor phase liquid nitrogen shipper designed to maintain temperatures below $-135\,°C$ (Figure 6.9). In this type of shipper, all liquid nitrogen is adsorbed into the liner of the vessel before shipment, thus creating a vapor phase environment and reducing the safety hazards of liquid nitrogen (Figure 6.10).

  – Depending on the manufacturer and model, vapor phase shippers may contain either a hydrophobic adsorbent such as a special-treated colloidal silicon dioxide, or a nonhydrophobic adsorbent such as calcium silicate. In either case, they do not contain free LN_2 and are classified as 'non-hazardous' for shipment around the world.

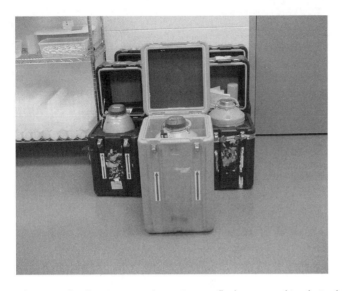

Figure 6.9 *A photograph of LN₂ vapor phase Dewar flasks secured in their shipping cases*

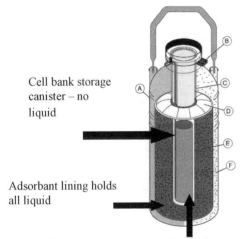

Cell bank storage canister – no liquid

Adsorbant lining holds all liquid

A Lightweight aluminum design reduces shipping costs

B All models come with locking tab for shipments

C Neck Tube– High strength neck tube reduces liquid nitrogen loss

D Advanced Chemical Vacuum Retention System provides superior vacuum performance for the life of the product

E Superior hydrophobic absorbent— repels moisture and humidity while maintaining a -150°C chamber environment

F Insulation — MVE's advanced insulation system provides maximum thermal performance

One model designed and approved to meet IATA and U.N. requirement for the shipment of infectious substances.

Shipping compartment will maintain temperature of - 180°C for up to 19 days (Eli Lilly validation data)

Figure 6.10 *Schematic drawing showing the inner components of a vapor phase shipping Dewar flask (Reproduced by permission of MVE/Chart Industries, Inc)*

– Additionally, it is advisable to include some type of temperature-logging thermocouple device in the shipping container to record actual temperature conditions during transit. Data stored in these devices can be downloaded to a personal computer and hard-copy printouts can be generated and filed for regulatory and quality review (Figure 6.11).

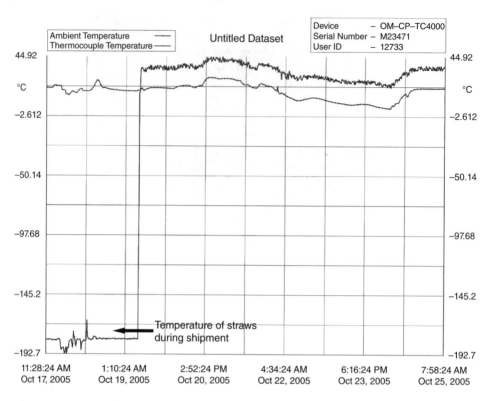

Figure 6.11   *Example of a graph of showing LN₂ vapor phase shipper temperature during transport. This plot is generated by a data-logging/thermocouple device*

6.4.6   Cryovial Thawing

- Generally, best recovery of frozen cell cultures is achieved by removing the cryovials from liquid nitrogen storage, and thawing them as quickly as possible in a water bath or incubator pre-warmed to about 37–40 °C. Cryovials should be gently agitated during thawing to facilitate uniform and rapid thawing of the frozen cell suspension. In a water bath, frozen cell cultures of 1–2 mL in plastic cryovials can be thawed in less than a minute or two. Cryovials should be removed from the water bath or incubator when a small sliver of ice remains in the liquid sample and transferred immediately to a laminar airflow biological safety cabinet for dilution of cells into an appropriate growth or thaw medium.
- For those cryovials thawed in a water bath, disinfection of the outside of the thawed cryovials with sterile 70% v/v isopropyl alcohol prior to opening helps to reduce the chance of culture contamination. Cryovials should be thawed in water that has been sterilized by autoclaving or filtration to minimize the risk of contamination. Alternatively, use high purity water, e.g. water for injection.
- Generally, growth medium should be added slowly to freshly thawed cells in order to minimize the osmotic shock resulting from dilution of the CPA (e.g., DMSO). Different cell types may be more or less affected by this sudden exposure to the hypotonic envi-

ronment of the growth or thawing media. In any case, it is usually beneficial to introduce the first few mL of growth medium slowly with gentle agitation to the freshly thawed cell suspension.

6.5   Microbial Cell Cultures

This section will review methods used to cryopreserve a wide range of microbial cells from bacteria to multicellular fungi. The organisms comprising microbial world exhibit a mind-boggling diversity of morphology, genetics, biochemistry, and physiology. These relatively simple organisms have developed mechanisms to exploit every imaginable habitat on the Earth. On the face of it, such biological diversity poses a formidable challenge to the laboratory practitioner responsible for preserving these life forms. Fortunately for us, there exists a significant degree of biological unity among all cells (see Section 6.2.2) and a fairly unified set of laboratory techniques have proven useful to mitigate the lethal effects of freezing. As will be discussed, there are common laboratory strategies for saving cells from an icy fate: use of CPAs, slow cooling, rapid warming, proper storage temperatures, and healthy cells at time of cryopreservation. First, let us step back and briefly review the time-honored methods of continuous subculture and freeze-drying. Of course, laboratory sophistication, in the form of cryopreservation, developed only after carefully observing cold-adapted animals in nature and learning how they had been doing it so successfully for millions of years!

6.5.1   Continuous Subculture

Probably the oldest and most traditional method of preserving microbial cultures has been the practice of serial transfer or continuous subculture on agar plates or slants. These media were stored at ambient temperatures on the laboratory bench top or at selected temperatures in incubators or refrigerators. Often, agar slants were covered with mineral oil to aid in protection and preservation of cultures by preventing desiccation. Although practised for many years in many laboratories, continuous subculture is fraught with many problems, both practical and technical. The most significant problems include the following:

- continued cell metabolism and division resulting in genetic drift and associated changes in phenotypes of sub-populations derived from parent culture;
- contamination with exogenous microorganisms introduced either by laboratory technicians or the laboratory environment;
- cross-contamination of preserved strains due to different live cultures being in close proximity to each other;
- loss of viability or death of cultures over time due to continued metabolic activity and exposure to a changing nutritional environment;
- lack of space to store and inventory large numbers of agar slants and plates.

As indicated above, this practice is unlikely to ensure that a source of high quality cultures is available for ongoing scientific investigation or commercial development. Although continuous subculture can be used to maintain cultures for short durations, e.g. days or weeks, it is not robust enough to be employed for longer periods of time.

Alternatively, cryopreservation or freeze-drying (lyophilization) techniques can either eliminate or minimize the problems associated with this age-old practice. For decades, freeze-drying and cryopreservation have proven to be reliable and robust methods for preservation of microbial cultures. Of course, no matter which method is used, well-trained, conscientious laboratory practitioners skilled in the art of aseptic techniques are still needed to maintain cell culture purity and viability! In the following section, lyophilization will be introduced as an important and time-honored method of preserving microbial cells and viruses.

6.5.2   Freeze-drying (Lyophilization)

The role of freeze-drying, or lyophilization, of microbial cells and viruses has been firmly established as a reliable means of preserving cultures for decades in the desiccated state. Freeze-drying is the process whereby water and other solvents are removed from a frozen aqueous sample by sublimation. Sublimation occurs when a frozen liquid changes directly to a gas without passing through the liquid phase [26]. The freeze-drying process consists of three steps including pre-freezing of the product, primary drying to remove most water, and secondary drying to remove residual or chemically bound water. The result is a dried product or 'cake' in the bottom of a glass vial. This freeze-dried cake must be reconstituted with a sterile, water-based medium before use. In the freeze-dried state, certain microbial cells and viruses remain relatively stable for many years. Freeze-dried stability depends on low residual moisture and inclusion of stabilizer compounds, or 'lyoprotectants', such as dextran, sucrose, trehalose, and mannitol, in the freezing matrix.

By contrast, the use of freeze-drying for long-term preservation of more complex biological systems (e.g., animal cells) has been largely unsuccessful. Even for microbial cells, size and type of cell can affect recovery following freeze-drying. For example, Gram-positive bacteria are easier to freeze-dry than are Gram-negative cells, due to the differences in the composition of their cell wall structures [27]. However, this technique has found universal success in preservation of many human and veterinary vaccines of both microbial and viral origin.

Freeze-drying may have some undesirable side effects that are not observed with cryopreservation. These include potential for genetic changes during freeze-drying due to DNA strand breakage and selection of mutants [28], denaturation of sensitive proteins [29], and decreased viability for many bacterial cell types [30]. It has been reported that mutation was induced in freeze-dried *E.coli* during repair of damaged DNA after rehydration [28].

Freeze-drying is a rather labor-intensive and time-consuming process. It requires complicated and expensive equipment using empirically derived protocols. From a practical standpoint, freeze-drying has only found a niche in preservation of simple life forms like viruses, bacteria, and yeast. More complex eukaryotic cells such as those derived from mammals and insects cannot be freeze-dried. However, it should be mentioned that investigations into unlocking the secrets of how to freeze-dry or desiccate eukaryotic cells are ongoing. Not surprisingly, investigators have again looked to the natural world for guidance. For example, we now understand that the disaccharide trehalose, in particular, is accumulated in freeze-tolerant insects and dehydration-tolerant animals (e.g., nematodes) and some plants. Bakers' yeast (*S. cerevisiae*) is a well-known anhydrobiotic organism, or one that is highly tolerant to desiccation through the accumulation of both intra- and extracellular trehalose. The use of trehalose as a lyoprotectant is being studied in desiccation of bacterial and mammalian cells [31, 32]. On the other hand, cryopreservation techniques are simpler and universally applicable to cells from all branches of the evolutionary tree of life.

6.5.3 Bacteria

It is important for the laboratory practitioner to consider the metabolic state of bacterial cells at the time of cryopreservation. For maximal recovery of colony forming units (CFUs), cultures should be grown under conditions that minimize metabolic adjustment, or lag phase, after inoculation into liquid media and continue to support optimal growth of cells into late exponential phase or very early stationary phase [27, 33, 34]. Generally, cells harvested from late exponential or early stationary growth phases are more resistant to freeze – thaw damage and provide the highest post-thaw recovery of viable cells and CFUs. It has been demonstrated that aerated and shaken cultures are more resistant to freezing and thawing than are cultures grown under static conditions on nutrient agar plates [27].

At Eli Lilly and Company, recombinant *E. coli* cultures harvested for cell banking are typically propagated in shake flasks vented to ambient air (for oxygen exchange) and incubated in integrated shaker-incubators. It is important to determine the optimal volume of medium for a given size and type of shake flask in order to provide adequate liquid surface area and culture 'head space' for exchange of gases. A simple 2-liter wide-mouth glass Erlenmeyer flask double wrapped with sterilization paper (e.g., Bio-Shield® wrap) serves this purpose quite well. Alternatively, disposable pre-sterilized polycarbonate Erlenmeyer flasks with $0.2\,\mu$m membrane-vented screw caps supplied by Corning® or an equivalent product from another manufacturer can be used. It is important to confirm that this type of flask does not affect growth of bacterial cells due to a change in dynamics of gas exchange through the vented cap. In any case, the oxygen demands of any given bacterial strain must be considered.

After optimal incubation time, broth cultures are harvested and cells pelleted by centrifugation at 5000–6000 $\times$ *g* for 10min. Cells are resuspended in cryopreservation medium to a final concentration in the range of 10^8 to 10^{10} cells/mL. For maximal recovery of CFUs, a minimum cell concentration of 10^7 cells/mL should be frozen [35]. Typical cryopreservation media are formulated with either 5–10% v/v glycerol or 5–10% v/v DMSO. A glycerol concentration exceeding 20% v/v in the freezing medium resulted in reduced post-thaw colony forming units of a recombinant *E. coli* production strain. It is recommended not to exceed concentrations of DMSO of 15% v/v [13]. Using the growth medium formulation as a base with the addition of CPAs is a logical starting point for optimizing the cryopreservation medium. For recombinant *E. coli* strains, the cryopreservation medium may be as simple as a peptone and a single CPA. However, as reviewed exhaustively by Hubálek [13], there is a plethora of CPAs used in microbiology, including glycerol, DMSO, ethylene glycol, mannitol, glucose, dextran, glycine, peptones, serum, and skim milk.

The optimal cooling rate for most bacterial cell cultures is in the range of 1 to 5 °C/min down to about −40 °C, followed by an increased rate of 10 to 30 °C/min down to a temperature of about −100 °C or lower. The cryovials are then transferred for storage in the vapor phase of liquid nitrogen. Although, it may be common to store bacterial cultures in a mechanical freezer at a temperature range of about −70 °C to −80 °C, long-term stability (i.e., years) can be affected, particularly when freezer doors are continuously opened and closed (see Section 6.2.7). To ensure decades-long genetic and phenotypic stability, all bacterial cell banks should be stored in the vapor phase of liquid nitrogen.

6.5.4   Actinomycetes

This group of aerobic Gram-positive microorganisms has been extensively studied for the biosynthesis of secondary metabolites, primarily molecules with antimicrobial properties. These bacteria produce characteristic branched, filamentous hyphae and reproduce either by spore formation or by hyphal fragmentation. The actinomycetes produce both aerial and submerged spores and exhibit distinctive morphological features such as spore chain structures, which can be useful in characterization and identification [36]. The aerobic actinomycetes are ubiquitous in nature and have been isolated worldwide from soil, freshwater, marine water, and organic matter. Generally, these are saprophytic soil organisms, primarily responsible for decomposition of organic plant material [37]. Although more than 40 genera of actinomycetes are currently recognized, *Streptomyces* is the genus of primary interest due to its propensity for producing antibiotics of clinical value. There are over 500 species identified in the genus *Streptomyces,* primarily inhabiting soil. Almost 50% of all *Streptomyces* species produce antibiotics and more than 60 of these compounds have found applications in human and veterinary medicine, agriculture, and industry [38]. Laboratory practitioners must be equipped to preserve these useful organisms properly in order to continue to exploit their important biological properties.

Streptomyces species are propagated for cryopreservation either on the surface of agar media or submerged in liquid media and agitated in flasks. *Streptomyces* species can produce a wide array of pigments that are displayed during growth of aerial hyphae, or mycelia, on agar media [38]. However, many strains do not produce aerial hyphae. Liquid medium may be as simple as a solution of Trypticase™ Soy Broth, or can be more complex involving combinations of yeast and malt extracts, various sources of carbohydrate, and even soybean flour. An example of a more complex medium is one formulated with malt extract, yeast, yeast extract, and glucose. The choice of medium for any particular *Streptomyces* strain is determined either empirically through repeated laboratory growth studies or, as is often the case, by replicating historical successes. In any case, the *Handbook of Microbiological Media*, by Ronald M. Atlas [39], is a comprehensive reference of existing media formulations used for a wide range of microbiological organisms. Typically, cultures are harvested after 24–48 h and then resuspended in a freezing medium containing 10–20% v/v glycerol as a cryoprotectant. Alternatively, some cultures have been grown on agar, harvested as plugs in pre-sterilized plastic straws, and frozen directly in the vapor phase of liquid nitrogen or in a mechanical freezer at −70 °C [36]. In those cases, it seemed that the agar medium contributed some cryoprotective or cryostabilizing properties to the frozen cultures.

At Eli Lilly and Company, the same methods used in cryopreservation of bacterial cell cultures described in the previous section have been successfully applied to mutant strains derived from various species of *Streptomyces*. It is advisable, however, to allow enough time for diffusion of glycerol into the cytoplasm of cells in order to ensure formation of the glassy state during freezing. Depending on the strain and density of biomass at harvest, 30–40 min or more may be needed. The unique composition of the cell wall in species of *Streptomyces* may affect the diffusion rate of glycerol.

Interestingly, there is at least one production strain of *Streptomyces* that has been cryopreserved successfully for several decades at Eli Lilly and Company in the absence of glycerol or any other cryoprotective agent. It should be noted that this production strain was cryopreserved in medium conditioned by cell growth for 24–48 h. In a situation analogous to the use of the conditioned agar medium, the conditioned liquid medium may have played a cryoprotective role in the process of freezing and thawing.

6.5.5   Filamentous Fungi

The filamentous fungi comprise a large and diverse group of spore-forming organisms that have adapted to exploit a wide range of ecological niches. These organisms have been utilized for the commercial production of enzymes, antibiotics, alcohols, and other industrial chemicals such as citric acid [40]. Generally, it has been reported that freezing and storage in liquid nitrogen is the preferred method of culture preservation. The advantage of cryopreservation is that is has been used successfully to preserve both sporulating and nonsporulating cultures. In contrast, the application of lyophilization has been limited primarily to preservation of sporulating cultures that are robust enough to survive the combined freezing and drying process; even so, not all types of fungal spores will withstand the freeze-drying process.

As with *Streptomyces* species, 10% v/v glycerol is the CPA of choice. At the International Mycological Institute (IMI) in Egham, UK, over 4000 species belonging to 700 genera have been successfully preserved in medium containing 10% v/v glycerol [40]. A proven method of cryopreservation at IMI has been controlled cooling at a rate of 1 °C/min to approximately −35 °C, followed by rapid uncontrolled freezing in a liquid nitrogen storage vessel. However, along with glycerol, DMSO has demonstrated its effectiveness as a CPA. DMSO has been used successfully at concentrations of 5 to 15% v/v. Also, 10% v/v DMSO has been used in combination with 8% v/v glucose to preserve nine strains of Deuteromycetes [36]. Additionally, there is evidence to demonstrate that DMSO is substantially more effective than glycerol for preservation of eight strains of mycelial cultures [40]. It was reported that the two strains surviving the longest were preserved in DMSO.

Again, slow cooling and rapid warming generally provide the highest recoveries of viable cultures. It has been reported that 'pre-growth' (i.e., cold conditioning) of cultures in the refrigerator at 4–7 °C can improve post-thaw viabilities of some fungi [40]. This practice is not unlike the cold hardening or pre-conditioning of plant cultures (or plant tissues) as a prelude to cryopreservation (see Section 6.6).

6.5.6   Yeasts

Yeasts are unicellular, eukaryotic cells that reproduce primarily by producing blasto-conidia or buds. Microscopically, they appear as round to oval cells, sometimes elongated and irregular in shape. Certain genera, particularly *Saccharomyces* and *Pichia*, are manipulated genetically to produce a wide range of recombinant proteins and peptides. These recombinant strains are amenable to existing bacterial fermentation technologies, and can be propagated to the high volumes and cell densities required for industrial production.

Historically, a wide variety of strains have been preserved by freeze-drying. However, some strains including pseudomycelium-forming cultures, cannot be lyophilized. The National Collection of Yeast Cultures (NCYC), Norwich, UK, has found that post-thaw viabilities of their repository strains are much higher when cryopreserved [41]. Additionally, the American Type Culture Collection (ATCC) has reported reduced viability of two plasmid-bearing strains of *S. cerevisiae* by two to three orders of magnitude after lyophilization [42]. The NCYC [41] has found no significant loss of viability or genetic stability of cultures cryopreserved for up to 10 years. The standard method of cryopreservation at NCYC includes the use of 10% v/v glycerol as the CPA, uncontrolled cooling of filled polypropylene cryovials or straws by transfer to a −30 °C methanol bath for 2 hours, and final storage in liquid nitrogen. They also reported that varying the primary freezing temperature of the methanol bath from −20 °C to −40 °C did not affect cell culture viability. The authors did not describe their exact method for freezing of cryovials in the methanol bath, but one can imagine that this freezing method exerted some degree of control over cooling rate that helped to preserve cell viability.

In contrast, a study of plasmid expression in recombinant *S. cerevisiae*, indicated that cooling rates of greater than 8 °C/min resulted in a marked decrease in post-thaw cell viabilities as well as permanent loss of plasmid DNA [43]. Results from another study demonstrated that the optimal cooling rate of *S. cerevisiae* was in the range of 30 °C to 10 °C/minute [44]. In this same study, using cryomicroscopy, they found that the incidence of intracellular ice formation increased markedly at cooling rates greater than 10 °C/min.

The condition of cells prior to cryopreservation was found to have a profound effect on the response of *S. cerevisiae* cells to freezing and thawing [44]. Post-thaw viability was significantly higher when cell concentrations were greater than 5×10^8/mL. In addition, it was reported that cells obtained from early stationary phase growth were more resistant to freezing damage.

At Eli Lilly and Company, methods used in the cryopreservation of recombinant *Pichia pastoris* production cultures (e.g., MCB and WCB) include the use of a filter-sterilized freezing medium composed of a plant-derived peptone and 10% v/v synthetic glycerin. Cells are harvested from shake flasks in early to middle stationary-phase growth and filled into straws at cell concentrations in the range of 10^7 to 10^8 viable cells/mL. Cell bank straws are cooled at a relatively slow rate of 3–7 °C/min in a controlled-rate freezer to approximately −110 °C, and then transferred to the vapor phase of liquid nitrogen.

6.6   Plant Cell Cultures

Plants comprise a significant part of the Earth's biosphere and are responsible for maintaining the ecological balance of our planet. They play a central role in the balance of nature so their economic importance to humankind cannot be overemphasized. Plants are sources of food, construction materials, fabrics, paper, and fuel. Additionally, they are important sources of pharmaceuticals, flavors, dyes, pigments, resins, enzymes, waxes, and agricultural chemicals. These plant-derived chemicals are referred to as secondary metabolites, meaning that these compounds are not essential to the survival of the plant but, fortunately, can serve as useful products for human use or consumption. Fowler and Scragg [45] reviewed natural products derived from higher plants and identified some commonly known pharmaceuticals including morphine (*Papaver somniferum*), digoxin (*Digitalis lanata*), theophylline (*Camellia sinensis*), vinblastine (*Catharanthus roseous*), quinine (*Cinchona* spp.), and codeine (*Papaver* spp.). Plants are a veritable treasuretrove of chemical gems just waiting to be discovered. However, in order for humankind to continue to reap the benefits, we must use our best technology to preserve the extant global biodiversity of plants.

Plant tissue culture technology has found application in a number of economically important areas including crop improvement, mass plant production, secondary metabolite production, and plant genetic conservation [46]. All these areas of endeavour require the preservation of plant cell phenotypes and genotypes to ensure consistent and predictable production of cells, tissues, and whole plants. Cryopreservation of plant cells and tissues is the fundamental technology underlying the current progress in the emerging field of plant biotechnology. For example, new developments in this field hold great promise for synthesis in transgenic plants of recombinant products for human therapeutic use [47]. As Gruber and Theisen [47] pointed out, many potentially therapeutic proteins have been successfully expressed in transgenic plants including antibodies, vaccine antigens, growth factors, hemoglobin, and hormones. As eukaryotic organisms, plants possess all the molecular and cellular machinery needed not only for synthesis of proteins, but also for extensive post-translational glycosylation, which is so important to biological activity in humans. In fact, Gruber and Theisen indicated that they are working on the generation of plant cell lines with humanized glycan processing machinery.

Totipotency is a remarkable biological property that allows an entire plant to be regenerated from a single embryonic or undifferentiated cell [48]. Unlike a fully differentiated cell or tissue from an adult animal, a mass of embryonic plant cells, or callus, possesses the intrinsic potential to generate all specialized cell types and structures that comprise an entire plant. Therefore, by appropriate laboratory manipulation of culture conditions, these callus cells can be induced to form a variety of plant tissues, including roots, stems, and leaves. This biological phenomenon has been exploited in an effort to advance progress in genetic improvement of plants. One study demonstrated the successful application of cryopreservation methods for regeneration of fertile maize (*Zea mays*) plants from protoplasts of elite maize inbreds [49]. Successful application of cryopreservation techniques to such research could have significant impact on genetic improvement of maize and other agricultural crops that are grown worldwide. Generally, callus tissue provides an important source of totipotent cells that can be genetically altered to produce new and improved plant strains.

Although some success has been achieved in the freeze-drying of pollen, this preservation technique has not been successfully applied to other plant tissue [46]. The first report by Quantrano [50] of successful cryopreservation and short-term storage of a plant cell suspension (i.e., flax cells) was published in 1968. Quantrano's initial work using DMSO opened the door to development and refinement of cryopreservation techniques for a variety of plant materials. A number of CPAs have since been used successfully in cryopreservation of plant cells and tissues. Some of the commonly used CPAs include DMSO, glycerol, sucrose, mannitol, trehalose, sorbitol, and proline. As with other cell types, empirical testing was necessary to determine those combinations and concentrations of CPAs that provided optimal post-thaw cell viability and plant regeneration potential. Owen [51] provides an extensive list of cryopreservation methods for various plant materials. A reading of that list reveals the repeated use of 5 to 15% v/v DMSO, typically in combination with 5 to 15% v/v glycerol, and 0.5 to 1.0 M sucrose as CPAs.

In contrast to other cell types, the pre-freeze acclimation cells and tissues to low temperatures (i.e., cold-hardening) is a technique that has found an important niche in the cryopreservation of plant cells and tissues. This transient stage referred to as 'pre-growth' can increase the freeze tolerance of cells and tissues by adjusting culture conditions, e.g. reduced temperature and presence of CPAs, prior to freezing without inducing phenotypic or genotypic selection [46]. This phenomenon of 'cold-hardening', or low temperature acclimatization, is more critical for cryopreservation of organized plant tissues such as shoot/meristem tip cultures or callus cultures. Increased freeze tolerance may be induced by changes in the proportions of lipid components in plant cell membranes, particularly in the composition of sterols and phospholipids, when cells are exposed to low, but non-freezing temperatures [52]. Steponkus and Lynch [52] demonstrated that the alterations in lipid composition of cell membranes during exposure to low temperatures (e.g., 0 to $-5\,°C$) were correlated with increases in plasma membrane stability during freezing and associated freeze tolerance.

As with other cell types, the most successful cryopreservation methods involve slow, step-wise cooling rates during freezing, followed by rapid warming rates during thawing. Once again, development of optimal cryopreservation procedures for a given plant cell or tissue remains an empirical process; nonetheless, some generalizations can be made regarding successful methods. Generally, a cooling rate in the range of $0.25\,°C$ to $2.0\,°C$/min down to a temperature of $-30\,°C$ to $-40\,°C$, followed by rapid cooling to liquid nitrogen temperature has proven to be effective for many species and different types of plant materials. Typically, a cooling rate of $1.0\,°C$/min (to $-40\,°C$) is used for cell suspensions, protoplast cultures, callus cultures, and immature embryos [18, 46, 51, 53, 54]. These frozen plant materials are stored either in vapor phase or liquid phase of liquid nitrogen and rapid thawing of cryovials in a pre-warmed water bath at $35\,°C$ to $40\,°C$ are proven methods for the successful regrowth of a majority of frozen plant cultures and materials.

A generally applicable protocol for cryopreservation of plant cells and tissues is based on a model developed for cell suspension cultures [46]. However, it must again be emphasized that such a model protocol will not be universally applicable, forcing the use of an empirical approach, particularly for optimizing CPA 'cocktails' (Table 6.1).

For a detailed review of methods optimized for a variety of plant materials including shoot tip (i.e., meristem), callus, protoplast, and embryo cultures, the reader is referred to a review by Withers [46]. Additionally, for laboratory protocols specific for cryopreservation of protoplasts and cell suspensions, the reader is referred to Grout (54) and Schrijnemakers and Van Iren (18), respectively.

In addition to the traditional cryopreservation procedures using controlled rates of cooling, newer techniques are being developed based on the principle of vitrification (see Section 6.2.3). Vitrification involves the use of a highly concentrated (e.g., 7.8 M) glycerol-based CPA solution to dehydrate cells prior to ultrarapid freezing [15]. The most

Table 6.1 *Model protocol for cryopreservation and thawing of plant cells*

---

(1) Use cells harvested in early exponential growth phase (cell division at maximum rate) when cells are small with a relatively high ratio of cytoplasm to vacuoles as opposed to large, highly vacuolated cells found at later growth stages.
(2) Use a cryprotectant solution containing 1 M DMSO, 1 M glycerol, and 2 M sucrose in standard culture media.
(3) Acclimatize cell suspension in cryoprotectant solution at 4 °C for 1 hour.
(4) Use a cooling rate of 1 °C/min to −35 °C, hold for 40 minutes, and then plunge into vapor phase of liquid nitrogen.
(5) Store ampoules directly in liquid nitrogen (if using appropriate container), or in vapor phase above liquid.
(6) Rapidly warm frozen ampoules in a 40 °C water bath.
(7) Transfer cells to semisolid media for recovery (2–4 weeks) followed by inoculation into liquid media and cultivation in shake flasks.

---

critical step in this process is pre-freeze cellular dehydration to prevent any intracellular ice formation. Engelmann [15] contends that vitrification-based techniques are more appropriate for cryopreservation of complex tissues like shoot tips and embryos, and these have proven successful for a broad range of plant materials. Of course, one practical advantage of using vitrification is that a controlled rate freezer is not needed.

6.7 Mammalian Cell Cultures

Mammalian cell lines are the cellular substrates of choice for the production of complex protein molecules. Mammalian cells possess the intrinsic biological machinery required for post-translational glycosylation of proteins that is critical for stability and bioactivity in humans. Mammalian cells have been cultured *in vitro* since early in the Twentieth century. Early tissue culture was limited by the availability of crude extracts derived from chicken plasma, human serum, bovine embryos, and human placental cord serum. It was not until 1955 that Harry Eagle, working at the National Institutes of Health, published his seminal paper on nutritional needs of mammalian cells in tissue culture and opened the door that helped to usher in the modern era of cell culture. He developed a chemically defined basal medium formulated with 13 amino acids, eight vitamins, a mixture of salts, glucose, and antibiotics. Due to the fastidious nature of mammalian cells, for stock cultures he recommended the addition of either 5% whole horse serum or 10% whole human serum. Harry Eagle's synthetic basal medium remains one of the fundamental and universal cell culture medium formulations in use today and is sold by various vendors as Eagle's Minimum Essential Medium (EMEM). Despite ongoing efforts to development chemically defined, serum-free, media formulations, today's cell culturist still relies extensively on the use of fetal bovine serum (FBS) for nutritional supplementation of media, both for cell propagation, and in particular, for cryopreservation.

The cellular architecture of mammalian cells is distinctly different than their more primitive microbial and plant counterparts. One of the fundamental differences is, of course, that mammalian cells possess only a single phospholipid bilayer to protect their cytoplasmic contents from the rigors of the external environment. They have no cell wall to afford them additional protection. In this way, they are more susceptible to changes, e.g. osmolarity and toxicity, in their extracellular environment than their microbial and plant relatives. Despite this important structural difference, the basic methods employed for cryopreservation of mammalian cell cultures are a reflection of those already described. Again, the essential elements of cryopreservation remain the same:

- healthy cells;
- a permeable CPA;
- slow cooling;
- rapid warming;
- storage at liquid nitrogen temperatures.

DMSO is the exclusive CPA used with mammalian cells today. Typically, DMSO is used at a concentration of 7.5% or 10% v/v, but it has been used successfully at concentrations as low as 5% v/v for many cell lines (personal observations [55]). Using a concentration of less than 5% v/v is not recommended, unless laboratory data are generated to support the practice. When preparing cryopreservation medium, DMSO should be added slowly to cold medium, e.g. 4–8 °C, while mixing due to the heat energy released by hydrogen bonding with water molecules. The heat generated can increase the temperature of the medium and potentially denature some proteins. However, this should only be a problem when using the extremely high concentrations of DMSO required in vitrification protocols, e.g. 50% v/v. In order to mitigate the potential toxicity of DMSO at higher temperatures, cryopreservation medium must be used cold (e.g., 4–8 °C). During the process of filling cryovials, the duration of exposure of cells to DMSO must be minimized to reduce the detrimental effects of increased osmolarity and reduced pH of the cryopreservation medium.

Ironically, DMSO can exhibit both hydrophobic and hydrophilic properties depending on the temperature [56]. The toxic effects of DMSO at higher temperatures (e.g., physiological temperatures) are due to the fact that it exhibits a hydrophobic character and preferentially binds to proteins leading to denaturation. Conversely, at low temperatures (e.g., below 0 °C), DMSO exhibits a hydrophilic quality and is preferentially excluded from the surface of proteins, leading to stabilization of the folded proteins.

Mammalian cells should be harvested for cryopreservation during mid-to-late exponential phase growth when cells are actively dividing, and the mitotic index, i.e. percentage of cells in mitosis phase, is high. Cells that have entered a stationary or quiescent phase (i.e., G_0) should not be cryopreserved. The effect of pre-freeze cell cycle on post-thaw recovery of cells has been investigated. One study reported that Chinese hamster cells were most resistant to the stresses of freezing and thawing when harvested in the M (mitosis) and late S (DNA replication) phases of the growth cycle but least resistant in G_2 [57]. Another study reported similar results for HeLa S3 cells, where the highest recovery rate was found when cells were frozen in the mid- to- late S phase and lower in G_2 [58]. On a more practical level, it is always important to harvest cells with high viability. As a general rule, cultures with viability measurements of less than about 90% should not be used for cell banking purposes, particularly for MCB and WCB. In any case, both cell viability and cell cycle or growth phase are crucial factors in successful cryopreservation, and viability criteria will depend on the application.

After an appropriate incubation time, cells are harvested and centrifuged. The resulting cell pellets are gently suspended in cold cryopreservation medium using a pipet (e.g., 10–25 mL) with a bore size large enough to avoid excessive shear forces on cell membranes. It is important to remember that DMSO alters membrane fluidity and permeability. Ultimately, the goal is to prepare a uniform single-cell suspension. The viable cell density used for filling cryovials should fall within the range of about $5 \times 10^6/mL$ to $2 \times 10^7/mL$ (personal observations, [59, 60]).

It is universally accepted that the optimal cooling rate for mammalian cells is close to 1 °C/min. If in doubt about what is optimal for a cell culture, start with this cooling rate and use the empirical approach only if cell viability does not meet expectations. However, acceptable cooling rates in the ranges of 1 to 5 °C/min [25] or 1 to 3 °C/min [59, 60] have been reported with the optimal rate being determined empirically. At the European Collection of Animal Cell Cultures (ECACC), a cooling rate of 3 °C/min in a programmable freezer is used for the majority of cells [60]. The ECACC has observed post-thaw viabilities that typically exceed 85% using their controlled freezing process [59]. Typically, cryovials are cooled at this slow rate down to a temperature of about −60 °C, when the cooling rate is increased to 10 to 30 °C/min, until a final hold temperature of about −100 °C is reached. Cryovials are removed from the freezing chamber and immediately transferred into the vapor phase of liquid nitrogen; if there is any delay during transport of cryovials, then use dry ice in the interim.

Eli Lilly and Company have optimized and validated a controlled-rate freezing program in which the initial cooling rate is 1.0 °C/min down to −60 °C, followed by a more rapid cooling rate of 10 °C/min down to a final holding temperature of −110 °C before transfer to liquid nitrogen vapor phase for cryogenic storage.

Typically, frozen cells are thawed as rapidly as possible in a water bath warmed to 37–40 °C. This is the simplest way to effect a relatively high rate of warming. Cryovials should be agitated gently during warming in an attempt to induce more uniform thawing. Using this method, cryovials containing 1–2 mL of culture can be thawed consistently in less than 3 minutes. Once thawed, cells should be immediately diluted with fresh growth medium at either ambient or refrigerator temperatures. It has been reported that dilution of freshly thawed cells with medium at room temperature may be less stressful and facilitate higher cell recovery [61, 62]. Additionally, it is recommended that the cryoprotectant from freshly thawed cells be slowly diluted in order to reduce the osmotic gradients that cause changes in cell volumes [24, 60, 62]. After dilution in growth medium, the cell is exposed to a hypotonic extracellular environment due to its cytoplasmic concentration of DMSO. Consequently, the cell will initially swell as water diffuses into the cytoplasm as DMSO equilibrates across the plasma membrane. If the swelling is too fast or too extensive, the plasma membrane can be irreparably damaged and cells will die. Fortunately, not only is DMSO itself highly permeable to cell membranes but it also enhances permeability of the membrane itself. These properties facilitate relatively rapid diffusion of solutes across membranes in order to reach equilibrium concentrations. Also, slowly adding fresh medium drop wise to the thawed cell suspension with gentle agitation should reduce the osmotic shock to cells, as the extracellular milieu is diluted more slowly and thus becomes hypotonic (relative to the cytoplasm) at a slower rate.

There seems to be no clear consensus as to whether or not to add cold, ambient, or warm medium to dilute the freshly thawed cell suspension. Immediately upon thawing, cells are in osmotic equilibrium with their cryopreservation medium. At lower temperatures immediately after thaw – somewhere between 0 °C and ambient – the toxicity of DMSO should not be significant, since its relatively low intrinsic toxicity is reduced further at lower temperatures. Also, at the low concentrations used in cryopreservation, DMSO is generally well tolerated by most cells. In actual practice, however, the goal is to dilute any effects of DMSO as soon as reasonably possible. The question still remains: What medium temperature is best? Cold medium will reduce any intrinsic toxicity of DMSO, but will slow down the diffusion process due to decreased permeability of membranes. Warm medium will tend to increase membrane permeability and increase the rates of diffusion, but will increase any intrinsic toxicity of DMSO. Ambient temperature medium is somewhere in between and for this reason it may be the best choice. In actual practice, there may not be a discernable difference in post-thaw viability regardless of which method is chosen.

Post-thaw centrifugation of cells should be done at a minimal g-force (e.g., $100 \times g$) in order to reduce shear forces on cell membranes that have been altered by the presence of DMSO. However, the centrifugation g-force and time must be adequate effectively to pellet all (or most) of the intact cells or a significant loss of viable cells may result (personal observations). For example, the author typically uses $100 \times g$ for 10 min in a refrigerated centrifuge. Table 6.2 summarizes the key elements of a typical protocol for cryopreservation of mammalian cells.

As for all other cell types, it is a universally recommended practice to store mammalian cells in the vapor phase of liquid nitrogen at temperatures below than about −135 °C (i.e., glass transition temperature of water). One issue of at least theoretical importance

Table 6.2 *Model protocol for cryopreservation and thawing of mammalian cells*

(1) Use cells harvested in early-to-middle exponential growth phase, i.e. when cell division or mitotic index is highest.

(2) Use DMSO as the CPA, prepared to a 10% v/v concentration in standard culture growth medium.

(3) Adjust final viable cell density to within 5×10^6/mL to 2×10^7/mL.

(4) Keep final cell suspension cool by using plastic cold packs around or under the harvest flask or vessel.

(5) As cryovials are filled, transfer to refrigerator to acclimatize cells in cryoprotective medium at 4–8 °C before freezing.

(6) If available, use a controlled-rate freezer programmed to a cooling rate of 1 °C/min to −40 °C, followed by a more rapid rate of 10 °C down to a final holding temperature of about −100 °C. If a freezer is not available, then use one of the alternatives discussed in Section 6.4.

(7) Store cryovials in the vapor phase of liquid nitrogen within a cryogen or storage Dewar Flask that is monitored routinely.

(8) Rapidly warm frozen cryovials in a 37–40 °C water bath.

(9) Transfer thawed cells to growth medium adjusted to ambient temperature followed by cultivation in shake flasks or static tissue flasks, depending on whether cells are attachment dependent or adapted to suspension growth.

(10) Cell seeding density out-of-thaw is critical and determined by specific cell line characteristics. For most suspension cell lines, seeding densities in the range of 2×10^5/mL to 4×10^5/mL should facilitate good cell growth.

of storage of mammalian cell cultures directly in liquid nitrogen is the possibility of contamination with viruses. Viruses could be introduced through contaminated LN_2, from the cryogen itself, or from personnel. Transmission of two bovine viruses – bovine viral diarrhea virus (BVDV) and bovine herpesvirus-1 (BHV) – to frozen bovine embryos was demonstrated during storage directly in liquid nitrogen [63]. The liquid nitrogen was experimentally inoculated with viruses and unsealed containers of embryos were plunged and stored in the liquid phase. After 3–5 weeks of storage, 21.3% of embryos tested positive for viral contamination. Conversely, all control embryos sealed in cryovials were free from contamination. This study illustrates the importance of maintaining cryovials in the vapor phase at all times during storage. As indicated in Section 6.4, plastic cryovials used for cryopreservation are not designed for storage in liquid and are prone to leakage if exposed to liquid nitrogen. This is particularly critical for storage in Dewar flasks filled manually, where liquid nitrogen levels may not be monitored carefully and cryovials may be immersed in liquid nitrogen for varying lengths of time.

The universal dogma of cell culture has always been that serum, particularly FBS, exerts some sort of protective effect on cells during cryopreservation. Presumably, as we have so diligently believed, FBS provides a source of proteins and other factors to enhance the cryopreservation process. However, the mode of action of serum in protection of cells during the freezing and thawing process is not known, or at least, has not been described in the literature, although, it has been inferred that serum proteins may form a protective coating over the cell membrane. In any case, FBS is readily recognized, anecdotally at least, by laboratory practitioners worldwide for its protective effects on cells during cryopreservation (personal observations, [64]. Even today, cell culturists remain dependent on the use of FBS to propagate many different cell types and cell lines. Typically, it is used as a nutritional supplement in growth media at concentrations ranging from 5–10% v/v. In the author's past experience, media used for cryopreservation were identical to the media used to grow the cells. This approach always seemed to make sense, and was easier than formulating and preparing special media used only for cryopreservation. Interestingly, FBS is often added to cryopreservation media at concentrations as high as 90% with DMSO comprising the remaining 10% of the volume. This practice may have been carried over from cryopreservation of hybridoma cell cultures (i.e., early post-fusion cultures) and certain other particularly sensitive diploid cell lines or primary cell cultures. Or, maybe the thought was (is) that if a little is good, more must be better? Honestly, this author doesn't know the answer, but, certainly, there is an established trend to eliminate FBS from both growth and cryopreservation media.

The development and use of serum-free and protein-free media for propagation of cells is a universal effort. Historically, this has been due in large part to concerns about high cost and limited availability of FBS. However since the discovery of 'mad cow disease' or bovine spongiform encephalopathy (BSE) in the mid-1980s, there has been growing concern about exposure of cell lines to bovine-sourced raw materials, particularly FBS. It is now recognized that BSE and other transmissible encephalopathies are caused by prions, or so-called 'infectious proteins', that induce abnormal folding of proteins and subsequent degeneration of neurological tissues. In contrast, the development and use of serum-free cryopreservation media has received much less attention and so the use of FBS continues. However, as global regulatory pressure has increased to remove all animal-sourced raw materials from biopharmaceutical processes, elimination of FBS from cryopreservation media formulations has become more commonplace, especially for those cell lines already adapted to grow in serum-free medium.

One early study in this area found that the addition of 0.1% w/v methylcellulose to freezing medium containing only 10% v/v DMSO and basal Minimal Essential Medium (MEM) increased almost twofold the post-thaw viability of a mouse L cell line (L.P3) over the same freezing medium without methylcellulose [65]. Additionally, Ohno *et al.* [65] reported the successful cryopreservation of HeLa cells and various hybridoma cell lines using serum-free freezing medium containing 0.1% w/v methylcellulose. Merten, Petres and Couvé [66] reported the successful replacement of 10% v/v fetal calf serum with either 0.1% w/v methylcellulose or 3.0% w/v polyvinyl pyrrolidone in freezing media for preservation of Vero (monkey) and BHK-21 (hamster) cell lines. In all studies cited above, the cell lines were propagated in serum-free media prior to freezing. All cryopreservation media in those same studies contained either 5% or 10% DMSO.

Other CPAs of nonanimal origin used in cryopreservation media to enhance recovery of mammalian cells include trehalose and *S*-adenosylmethionine; membranes are not normally permeable to either compound. Trehalose has been introduced through the membranes of human pancreatic islet cells using DMSO to increase membrane permeability [67]. Another group has used a genetically engineered alpha-hemolysin to create pores in cell membranes of 3T3 mouse fibroblasts and human keratinocytes to induce uptake of trehalose into cells without the synergistic action of DMSO [68]. Trehalose exerts its cryoprotective effects through stabilization of membranes and proteins both intracellularly and extracellularly during freezing and thawing The addition of *S*-adenosylmethionine to cryopreservation media containing DMSO proved to increase both viability and metabolic activity of thawed rat hepatocyte cultures [69].

Despite all the elegant empirical work to find alternatives to the use of FBS in cryopreservation media, in many cases the solution is simpler than one might expect. As indicated earlier, the key to elimination of FBS in cryopreservation media is elimination of FBS in growth media. In the author's experience, both recombinant and nonrecombinant Chinese hamster ovary (CHO) cell lines adapted to growth in FBS-free medium, can be successfully cryopreserved in their respective growth medium with 10% DMSO as the only additive.

6.8   Facilities for Preparation and Storage of Cell Banks for Commercial Use

Laboratory facilities used for the preparation and storage of MCB and WCB to support commercialization of biopharmaceutical products must be designed and operated to meet all current Good Manufacturing Practices (cGMP). As such, these facilities are open to inspection by regulatory agencies such as the FDA. Cell banking facilities that produce MCB and WCB for cGMP manufacturing processes must be designed and operated to preserve the purity (i.e., sterility or axenicity) of the cell cultures during propagation, harvest, and filling processes, and to maintain the integrity of frozen cultures during long-term storage in liquid nitrogen vessels. At Eli Lilly and Company, the cGMP cell banking facility is operated as a cleanroom environment requiring the following design features:

- HEPA-filtered air supply with no recycling within cell culture suites;
- air pressure differentials between rooms;
- airlocks to provide biological containment;
- gowning/de-gowning rooms;
- cleanable surfaces, e.g. epoxy-coated floors and coved walls and ceiling;
- physical segregation of operations, e.g. separate laboratories for different cell types;
- unidirectional traffic flow.

The cell banking facility is designed for unidirectional traffic flow of both personnel and equipment (Figure 6.12). One of the key features of this facility is physical segregation of different activities. The core cell culture laboratories lead to the cryogenic storage areas by way of de-gowning rooms and airlocks. Each cryogenic storage room is equipped with automatic-fill liquid nitrogen cryogens supplied from an outside bulk storage tank through a vacuum-jacketed insulated piping system. Cell banks are stored in the vapor phase at approximately −150 °C or lower. Each storage cryogen is continuously monitored and alarmed for temperature through a validated computerized building alarm system. Access to the cryogenic storage rooms is strictly controlled through the use of multiple card readers and each cryogen is locked.

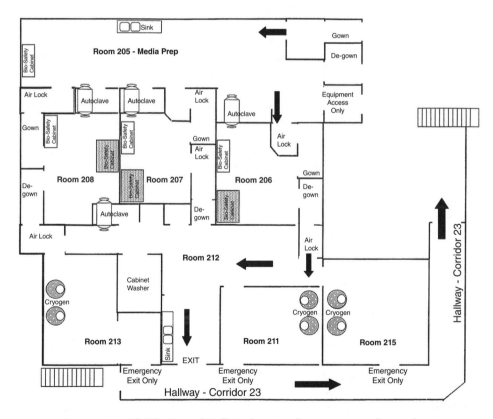

Figure 6.12   *Eli Lilly Central Cell Banking Facility Layout – Indianapolis, IN*

Each cryogen has been validated to maintain required temperatures (i.e, $<-135\,°C$) throughout the vapor phase storage compartment. The validation studies were performed using a calibrated temperature logging instrument (i.e., Validator® 2000, Kaye Instruments, Inc.) with 12 thermocouples distributed throughout the cryovial storage racks to test all positions from top to bottom within the vapor phase storage areas of the cryogen (Figure 6.8). Those studies demonstrated the capability of the cryogens to maintain temperatures below $-145\,°C$ at all locations and levels tested in the vapor phase storage compartments even when lids were opened for 2 minutes. Cryogens should be validated, or at least tested, to confirm that the required vapor phase temperature of $<-135\,°C$ in all locations where cryovials are stored.

To safeguard personnel, it is essential that liquid nitrogen storage vessels be located in a well-ventilated room to prevent depletion of oxygen by nitrogen gas, and that oxygen monitors and alarms should be located within the room. For example, at the ECACC an oxygen monitor has been connected to an automatic ventilation system that operates when the oxygen level decreases to 18.5% v/v [60]. Similarly, at Eli Lilly and Company each cryogenic storage room is equipped with two oxygen monitors and associated audible and visual alarms to alert personnel to reduced oxygen levels.

6.9   Current Quality/Regulatory Considerations

The elimination of animal-sourced or animal-derived raw materials in cell banks is an emerging issue that impacts global licensing and marketing of biotechnology products generated through fermentation of cell cultures. Due to escalating concerns about transmissible spongiform encephalopathy (TSE) and bovine spongiform encephalopathy (BSE)-related diseases, global regulatory agencies are closely scrutinizing the use of animal-sourced raw materials used in manufacturing processes. At Eli Lilly and Company, whenever technically feasible, all animal-derived raw materials are removed from our cell banking process. For microbial MCB and WCB, animal-sourced peptones and glycerol have been replaced with plant-derived or synthetic alternatives for both growth and cryopreservation.

Table 6.3 includes some important quality and regulatory guidance on control and testing of MCB and WCB used for clinical trials or commercial production:

Interested readers are referred to key guidance documents [70, 71] for additional information on current international regulatory and quality standards applied to biotechnology products derived from cell cultures.

6.10   Interesting and Emerging Areas for Further Development

The introduction of technology for the transgenic plant-based production of biopharmaceuticals provides a promising avenue for avoidance of the regulatory and quality risks associated with use of animal-derived raw materials and use of animal cell lines. This alternative approach to the production of recombinant proteins has been developed using transgenic plants as so-called '*in vivo* bioreactors'[47]. Although plants possess cellular machinery needed to accomplish post-translational glycosylation of proteins, there are

Table 6.3   *Quality and regulatory guidance for biotechnology*

---

(1) The origin, history of laboratory manipulations, and biosafety testing of cells used to manufacture biotechnology products must be included in the product registration application package. Some commonly used host cells, e.g. CHO and recombinant *E. coli* K12 strains, already have established safety profiles and are more easily accepted by regulatory agencies.

(2) All incoming cultures for banking should be obtained from a pre-approved or reputable source with appropriate documentation. Incoming cells should be segregated into a designated quarantine area until appropriate test results are confirmed (e.g., sterility and mycoplasma for mammalian cells).

(3) Animal-free raw materials for media and laboratory reagents should be used. Any bovine-derived products, e.g. FBS, should be purchased from a vendor who can document sourcing from BSE-negative countries such as New Zealand or Australia, or from known BSE-negative herds. All FBS should be gamma-irradiated by a reputable vendor to reduce further the risk of viral contamination.

(4) The maintenance of accurate and detailed laboratory records is required. All records, including approved laboratory notebooks, should be filed to enable easy retrieval for use in regulatory submissions, inspections, and quality audits.

(5) Cell cultures must be tested both before and after banking. Pre-bank testing may include the following, as appropriate, (i) sterility or purity; (ii) mycoplasma (for mammalian or insect cells only); (iii) viral assays specific for the host cell species; (iv) bacteriophage; (v) phenotyping. Post-bank testing may include all of the above plus the following: (i) additional viral assays, both *in vivo* and *in vitro*, as required for a given cell type; (ii) DNA sequencing; (iii) plasmid copy number; and (iv) restriction enzyme analysis.

(6) Testing must be performed on cells obtained from the end-of-fermentation, or so-called 'end-of-production cells'. Genetic stability must even be verified on cells propagated beyond their final harvest from the bioreactor/fermenter, or 'beyond the *in vitro* cell age'.

(7) Stability of cell banks during storage should include testing for post-thaw viability (e.g., trypan blue or fluorescein diacetate staining), population doubling time, and consistent expression of product (e.g., yields, proper glycosylation patterns).

---

differences between glycan structures observed in plant and mammalian cells. Nevertheless, this promising technology has generated several potential therapeutic products that are now in clinical trials.

One fascinating application of this technology is in the area of human vaccine development. Huang *et al.* [72] have successfully expressed both hepatitis B surface antigen (HbsAg) and Norwalk virus capsid protein in suspension cultures of tobacco and soybean cells, as well as in potato tuber and leaf tissues. In a promising first step, these investigators were able to demonstrate that ingestion of transgenic fresh (uncooked) potato tuber stimulated immunoglobulin responses against HbsAg in both mice and humans. Although raw potatoes are not very palatable, it is less traumatic than administration of vaccine via a needle!

Genetic stability is the hallmark of cryopreservation. However, genetic responses to cold and freezing temperatures are not well understood. The induction of heat shock proteins and apoptosis enzymes have been observed immediately after thawing of mammalian cells [73]. Similarly, Haung and Tunnacliffe [74] investigated gene expression in a human embryonic kidney cell line when exposed to desiccation stresses. It is hoped that studies like this one will provide fundamental scientific insights into anhydrobiosis, i.e. preservation by desiccation, of mammalian cells.

On a more practical level, demonstration of genetic stability is a critical component of the overall testing program required fully to qualify recombinant MCB and WCB for commercial use. For example, studies to determine DNA sequences and plasmid copy number are required to ensure stability and fidelity of the genetic construct not only during the rigors of cryopreservation but also throughout the cell generations (i.e., *in vitro* cell age) needed for scale-up and final fermentation. Interested readers are referred to a practical overview of the subject [75] and a key international regulatory document providing guidance in this area [76].

6.11   Final Thoughts

Since all cells are not created equal, it is incumbent on the laboratory practitioner to continue to use an empirical approach in finding the best method of cryopreservation for their particular cells or tissues. However, regardless of the dissimilarity in their evolutionary status, all cells share enough fundamental characteristics to make the approaches to cryopreservation surprisingly similar. There are universal physical and chemical forces acting on all cells and tissues during freezing and thawing. Early cryobiologists, faced with these unforgivable natural forces, first looked to nature for answers and found many intriguing examples of successful adaptation to cold and freezing environments. For example, use of CPAs evolved millions of years ago in insects to enable overwintering. Although, for most laboratory practitioners the day-to-day challenges of cryopreservation have been overcome, the quest for the 'holy grail' goes on as cryobiologists continue to search for ways to preserve more complex multicellular structures such as embryos, tissues, and even organs. Cryopreservation of tissues and whole organs now seems possible with the development of new vitrification techniques. As reported by Fahy [77] at the 2005 Annual Meeting of the Society of Cryobiologists, a number of diverse tissues such as mouse ova, human corneas, and rat brain slices have now been successfully cryopreserved through the use of new vitrification solutions with lower toxicity. He indicated that these investigations are opening the door for potential cryopreservation of mammalian kidneys.

Fortunately, most laboratory practitioners reading this book work with cells adapted to grow in suspension either singly or as relatively small cell aggregates. The basic blueprint for cryopreservation of typical cell cultures used in biotechnology has largely been defined. It is up to each laboratory practitioner to use and modify this blueprint accordingly in order to create their own cell banks that will truly become the building blocks upon which to develop a fruitful fermentation process.

Acknowledgments

I wish to thank the editors for providing me with the opportunity to contribute this chapter. I am especially grateful to Dr K. Roger Tsang for his invaluable scientific mentorship over the years and for sharing his expertise in the art of cell culture.

Selected Reading List for Practitioners of the Art

Methods in Molecular Biology, Vol. 38: *Cryopreservation and Freeze-drying Protocols* (Day, J.G. and McLellan, M.R., eds), Humana Press, Inc., New Jersey.

ATCC Preservation Methods: Freezing and Freeze-drying, second edition, 1991, American Type Culture Collection, Rockville, MD.

Maintaining Cultures for Biotechnology and Industry (Hunter-Cevera, J.C. and Belt, A., eds), Academic Press, Inc., San Diego, CA.

Hubálek, Z. (2003) Protectants used in the cryopreservation of microorganisms. *Cryobiology* **46**, 205–229.

ICH Topic Q 5 D, Quality of biotechnological products: Derivation and characterization of cell substrates used for the production of biotechnological/biological products. The European Agency for the Evaluation of Medicinal Products, Human Medicines Evaluation Unit, 1997.

Maintenance of Microorganisms and Cultured Cells – A Manual of Laboratory Methods, Second edition, 1991 (Kirsop, B.E., and Doyle, A., eds), Academic Press Limited, London.

Withers, L.A. (1990) Cryopreservation of plant cells, in *Methods in Molecular Biology,* Vol. 6: *Plant and Tissue Culture* (Pollard, J.W. and Walker, J.M., eds), The Humana Press, New Jersey.

References

1. Fuller, B. and Paynter, S. (2004) Fundamentals of cryobiology in reproductive medicine. *Reproductive BioMedicine Online* **9**, 680–691.
2. Georlette, D., Blaise, V., Callins, T., and D'Amico, S. (2004) Some like it cold: biocatalysis at low temperatures. *FEMS Microbiology Reviews* **28**, 25–42.
3. Devireddy, R.V., Barratt, P.R., Storey, K.B., and Bishof, J.C. (1999) Liver freezing response of the freeze-tolerant wood frog, *Rana sylvatica,* in the presence and absence of glucose. I. Experimental measurements. *Cryobiology* **38**, 310–326.
4. Pegg, D.E. (1994) Cryobiology: life in the deep freeze. *Biologist* **41**, 53–56.
5. Polge, C., Smith, A.U., and Parkes, A.S. (1949) Revival of spermatozoa after vitrification and dehydration at low temperatures. *Nature* **164**, 666.
6. Lovelock, J.E. and Bishop, M.W.H. (1959) Prevention of freezing damage to living cells by dimethyl sulphoxide. *Nature* **183**, 1394–1395.
7. Matthes, G. and Hackensellner, H.A. (1981) Correlations between purity of dimethyl sulfoxide and survival after freezing and thawing. *Cryo-Letters* **2**, 389–392.
8. Withers, L.A. (1990) Cryopreservation of plant cells, in *Methods in Molecular Biology,* Vol. 6: *Plant and Tissue Culture* (Pollard, J.W. and Walker, J.M., eds), The Humana Press, New Jersey, pp. 39–48.
9. Grout, B.W.W. (1995) Cryopreservation of plant protoplasts, in *Methods in Molecular Biology,* Vol. 38: *Cryopreservation and Freeze-drying Protocols* (Day, J.G. and McLellan, M.R., eds), Humana Press, Inc., New Jersey, pp. 91–101.
10. Fuller, B. (2004) Cryoprotectants: The essential antifreezes to protect life in the frozen state. *CryoLetters* **25**, 375–388.
11. Yu, Z.-W. and Quinn, P.J. (1994) Dimethyl sulfoxide: A review of its applications in cell biology. *Biosciences Reports* **14**, 259–281.
12. Meryman, H.T. (1971) Cryoprotective agents. *Cryobiology* **8**, 173–183.
13. Hubálek, Z. (2003) Protectants used in the cryopreservation of microorganisms. *Cryobiology* **46**, 205–229.
14. Fahy, G.M. (1986) The relevance of cryoprotectant 'toxicity' to cryobiology. *Cryobiology* **23**, 1–13.
15. Engelmann, F. (2004) Plant cryopreservation: progress and prospects. *In Vitro Cellular and Developmental Biology-Plant* **40**, 427–433.

16. Mazur, P. (1984) Freezing of living cells: mechanisms and implications. *American Journal of Physiology* **247**, C125–C142.

17. Wolfe, J., Bryant, G., and Koster, K. (2002) What is 'unfreezable' water, how unfreezable is it and how much is there? *CryoLetters* **23**, 157–166.

18. Schrijnemakers, E.W.M. and Van Iren, F. (1995) A two-step or equilibrium freezing procedure for the cryopreservation of plant cell suspensions, in *Methods in Molecular Biology,* Vol. 38: *Cryopreservation and Freeze-drying Protocols* (Day, J.G. and McLellan M.R., eds), Humana Press, Inc., New Jersey, pp. 103–111.

19. Grout, B.W.W. and Morris, G.J. (1987) Freezing and cellular organization, in *The effects of low temperature on biological systems* (Grout, B.W.W. and Morris, G.J., eds), Edward Arnold, London, pp. 147–174.

20. Ashwood-Smith, M.J. and Friedmann, C.B. (1979) Lethal and chromosomal effects of freezing, thawing, storage time, and X-irradiation on mammalian cell preserved at $-196\,°C$ in dimethyl sulfoxide. *Cryobiology* **16**, 132–140.

21. Mazur, P. (1970) Cryobiology: the freezing of biological systems. *Science* **168**, 939–949.

22. Simione, F.P. and Brown, B.S. (eds) (1991) Freezing methods, in *ATCC Preservation Methods: Freezing and Freeze-drying,* second edition, American Type Culture Collection, Rockville, MD, pp.5–46.

23. Simione, F.P. (Jr.), Daggett, P.M., McGrath, M.S., and Alexander, M.T. (1977) The use of plastic ampoules for freeze preservation of microorganisms. *Cryobiology* **14**, 500–502.

24. Doyle, A., Morris, C.B., and Armitage, W.J. (1988) Cryopreservation of animal cells, in *Advances in Biotechnological Processes,* Vol. 7, *Upstream Processes: Equipment and Techniques,* Alan R. Liss, UK, pp.1–17.

25. Álamo, D., Batista, M., Gonzales, F., and Rodriguez, N. (2005) Cryopreservation of semen in the dog: use of ultra-freezers of $-152\,°C$ as a viable alternative to liquid nitrogen. *Theriogenology* **63**, 72–82.

26. Simione, F.P. and Brown, B.S., eds (1991) Principles of freezing and freeze-drying, in *ATCC Preservation Methods: Freezing and Freeze-drying,* second edition, American Type Culture Collection, Rockville, MD, pp.1–4.

27. Heckly, R.J. (1978) Preservation of microorganisms, in *Advances in Applied Microbiology,* Vol. 24, Academic Press Inc., San Diego, CA, pp. 1–53.

28. Tanaka, Y., Yoh, M., Takeda, Y., and Miwatani, T. (1979) Induction of mutation in *Escherichia coli* by freeze-drying. *Applied Environmental Microbiology* **37**, 369–372.

29. Carpenter, J.F., Crowe, L.M., and Crowe, J.H. (1987) Stabilization of phosphofructokinase with sugars during freeze-drying: characterization of enhanced protection in the presence of divalent cations. *Biochimica Biophysica Acta* **923**, 109–115.

30. MacKenzie, A.P. (1977) Comparative studies on the freeze-drying survival of various bacteria: Gram type, suspending media and freezing rate. *Developments in Biological Standardization* **36**, 263–277.

31. Tunnacliffe, A., García de Castro, A., and Manzanera, M. (2001) Anhydrobiotic engineering of bacterial and mammalian cells: is intracellular trehalose sufficient? *Cryobiology* **43**, 124–132.

32. Puhlev, I., Guo, N., Brown, D.R., and Levine, F. (2001) Desiccation tolerance in human cells. *Cryobiology* **42**, 207–217.

33. Perry, S.F. (1995) Freeze-drying and cryopreservation of bacteria, in Advances in Applied Microbiology, Vol. 24, Academic Press, Inc., San Diego, CA, pp. 21–30.

34. Nakamura, L.K. (1996). Preservation and maintenance of eubacteria, in *Maintaining Cultures for Biotechnology and Industry* (Hunter-Cevera, J.C. and Belt, A., eds), Academic Press, Inc., San Diego, CA, pp. 65–84.

35. Simione, F.P. and Brown, B.S., eds (1991) Bacteria and bacteriophages. *Biochimica Biophysica Acta* **923**, 14–16.

36. Dietz, A. and Currie, S. (1996) Actinomycetes, in *Maintaining Cultures for Biotechnology and Industry* (Hunter-Cevera, J.C. and Belt, A., eds), Academic Press, Inc., San Diego, CA, pp. 85–99.

37. Brown, J.M. and McNeil, M.M. (2003) *Nocardia, Rhodococcus, Gordonia, Actinomadura, Streptomyces,* and other aerobic actinomycetes, in *Manual of Clinical Microbiology, eighth edition* (Murray, P.R., Baron, ellen Jo, Jorgensen, J.H., Pfaller, M.A., Yolken, R.H., eds), ASM Press, Washington, DC, pp. 502–531.

38. Madigan, M.T. and Martinko, J.M. (2006) Prokaryotic diversity: the bacteria, in *Brock Biology of Microorganisms,* eleventh edition, Pearson/Prentice Hall, NJ, pp. 390–394.

39. Atlas, R.M. (1997) *Handbook of Microbiological Media,* second edition (Parks, L.C., ed.), CRC Press, Inc., Boca Raton, FL.

40. Smith, D.S. and Kolkowski, J. (1996). Fungi, in *Maintaining Cultures for Biotechnology and Industry* (Hunter-Cevera, J.C. and Belt, A., eds), Academic Press, Inc., San Diego, CA, pp. 101–132.

41. Bond, C.J. (1995) Cryopreservation of yeast cultures, in *Methods in Molecular Biology,* Vol. 38: *Cryopreservation and Freeze-drying Protocols* (Day, J.G. and McLellan M.R., eds), Humana Press, Inc., New Jersey, pp. 39–47.

42. Nierman, W.C. and Feldblyum, T. (1985) Cryopreservation of cultures that contain plasmids. *Developments in Industrial Microbiology* **26**, 423–434.

43. Pearson, B.M., Jackman, P.J.H., Painting, K.A., and Morris, G.J. (1990) Stability of genetically manipulated yeasts under different crypreservation regimes. *Cryo-Letters* **11**, 205–210.

44. Morris, G.J., Coulson, G.E., and Clarke, K.J. (1988) Freezing injury in *Saccharomyces cerevisiae*: the effect of growth conditions. *Cryobiology* **25**, 471–482.

45. Fowler, M.W. and Scragg, A.H. (1988) Natural products from higher plants and plant cell culture, in *Plant Cell Biotechnology* (Pais, M.S.S. ed.), Springer-Verlag, Berlin-Heidelberg, pp. 165–177.

46. Withers, L.A. (1991) Maintenance of plant tissue cultures, in *Maintenance of Microorganisms and Cultured Cells – A Manual of Laboratory Methods, second edition* (Kirsop, B.E., and Doyle, A., eds), Academic Press Limited, London, pp. 243–267.

47. Gruber, V. and Theisen, M. (2000) Transgenic plants in the production of therapeutic proteins. *Innovations in Pharmaceutical Technology* Vol.00, Issue 6, pp. 59–63.

48. Cooper, G.M. (2000). An overview of cells and cell research, in *The Cell, A Molecular Approach,* second edition, ASM Press, Washington, DC, pp. 3–39.

49. Shillito, R.D., Carswell, G.K., Johnson, C.M., DiMaio, J.J., and Harms, C.T. (1989) Regeneration of fertile plants from protoplasts of elite inbred maize. *Bio/Technology* **7**, pp. 581–587.

50. Quantrano, R.S. (1968) Freeze-preservation of cultured flax cells utilizing DMSO. *Plant Physiology* **43**, 2057.

51. Owen, H.R. (1996) Plant germplasm, in *Maintaining Cultures for Biotechnology and Industry* (Hunter-Cevera, J.C. and Belt, A., eds), Academic Press, Inc., San Diego, pp. 197–228.

52. Steponkus, P.L. and Lynch, D.V. (1989) Freeze/thaw-induced destabilization of the plasma membrane and the effects of cold acclimation. *Journal of Bioenergetics and Biomembranes* **21**, 21–41.

53. Withers, L.A. (1987) The low temperature preservation of plant cell, tissue and organ cultures and seed for genetic conservation and improved agricultural practice, in *The Effects of Low Temperatures on Biological Systems* (Grout, B.W.W. and Morris, G.J., eds), Edward Arnold, London, pp. 389–409.

54. Grout, B.W.W. (1995) Cryopreservation of plant protoplasts, in *Methods in Molecular Biology,* Vol. 38: *Cryopreservation and Freeze-drying Protocols* (Day, J.G. and McLellan M.R., eds), Humana Press, Inc., New Jersey pp. 91–101.

55. Shannon, J.E. and Macy, M.L. (1973) Freezing, storage, and recovery of cell stocks, in *Tissue Culture Methods and Applications* (Kruse, P.F. and Patterson, M.K., eds), Academic Press, Inc, New York, pp. 712–718.

56. Crowe, J.H., Carpenter, J.F., and Crowe, L.M. (1990) Are freezing and dehydration similar stress vectors? A comparison of modes of interaction of stabilizing solutes with biomolecules. *Cryobiology* **27**, 219–231.

57. Koch, G.J., Kruuv, J., and Bruckschwaiger, C.W. (1970) Survival of synchronized Chinese hamster cells following freezing in liquid nitrogen. *Experimental Cell Research* **63**, 476–477.

58. Terasima, T. and Yasukawa, M. (1977) Dependence of freeze-thaw damage on growth phase and cell cycle of cultured mammalian cells. *Cryobiology* **14**, 379–381.
59. Morris, C.B. (1995) Cryopreservation of animal and human cell lines, in *Methods in Molecular Biology,* Vol. 38: *Cryopreservation and Freeze-drying Protocols* (Day, J.G. and McLellan M.R., eds), Humana Press, Inc., New Jersey, pp. 179–187.
60. Doyle, A. and Morris, C.B. (1991) Maintenance of animal cells, in *Maintenance of Microorganisms and Cultured Cells*, second edition (Kirsop, B.E., and Doyle, A., eds), Academic Press Limited, London pp. 227–241.
61. Armitage, W.J. and Juss, B.K. (1996) The influence of cooling rate on survival of frozen cell differs in monolayers and in suspensions. *Cryo-Letters* **17**, 213–218.
62. Armitage, W.J. (1987) Cryopreservation of animal cells, in *Symposia of the Society for Experimental Biology*, Number XXXXI, *Temperature and Animal Cells* (Bowler, K. and Fuller, B.J., eds), The Company of Biologists Limited, Cambridge, UK, pp. 379–393.
63. Bielanski, A., Nadin-Davis, S., Sapp, T., and Lutze-Wallace, C. (2000) Viral contamination of embryos cryopreserved in liquid nitrogen. *Cryobiology* **40**, 110–116.
64. Doyle, A. and Morris, C.B. (1996) Cryopreservation, in *Cell and Tissue Culture: Laboratory Procedures* (Doyle, A., Griffiths, J.B., and Newell, D.G, eds), John Wiley & Sons, Chichester, pp. 4C:1.1–4C:1.7.
65. Ohno, T., Kurita, K., Abe, S., Eimori, N., and Ikawa, Y. (1988) A simple freezing medium for serum-free cultured cells. *Cytotechnology* **1**, 257–260.
66. Merten, O.-W., Petres, S., and Couvé, E. (1995) A simple serum-free freezing medium for serum-free cultured cells. *Biologicals* **23**, 185–189.
67. Beattie, G.M., Crowe, J.H., Lopez, A.D., Cirulli, V., Ricordi, C., and Hayek, A. (1997) Trehalose: a cryoprotectant that enhances recovery and preserves function of human pancreatic islets after long-term storage. *Diabetes* **46**, 519–523.
68. Eroglu, A., Russo, M.J, Bieganski, R., Fowler, A., Cheley, S., Bayley, H., and Toner, M. (2000) Intracellular trehalose improves the survival of cryopreserved mammalian cells. *Nature Biotechnology* **18**, 163–167.
69. Vara, E., Arias-Diiaz, J., Villa, N., Hernandez, J., Garcia, C., Ortiz, P., and Balibrea, J.L. (1995) Beneficial effect of *S*-adenosylmethionine during both cold storage and cryopreservation of isolated hepatocytes. *Cryobiology* **32**, 422–427.
70. *ICH Topic Q 5 D, Quality of Biotechnological Products: Derivation and Characterization of Cell Substrates Used for the Production of Biotechnological/Biological Products*. The European Agency for the Evaluation of Medicinal Products, Human Medicines Evaluation Unit, London 1997.
71. Annex 2, Manufacture of biological medicinal products for human use, in *Medicines Control Agency, Rules and Guidance for Pharmaceutical Manufacturers and Distributors 1997*, The Stationery Office, London pp. 107–114.
72. Huang, Z., Elkin, G., Maloney, B.J., and Beuhner, N. (2005) Virus-like particle expression and assembly in plants: hepatitis B and Norwalk viruses. *Vaccine* **23**, 1851–1858.
73. Fuller, B. (2003) Gene expression in response to low temperatures in mammalian cells: a review of current ideas. *Cryo-Letters* **24**, 95–102.
74. Huang, Z. and Tunnacliffe, A. (2005) Gene induction by desiccation stress in human cell cultures. *FEBS Letters* **579**, 4973–4977.
75. Kittle, J.D. and Pimental, B.J. (1997) Testing the genetic stability of recombinant DNA cell banks. *BioPharm* October, 48–51.
76. *ICH Topic Q 5 B, Quality of Biotechnological Products: Analysis of the Expression Construct in Cell Lines Used for Production of r-DNA Protein Products*. The European Agency for the Evaluation of Medicinal Products, Human Medicines Evaluation Unit, 1996.
77. Fahy, G. (2005) Vitrification as an Approach to Cryopreservation. Presented at the 42nd Annual Meeting of the Society for Cryobiology, Minneapolis, MN, July 24–27.

7

Modelling the Kinetics of Biological Activity in Fermentation Systems

Ferda Mavituna and Charles G. Sinclair

7.1 Introduction to Modelling

The aim of this chapter is to introduce the principles of modelling for constructing and manipulating mathematical expressions that describe biological activity. Although the chapter focuses mainly on microorganisms in bioreactors as a model system, the fundamental approach however, can be used to describe any biological activity in any system. The term system is used frequently in engineering and more increasingly in science, and it simply means something we want to study, understand, describe, predict, control and/or design (Figure 7.1).

The chapter starts with an introduction to principles of modelling; then, a brief section covers basic enzyme kinetics. After defining the kinetic rate expressions for basic biological activity, these are then used in material balances to describe the behaviour of microbial cultures in batch and steady-state continuous bioreactors. Finally, the dissolved oxygen limitation is considered in continuous and batch bioreactors.

7.1.1 Why Do We Need Models?

A model is a set of relationships between the variables of interest in the system being studied (Figure 7.2). The set of relationships, i.e. the model, can be a set of mathematical equations, graphs, tables, or unexpressed set of cause/effect relationships. The system being studied can be a bioreactor, a single cell, a microbial culture, an immobilised cell, an enzyme, any equipment of unit operations, such as a heat exchanger, a centrifuge, an

Practical Fermentation Technology Edited by Brian McNeil and Linda M. Harvey
© 2008 John Wiley & Sons, Ltd

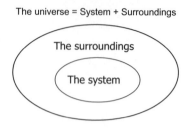

Figure 7.1   *The concept of the system, something we want to study*

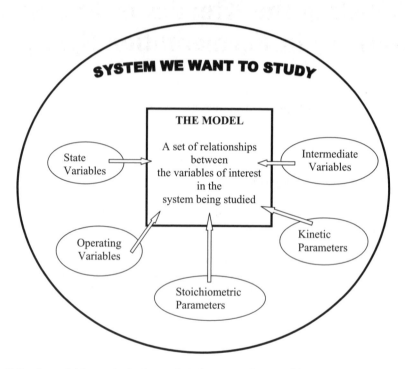

Figure 7.2   *A model is a set of relationships between the variables of interest in the system being studied*

HPLC column, etc. The variables of interest can be the feed rate, temperature, pH, the rate and mode of agitation, inoculum quality, and operational costs. Whether mathematical or not, a model therefore, describes how the system will behave in response to changes we make in the system or its environment.

In a mathematical model, we can have one or more equations that we can use to answer questions about the system: what will happen if we increase the temperature, decrease the concentration of glucose, decrease the spore counts in the inoculum, change from nitrate to ammonia as the nitrogen source, use a smaller bioreactor, a higher impeller speed, a lower aeration rate, use a genetically modified strain, etc. Other example ques-

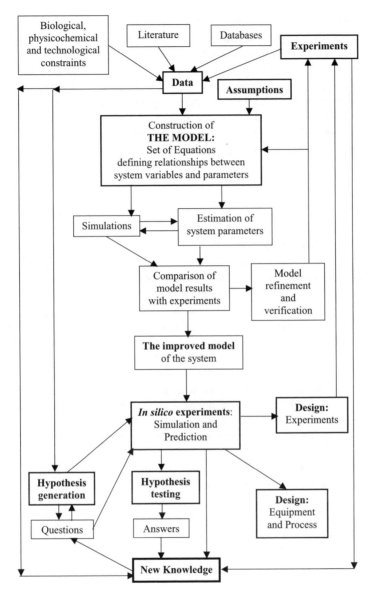

Figure 7.3 *The cyclic processes of model construction, verification and applications*

tions may involve the identification of the operating conditions for an economically breakeven point, for a given profit margin, etc.

Figure 7.3 shows schematically the cyclic processes of model construction, its verification and applications of the model in simulation, prediction and control, as well as the design of new experiments, equipment and processes, hypothesis generation and hypothesis testing, all leading to knowledge generation.

With advances in instrumentation, measurement, information technology, molecular biology and high-throughput techniques, there has been an explosion of quantitative and qualitative data and information in biotechnology and bioprocess engineering fields. In order to make any sense of this information, we need to look for relationships, connectivity amongst them. Various software tools are available for this purpose. Once we find a relationship, we have a model, however crude or complicated it may be. Information from experiments, observation and literature are the conventional starting points for modelling as shown in Figure 7.3. Since this figure has a cyclic nature, we do not have to start with the data and observations but we may choose to start with a hypothesis, construct a simple model based on the hypothesis, design some experiments based on the hypothesis and the model, and test it by comparison of the predictive model results and the experimental results.

We may start modelling by looking at a set of results from laboratory batch fermentation. Even when we plot changes in the concentration of a compound or cells against time, we are modelling. If we have different strains of a microorganism, including some genetically modified strains, we can compare their relative performance using quantitative definitions based on simple models, such as their specific growth rate or the specific production rate of a protein. Such a quantitative comparison can help us with our decision as to which strain to choose or which genetic modifications to perform in order to improve the desirable characteristics of the strain. Using model based definitions or terms, we can communicate and discuss with scientists and engineers from different backgrounds. A model therefore is a scientific communication language.

If we have a model that can predict how fast the cells will grow in a flask or a bioreactor as a function of the inoculum concentration in a given medium, we can plan our experiments knowing how much inoculum to use and when to start so that the experiment finishes before the weekend! If we have a model, we will also know what to measure or monitor and sometimes a model highlights the importance of factors that we did not think of as relevant or that we overlooked.

We therefore, use and need models in order to understand and describe some characteristics of the system as a function of other system components or the environment. This leads to new knowledge generation. We can use models in predictive mode and to design experiments, equipment and bioprocesses. For example, in order to produce a required amount of amino acid over a given period of batch time, using a given strain and using a model of the culture activity, we may have to choose the volume of the liquid medium in the bioreactor, its agitation and aeration, inoculum level, initial concentrations of the carbon-and-energy source and the nitrogen source. These applications of the models justify the importance of modelling.

In order to use models, we need to create them; hence the cyclic processes of Figure 7.3. The fundamentals of modelling will be introduced in the following sections. We need to mention that the computational software available often encourages the beginner to use curve-fitting approaches to modelling. Although this can be part of the modelling, one should try to associate biological, chemical and physical meaning with the equations rather than simply curve-fitting using a polynomial equation. Furthermore, although estimation of the values of the model parameters will be an important part of modelling, in the correct approach to modelling, a model should be tested by its ability to predict results from an independent set of experiments; that is, a set different from the experiments from which the values of model parameters were estimated. It should also be mentioned that the experimental errors should be considered in model verification and prediction. Sometimes, it is necessary to omit data that have a high degree of experimental error from modelling. It is therefore important to replicate the experiments, sampling and analyses. One should also remember that in the real world, negative concentrations do not exist, just in case the software gives a negative value as the solution. Another important point to remember is the constraints of the system. Constraints mean technical, biological, chemical and physical upper and lower limits to the range of values a system variables and parameters can take on.

7.1.2   The Components of Modelling

The components of modelling include the control volume, variables, parameters and the equations. Assumptions are also a very important and indispensable component of modelling.

The Control Region (Volume)

This is a very important concept, and the success or failure of the model may depend on the correct choice of it. The term control region or volume is used interchangeably. The control region is a space in the system we want to model, chosen by the modeller in such a way that all the variables of interest (concentration, temperature, pH, pressure, etc.) are uniform everywhere in the control region. This means that the concentration of a compound, for example, is the same everywhere within a chosen control region. This does not mean that the concentration in the control region is constant with time; the concentration can be constant or it can change with time so long as at any point in time, it is the same everywhere in the control region. Such a definition of the control region simplifies the modelling of complex systems but necessarily implies that sometimes, the control region may have to be chosen by the modeller as an imaginary space within a system since most real systems have a heterogeneous nature. In order to account for the heterogeneity of the system, we may define several control regions within the system we want to model. In Figure 7.4, the bioreactor on the left is very well mixed and the concentration

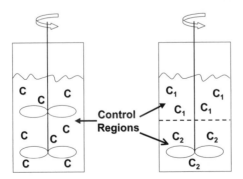

Figure 7.4 *The control region is a space in the system we want to model, chosen by the modeller in such a way that all the variables of interest (concentration, temperature, pH, pressure, etc.) are uniform everywhere in the control region*

is uniform everywhere in the bulk liquid; the bulk liquid is the control region. In the bio-reactor on the right, the impeller mixing characteristics may create two sections in the bulk liquid with different but uniform concentrations; the bulk liquid has two control regions. There may be exchange of matter, energy and momentum between different control regions. The volume of the control region may be constant or it may vary. The volume of the control region may be finite or infinitesimal.

The boundaries of the control region may be:

• phase boundaries across which no exchange takes place;
• phase boundaries across which an exchange of mass and/or energy takes place;
• geometrically defined boundaries within one phase across which exchanges take place either by bulk flow or by molecular diffusion.

Choosing the control volume is crucial to successful modelling, and although it may seem easy, several factors need to be considered for a complicated system. As summarised in Figure 7.5, the system we want to model must be decided first. Then, the consideration of the mode of operation or activity leads to the assumptions of either steady state or unsteady state, that is whether the system properties change with time or not. This, and the following consideration of the heterogeneity of the system, may lead to the choice of finite or infinitesimal control regions.

Variables

State variables.   These define the state of the process and there is one for each extensive property, for example:

$$x_v \quad \text{viable cell concentration;}$$
$$x_d \quad \text{nonviable cell concentration;}$$
$$S \quad \text{outlet and bioreactor limiting substrate concentration;}$$
$$P \quad \text{outlet and bioreactor product concentration.}$$

Operating variables.   These are variables the values of which can be set by the operator of the process, for example:

$$D \qquad \qquad \text{dilution rate;}$$
$$F \qquad \qquad \text{volumetric feed flow rate;}$$
$$S_i, x_{vi}, x_{di}, P_i \quad \text{inlet concentrations of the four conserved quantities.}$$

Intermediate variables.   These are all the volumetric rates r_x, r_d, r_{Sx}, r_{Sm}, r_{Sp} and r_P, which can all be expressed in terms of the state variables listed above.

Parameters

Kinetic parameters.   These are constants that are associated with the kinetic rate expressions for the system, such as μ_{max}, K_S, k_d, m_S, α, β, etc.

Stoichiometric parameters.   These define the stoichiometric relationships in the reactions or biological activity, such as yields: $Y_{P/S}$, $Y_{x/S}$, etc.

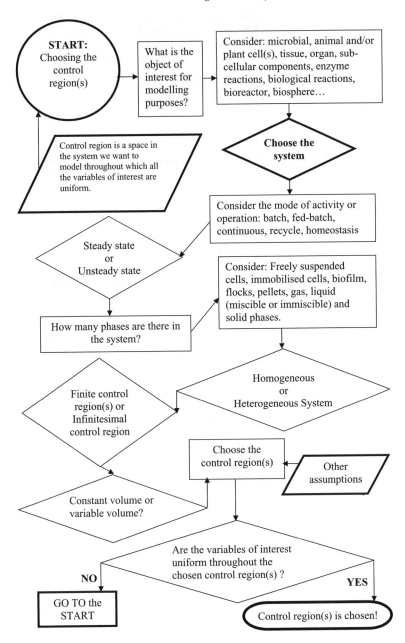

Figure 7.5 *The process of choosing the control region for modelling purposes*

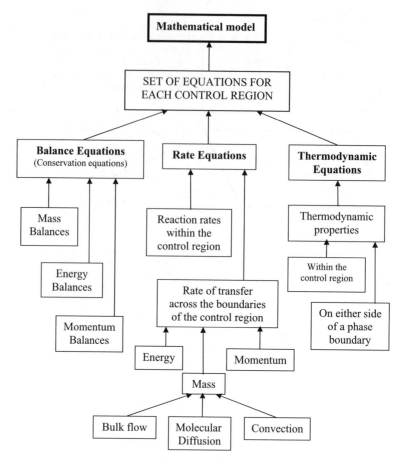

Figure 7.6 *Different types of mathematical equation that can be used in the construction of the models*

Equations

A conventional mathematical model is constructed by writing a set of equations for each control region. This set of equations consists of (Figure 7.6):

(a) balance equations for each extensive property of the system;
(b) rate equations:
   – rates of reaction, that is, generation or consumption of individual species within the control region;
   – rates of transfer of mass, energy, momentum across the boundaries of the control region;
(c) thermodynamic equations.

7.1.3 Mass Balances

Balance equations are written for every extensive property of interest in each control region.

- Extensive properties are those that are additive over the whole of a system. Thus mass or energy are extensive properties whereas concentration or temperature are not (they are intensive properties).
- Each balance equation should be linearly independent of the others. Linearly independent means that no balance equation can be formed by adding together any combination of the others.

$$\begin{bmatrix} \text{Rate of accumulation} \\ \text{or} \\ \text{Rate of depletion} \\ \text{in the control region} \end{bmatrix} = \begin{bmatrix} \text{Rate of input} \\ \text{to control region} \end{bmatrix} - \begin{bmatrix} \text{Rate of output} \\ \text{from control region} \end{bmatrix} \quad (7.1)$$

Input (positive) and output (negative) terms can describe both mass transfer and reaction phenomena:

- bulk flow across geometric boundaries;
- diffusion across geometric boundaries;
- transfer across phase boundaries;
- generation (input) and consumption (output) due to reaction within the control region.

Accumulation or depletion is simply the rate at which the amount of the extensive property within the control region changes with respect to time. Whether an extensive property accumulates or depletes will be determined by the numerical values, or the relative magnitude, of the input and output terms, considering all the input terms as positive and output terms as negative. Hence, if the total value of the input terms is larger than those of the output, the overall summation value of Equation (7.1) is positive and that particular extensive property accumulates in the control region. If the total value of the output terms is larger than those of the input terms, the overall summation value of Equation (7.1) is negative and that particular extensive property depletes in the control region.

7.2 Biological Reaction Rates for Basic Biological Activity

A living cell is capable of performing about two thousand metabolic reactions although not all may occur at any given time. The manifestation of the end result of all these metabolic reactions is the cell physiology (Figure 7.7). With the analytical and measurement instruments and techniques available, we can observe and measure (hence quantify) cell physiology in terms of cell division, growth, death, cell lysis, extent of viability, cell numbers, cell morphology such as size, shape and sporulation, substrate uptake, including dissolved oxygen for the respiration of aerobic cultures, effect of starvation and product formation. In order to illustrate the mathematical modelling of biological activity, we shall treat each of these complex activities as an individual biological reaction. Hence,

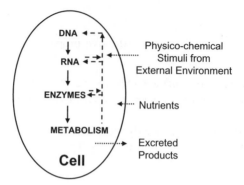

Figure 7.7 *The complex interrelationship between the activity of the DNA (transcription), RNA (translation), enzymes and metabolism in response to external stimuli and nutrients*

we shall use simple mathematical rate expressions for cell growth, death, lysis, substrate uptake and product formation. We shall also include cell maintenance or endogenous respiration among the basic cell activities. Although we shall use microbial reactions for the purposes of illustration, the same approach can be used to model the biological activity of plant and animal cells.

In order to simplify the modelling, we shall assume that in a microbial culture, all the living cells are of the same size, shape, mass and age, demonstrating the same physiology (a synchronous culture) even if this is rarely the case in reality. If we assume this, we can treat the whole living cells in the microbial culture as one uniform biomass. We shall treat the dead cells as a collection of a uniform mass as well. We shall therefore use the terms cell and biomass interchangeably. Segregated and structured models can be found in the literature that account for nonuniformity and/or heterogeneity in the cultures.

7.2.1 Definition of Volumetric and Specific Rates for Basic Microbial Activities

The volumetric rate of any biological reaction is defined as:

$$\text{Volumetric rate} \equiv \frac{\text{Amount of a compound produced or consumed}}{(\text{Unit volume})(\text{Unit time})} \tag{7.2}$$

The amount of something per unit volume is its concentration. The extent of any of the microbial activities, expressed as *volumetric rates*, depends on the concentration of viable biomass x_v in the control volume. Only in certain unusual cases may the volumetric rate depend on the concentration of dead cells. For example, a product that is normally kept in the cells may start appearing in the bioreactor bulk liquid when the cells are dead, due to the loss of membrane integrity. Although volumetric rates are useful to the process engineer for approximate design calculations, they do not allow us to compare the performance of cultures of the same or different microorganisms, for example genetically engineered with wild type strains, on a 'cellular' or physiological level. For this reason, we also define and use specific rates.

The specific rate of a microbial activity is equal to the volumetric rate for that activity divided by the cell concentration performing that activity. Specific rates are usually defined for growth, product formation and substrate uptake:

$$\text{Specific rate} \equiv \frac{\text{Volumetric rate}}{\text{Biomass concentration}} \tag{7.3}$$

$$\text{Specific rate} \equiv \frac{\text{Amount of a compound produced or consumed}}{\text{(Unit volume)(Unit time)(Concentration of biomass)}} \tag{7.4}$$

With these definitions, we use the following nomenclature for the basic microbial activities.

For growth:

$$r_x \equiv \text{Volumetric growth rate} \equiv \frac{\text{Amount of biomass formed}}{\text{(Unit volume)(Unit time)}} \tag{7.5}$$

Units of r_x are (kg live biomass) m^{-3} h^{-1}.

$$\mu \equiv \text{Specific growth rate} \equiv \frac{r_x}{x_v} \tag{7.6}$$

Units of μ are (kg live biomass) (kg live biomass)$^{-1}$ h^{-1} or simply h^{-1}.

Here, with x_v we denote the concentration of living cells as opposed to dead, and we make the distinction that growth is a biological activity performed by living cells.

For death:

$$r_d \equiv \text{Volumetric death rate} \equiv \frac{\text{Amount of dead cells formed}}{\text{(Unit volume)(Unit time)}} \tag{7.7}$$

Units of r_d are (kg dead biomass) m^{-3} h^{-1}.

$$k_d \equiv \text{Specific death rate} \equiv \frac{r_d}{x_v} \tag{7.8}$$

Units of k_d are (kg dead biomass) (kg live biomass)$^{-1}$ h^{-1}.

It should be noted that in the definition of the specific death rate we again use the living cell concentration since the process of dying is performed by living cells only. Once a cell is dead it does not die again to contribute to the process of dying.

For product formation:

$$r_P \equiv \text{Volumetric product formation rate} \equiv \frac{\text{Amount of product formed}}{\text{(Unit volume)(Unit time)}} \tag{7.9}$$

Units of r_P are (kg product) m^{-3} h^{-1}

$$q_P \equiv \text{Specific product fromation rate} \equiv \frac{r_P}{x_v} \tag{7.10}$$

Units of q_p are (kg product) (kg live biomass)$^{-1}$ h^{-1}.

Here, we assume that product formation is performed by the living cells, and hence live biomass concentration is used in the definition of the specific product formation rate. In some rare cases, product formation may be due to dead cells, for example, if the product is formed as a result of dead cells autolysing.

For substrate uptake:

$$r_S \equiv \text{Volumetric substrate uptake rate} \equiv \frac{\text{Amount of substrate consumed}}{(\text{Unit volume})(\text{Unit time})} \quad (7.11)$$

Units of r_S are (kg substrate) m^{-3} h^{-1}.

$$q_S \equiv \text{Specific substrate uptake rate} \equiv \frac{r_S}{x_v} \quad (7.12)$$

Units of q_S are (kg substrate) (kg live biomass)$^{-1}$ h^{-1}.

Again, the process of substrate uptake is normally performed by the living cells and therefore, the live biomass concentration is used in the definition of the specific substrate uptake rate.

7.2.2   Explicit Rate Expressions for Microbial Activity

In this section, we shall introduce commonly used explicit mathematical expressions for the volumetric rates of microbial activity. These basic forms of the kinetic rate expressions will be used later in the material balance equations for batch, fed-batch and continuous microbial cultures in bioreactors.

Growth

Requirements for growth.   The essential requirements for the growth of any microbial culture are as follows:

- a viable inoculum;
- a carbon source;
- an energy source;
- essential nutrients for biomass synthesis;
- suitable physicochemical conditions.

In order to divide and grow, a microbial culture needs, initially, at least one living and healthy cell. If the culture starts with only one cell, it takes a long time to obtain a measurable amount of cells. This is usually observed with microbial contaminations that may take a while before we can observe them (see Chapter 12 for more details). We therefore use an inoculum as a starter, or seed culture, which is a collection of healthy, viable cells in order to increase the volumetric rate. The rate of increase in live biomass concentration when all the above requirements are provided, will be proportional to the viable biomass concentration.

The specific growth rate can be expressed in two different ways: (i) substrate dependent and (ii) substrate independent.

Monod kinetics for growth.   Cell metabolism is made up of hundreds of sequential, branched and parallel biological reactions that are normally catalysed by enzymes. The production of these enzymes themselves is an important aspect of metabolism. We can assume that growth is the result of hundreds of such enzyme-catalysed reactions (Figure 7.8).

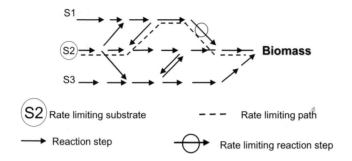

Rate limiting substrate — — — Rate limiting path

Reaction step $\ominus$ Rate limiting reaction step

Figure 7.8 *Growth is the result of hundreds of enzyme-catalysed metabolic reactions. One of these metabolic reactions may become the rate controlling step for growth*

The simplest of the expressions relating the enzyme reaction rate to the rate limiting substrate concentration is the Michaelis–Menten expression:

$$v = \frac{kES}{K_M + S} \tag{7.13}$$

in which v = the reaction velocity; k = the rate contant; E = the total amount of enzyme; K_M = the Michaelis–Menten (or saturation) constant; S = the substrate concentration.

The product kE is the maximum rate at which the enzyme reaction can proceed (when $S \gg K_M$) and is often written as v_M.

If we identify an enzyme reaction of the Michaelis–Menten type with the rate controlling step for growth (Figure 7.8), and if we assume moreover that the concentration of this rate controlling enzyme is proportional to the viable cell concentration, while the concentration of the substrate for the rate controlling step is proportional to the limiting substrate concentration in the nutrient medium, then we can write an analogous expression – the classical Monod equation for cell growth.

Normally the specific growth rate function $\mu(S)$ is simply abbreviated as μ, and it has the dimensions h^{-1}. Using the definition of the volumetric and specific rates for growth, from Equation (7.6) we have

$$r_x = \mu x_v \tag{7.14}$$

The relationship between the specific growth rate and limiting substrate concentration proposed by Monod states that:

$$\mu = \mu_{max} \frac{S}{(K_S + S)} \tag{7.15}$$

where r_x is the volumetric rate of cell growth, kg cells m^{-3} h^{-1}; μ_{max} is the maximum specific growth rate, h^{-1}; S is limiting substrate concentration, kg substrate m^{-3}; K_S is the saturation constant, kg substrate m^{-3}; x_v is the viable cell concentration, kg cells m^{-3}.

When the substrate concentration is not limited, that is when $S \gg K_S$ numerically, K_S can be ignored in Equation (7.15) and then S cancels each other, the specific growth rate approaches μ_{max} and the growth rate becomes independent of S and only proportional to the cell concentration. This is the zero order asymptote of the Monod expression, that is the specific growth rate is zero order with respect to substrate concentration:

$$\mu = \mu_{max} \qquad (7.16)$$

When the substrate concentration is lower than the numerical value of the Monod saturation constant, $S \ll K_S$, then S can be ignored in the denominator of Equation (7.15) and the specific growth rate becomes first order with respect to the growth limiting substrate concentration as

$$\mu = \frac{\mu_{max}}{K_S} S \qquad (7.17)$$

This is the first order asymptote of the Monod expression. For the substrate concentrations in between these two asymptotes, the specific rate is a function of the substrate concentration according to the Monod equation (Equation 7.15). In transition from zero order to Monod equation for the specific growth rate, we can define arbitrarily a critical substrate concentration, S_{crit}, in such as way that:

$$\mu = 0.99\mu_{max} \qquad (7.18)$$

Then inserting Equation (7.18) into Equation (7.15), we get:

$$S_{crit} = 99K_S \qquad (7.19)$$

It should be noted that Equation (7.19) will depend on the arbitrary definition we used for the critical substrate concentration; if we had defined it in Equation (7.18) as 90% instead of 99%, then we would get $S_{crit} = 90K_S$.

When the substrate concentration S is numerically equal to the saturation constant K_S then, the specific growth rate from Equation (7.15) is:

$$\mu = \frac{\mu_{max}}{2} \qquad (7.20)$$

For this reason, and considering the arbitrary definition of the critical substrate concentration, sometimes the Monod saturation constant, K_S is called the critical substrate concentration. The critical substrate concentration is useful in the design of the medium composition, indicating the concentration of a particular substrate when it becomes growth limiting.

Other growth rate expressions. Of course, there are several other mathematical expressions for the volumetric or specific growth rate. For example, the Monod equation for the correlation of specific growth rate and substrate concentration does not hold true when the intracellular substrate concentration is reduced during fast growth, even though adequate substrate is still available in the medium. In order to handle such situations, additional models have been developed. The logistic equation, which is a substrate independent method, is an alternative empirical function to the Monod equation. The logistic equation is also useful in processes where information on the limiting carbon substrate is not available, for instance, the limitation of an unidentified component in complex media. In the logistic equation, the specific growth rate can be expressed as:

$$\mu = \mu_{max}\left(1 - \frac{x_v}{x_{vm}}\right) \tag{7.21}$$

and for the cell growth:

$$r_x = \mu_{max}\left(1 - \frac{x_v}{x_{vm}}\right)x_v \tag{7.22}$$

where, μ_{max} is the maximum specific growth rate (h^{-1})and x_{vm} is the maximum viable biomass concentration.

A few examples of other forms of growth rate expressions are given below.

Tessier model:

$$\mu = \mu_{mas}\left(1 - e^{-s/K_s}\right) \tag{7.23}$$

Moser model:

$$\mu = \mu_{max}\left(1 + K_S S^{-\lambda}\right)^{-1} \tag{7.24}$$

where λ is a constant.

Contois model:

$$\mu = \mu_{max}\frac{S}{Bx + S} \tag{7.25}$$

where B is a constant implying that Bx is an apparent Monod constant that is proportional to biomass concentration x.

Growth inhibition.   Cell growth can be inhibited by various factors. The equations below give some examples of growth rate inhibition.

The substrate inhibition effect is often modelled by the expression:

$$\mu = \frac{\mu_{max}S}{K_S + S + \left(\dfrac{S}{k_i}\right)^2} \tag{7.26}$$

where k_i is the inhibition coefficient.

Inhibition by other substances can include the effect of poisons on cell growth, antibiotics, the accumulation of a product that becomes toxic above a threshold concentration, such as ethanol. Equations below describe such inhibition of growth by an inhibitor I.

$$\mu = \frac{\mu_m S}{K_S + S}(1 - k_i I) \tag{7.27}$$

$$\mu = \frac{\mu_m S}{K_S + S} \cdot \frac{k_i}{k_i + I} \tag{7.28}$$

$$\mu = \frac{\mu_m S}{K_S + S}\exp(-k_i I) \tag{7.29}$$

In these expressions I = inhibitor concentration and k_i = inhibition constant.

Death and Cell Lysis/Autolysis

During the course of fermentation some of the cells become nonviable, i.e. incapable of growth and reproduction. In this case, the cells cannot maintain their physiological activities, the cell membranes and wall lose their integrity and autolysis can occur. The rate cell death can be mathematically expressed as a first-order rate expression with respect to viable cell concentration:

$$r_d = k_d x_v \qquad (7.30)$$

where r_d is volumetric rate of conversion to nonviable form (kg cells m^{-3} h^{-1}); x_v is the concentration of viable cells (kg cells m^{-3}); k_d is the rate constant (kg cells kg $cells^{-1}$ h^{-1}).

Equation (7.30) means that the rate of conversion of viable cells to the nonviable form, the death rate, is assumed to be directly proportional to the concentration of viable cells. It should be noted that it is the viable cells that are contributing to the act of dying and not the dead cells; once cells are dead, they cannot die again to contribute to the death rate. Dead cells however, contribute to the volumetric rate of autolysis if they lose their cell membrane integrity. Normally lysis means that an external factor such as a change in the osmotic pressure of the cells' environment or a toxic chemical such as a detergent, disrupts the integrity of the cell membrane. Autolysis on the other hand, implies a self-inflicted disruption of the cell membrane integrity, for example, following death or starvation. Lysis as a consequence of viral infection is somewhere in between the two. Ideally therefore, autolysis should be a function of dead cell concentration whereas lysis can involve viable cell concentration as well as the concentration of the external lysis-causing factor. We shall not distinguish between autolysis and lysis in the treatment below for the sake of simplicity.

Cell autolysis can be expressed as a first order rate expression with respect to the dead biomass concentration:

$$r_l = k_l x_d \qquad (7.31)$$

where r_l is volumetric rate of cell lysis (kg cells m^{-3} h^{-1}); x_d is the concentration of dead cells (kg cells m^{-3}), and k_l is the rate constant (kg cells kg $cells^{-1}$ h^{-1}).

Product Formation

The cells themselves are the most important products of fermentation or biological activity. Since production of cells (or biomass) is covered under growth, here we consider other intracellular or extracellular products of fermentation or biological activity. The energy required to drive the cell processes is the chemical energy of ATP or similar substances. ATP and other energy currency compounds in the cell in most cases is provided either aerobically or anaerobically. Figure 7.9 is a schematic description of anaerobic and aerobic breakdown of carbon-and-energy substrates, and concomitant production of energy in the form of ATP and formation of various products. In aerobic cells, ATP is generated by the oxidation of substrate (usually a carbohydrate, the carbon and energy source) by molecular oxygen to CO_2 and water (oxidative phosphorylation). In anaerobic cells ATP (and other energy currency compounds) is generated by the degradation of substrate to simpler products such as ethanol, lactic acid, CO_2 and water, etc., which are excreted by the cell (substrate level phosphorylation).

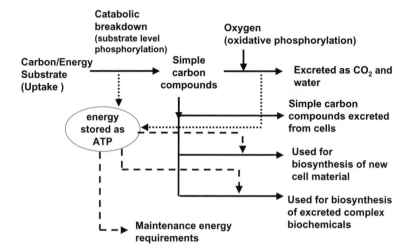

Figure 7.9 *A schematic description of anaerobic and aerobic breakdown of carbon-and-energy substrates, and concomitant production of energy in the form of ATP and its use in the formation of various products*

Other extracellular products include such compounds as:

- exoenzymes (for breaking down substrates that cannot pass through the cell wall);
- polysaccharides (for cell aggregation, avoidance of desiccation, binding metal ions, etc.);
- special metabolites (e.g., antibiotics).

There are substances produced in situations where the carbon substrate is in excess and other substrates, such as nitrogen or magnesium, are limiting. They include possible 'energy storage' compounds such as glycogen or lipids, etc., which are stored within the cell, or similar polysaccharides, etc., excreted by the cell. These products are considered by some to act as an 'energy sink'; excess ATP is produced so as to use the limiting substrate more efficiently, and the formation of energy storage products then dissipates the chemical energy of the excess.

The products of biological activity can be classified based on their production kinetics as follows:

Growth-associated

$$r_P = ar_x \tag{7.32}$$

Nongrowth associated

$$r_p = \beta x_v \tag{7.33}$$

Mixed kinetics

$$r_p = ar_x + \beta x_v \tag{7.34}$$

In these equations, α and β are constants. Equation (7.34) is called the Luedeking–Piret expression for product formation. Using the Monod equation for cell growth, it expands to:

$$r_P = \left[\frac{\alpha \mu_m S}{K_S + S} + \beta \right] x_v \qquad (7.35)$$

Production inhibition and degradation.   Product formation rate can be inhibited either by the product concentration itself, such as the production of ethanol, through its inhibitory action on cell metabolism, or it can be inhibited by another compound. Equation (7.36) is an example of the inhibition of product formation rate by the product concentration itself.

$$r_P = \left(1 - \frac{P}{P_m} \right)(\alpha r_x + \beta x_v) \qquad (7.36)$$

where P_m is the maximum attainable product concentration under inhibition conditions.

Maintenance and Endogeneous Respiration

The total energy required to maintain the concentration gradients that usually exist between the interior and the exterior of cells, and to drive turnover reactions in which labile cell components are continuously resynthesised is usually referred to as the maintenance energy requirement, being used only to maintain the cell in a viable state and not to produce cell material or excreted or energy storage products. Maintenance energy is also used in cell motility.

The volumetric rate of consumption of substrate to provide the energy for maintenance is written as r_{Sm} (kg substrate m^{-3} h^{-1}), and the generally accepted kinetic expression for the maintenance energy requirement is:

$$r_{Sm} = m_S x_v \qquad (7.37)$$

where m_S is the rate constant (kg substrate kg cells^{-1} h^{-1}).

When the external energy source is limited or exhausted, the cell switches to the use of internal energy storage compounds such as lipids, phospholipids, polysaccharides, glycogen, etc. This use of internal energy reserves, which is called endogenous respiration, results in a decrease in biomass concentration; cells are losing weight as a result of starvation!

The volumetric rate of endogenous respiration can be written as:

$$r_e = k_e x_v \qquad (7.38)$$

where r_e is volumetric rate of endogenous respiration (kg cell m^{-3} h^{-1}) and k_e is the rate constant (kg cell matter kg cells^{-1} h^{-1})

It should be apparent that, according to whether we use the maintenance energy concept or the endogenous respiration concept, the two consumption terms r_{Sm} and r_e will appear in different balance equations.

If we use the maintenance concept then r_{Sm} appears in the substrate balance, whereas if we use the endogenous respiration concept then r_e appears in the cell mass balance. However, there is no practical difference between the resulting sets of equations. We can convert from one to the other by making the substitution:

$$k_e = m_S Y'_{x/S} \qquad (7.39)$$

where $Y'_{x/s}$ is the yield of biomass on substrate, which is explained in the next section, except in the case where the external substrate supply is less than the maintenance requirement.

Stoichiometric Aspects of Metabolism

While cell composition may vary with cell type and physiological/environmental conditions, a typical cell can be assumed to contain protein, RNA, DNA, lipids, lipopolysaccharides, peptidoglycan, and glycogen.

The macromolecular composition of a typical *E. coli* cell can be: 70% water; 15% protein; 7% nucleic acids; 3% polysaccharides; 3% lipids; 1% inorganic ions, and 0.2% metabolites. Proteins are the most abundant organic molecules within the cell. Proteins usually contain around 50% C, 7% H, 23% O and 16% N. They also contain sulfur (up to 3%) for formation of S–S bonds.

Elemental balances.   Assuming that biomass consists of certain types of macromolecule (e.g. protein, RNA), it is possible to calculate an average elemental composition for biomass from the average content of the individual building blocks. The following are typical values based on the elemental composition of various cell components: protein $CH_{1.58} O_{0.31} N_{0.27} S_{0.004}$; DNA $CH_{1.15} O_{0.62} N_{0.39} P_{0.10}$; RNA $CH_{1.23} O_{0.75} N_{0.38} P_{0.11}$; carbohydrates $CH_{1.67} O_{0.83}$; phospholipids $CH_{1.91} O_{0.23} N_{0.02} P_{0.02}$; neutral fat $CH_{1.84} O_{0.12}$, and Biomass $CH_{1.81} O_{0.52} N_{0.21}$.

Single cell protein, or SCP, refers to proteinaceous materials that are dried cells of microorganisms. Example species that have been cultivated for use in animal or human foods include algae, actinomycetes, bacteria, yeasts, moulds and higher fungi. While human consumption of microbial protein is ancient in origin, more recent food products involve microbial growth in aerated bioreactors using substrates, such as natural gas and paraffins.

In order to illustrate the use of stoichiometry in the mathematical modelling of biological activity, we shall write a stoichiometric equation describing SCP production from methane. You may assume that the only metabolic products are carbon dioxide and water, and the nitrogen source is ammonia. Assume that the experimentally measured oxygen consumption is 1.35 mol oxygen per mol methane consumed.

The following reaction scheme can be assumed to describe the growth of cells on methane, oxygen and ammonia:

$$CH_4 + aO_2 + bNH_3 \quad \longrightarrow \quad cCH_{1.81} O_{0.52}N_{0.21} + dCO_2 + eH_2O$$

a, b, c, d and e are the stoichiometric coefficients indicating how many moles of a particular species are needed for the balanced reaction. The following stoichiometric balances can be written for the elements:

$$\text{Carbon:} \quad 1 = c + d \tag{7.40}$$

$$\text{Hydrogen:} \quad 4 + 3b = 1.81c + 2e \tag{7.41}$$

$$\text{Nitrogen:} \quad b = 0.21c \tag{7.42}$$

$$\text{Oxygen:} \quad 2a = 0.52c + 2d + e \tag{7.43}$$

Number of unknowns: five (a, b, c, d and e)
Number of equations: four (carbon, hydrogen, nitrogen and oxygen balance)
Degrees of freedom: one

With the degrees of freedom of one, one equation is still needed to describe the system fully. In this particular case, the missing equation comes from the experimental finding:

$$a = 1.35 \text{ mol oxygen used per mol methane used}$$

Solving the above elemental balance equations gives:

$$a = 1.35 \text{ (given)}$$
$$b = 0.13$$
$$c = 0.63$$
$$d = 0.37$$
$$e = 1.63$$

Substrate Utilisation

Uses of carbon in the cell.   Microorganisms require substrates for three main functions:

(i)   to synthesise new cell material;
(ii)   to synthesise extracellular products;
(iii)   to provide the maintenance energy necessary:
(a)   to drive the synthetic reactions;
(b)   to maintain concentrations of materials;
(c)   to drive recycling (turnover) reactions within the cell.

Thus growth, substrate utilisation, maintenance and product formation are all intimately related (Figure 7.8), and, as will be shown later, the various rate expressions are also mathematically related.

Yield coefficients and yield factors in rate expressions.   Substrate is used to form cell material and metabolic products, and therefore, the rate of substrate utilisation is related stoichiometrically to the rates of formation of these materials. In some cases it is possible to write the chemical equations; for example the equation for ethanol production from glucose is:

$$C_6H_{12}O_6 \quad \rightarrow \quad 2C_2H_5OH + 2CO_2$$

from which it is easy to calculate that 0.51 kg of ethanol will be formed from 1 kg of glucose. If for this case we write r_p as the rate of product (ethanol) formation and r_{SP} as the rate of substrate uptake for the product formation then:

$$r_{SP} = \frac{r_p}{0.51} \quad \text{or more generally:} \quad r_{SP} = \frac{r_p}{Y_{P/S}} \qquad (7.44)$$

where $Y_{P/S}$ is the yield coefficient for product on substrate and has the units kg product formed per kg substrate converted to product.

From generalised observation, anaerobic cell growth might be described by:

$$CH_2O + 0.2NH_4^+ + e^- \quad \rightarrow \quad CH_{1.8}O_{0.5}N_{0.2} + 0.5H_2O$$

where CH_2O represents the carbohydrate substrate, NH_4^+ is the nitrogen substrate, and $CH_{1.8}O_{0.5}N_{0.2}$ is an approximate formula for many cells (phosphorus and other elements that occur in much smaller proportions have been ignored).

From this we can readily calculate that $0.82\,kg$ of cells will be formed per kg of carbohydrate substrate utilised for cell formation, and $8.8\,kg$ of cells per kg of nitrogen. Thus we could write as above:

$$r_{Sx} = \frac{r_x}{0.82} \quad \text{or} \quad r_{Sx} = \frac{r_x}{Y_{x/S}} \tag{7.45}$$

$$r_{Nx} = \frac{r_x}{8.8} \quad \text{or} \quad r_{Nx} = \frac{r_x}{Y_{x/N}} \tag{7.46}$$

where

$\quad r_{Sx} \quad$ is the volumetric rate of carbohydrate utilisation for cell growth,
$\quad r_{Nx} \quad$ is the volumetric rate of nitrogen utilisation for cell growth,
$\quad r_x \quad$ is the volumetric rate of cell growth,
$\quad Y_{x/S} \quad$ is the yield coefficient for cells on carbohydrate,
$\quad Y_{x/N} \quad$ is the yield coefficient for cells on nitrogen.

The two material balances for carbon substrate (S) and nitrogen substrate (N) are therefore (using volumetric rates):

$$r_S = r_{Sx} + r_{Sp} + r_{So} \tag{7.47}$$

$$r_N = r_{Nx} + r_{Np} \tag{7.48}$$

This anaerobic utilisation of carbon-and-energy substrate is illustrated in Figure 7.10.

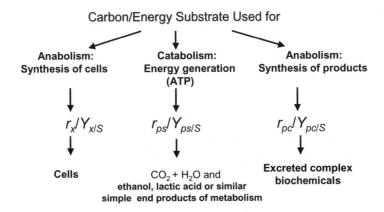

Figure 7.10 *Carbon-and-energy substrate utilisation in anaerobic metabolism and the use of yield coefficients*

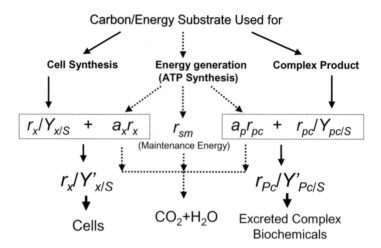

Figure 7.11   *Aerobic use of carbon-and-energy substrate and the use of yield factors*

It is important not to confuse these yield coefficients with 'yield factor' as normally reported in the literature. A yield coefficient is a stoichiometric constant that depends on the chemical equation relating the reactants and the products, whereas a yield is a ratio of one product to one reactant that may be entering into multiple reactions that form a variety of products. Thus in a normal fermentation process, with cell growth and product formation both consuming substrate, the 'yield' will be calculated on the total amount of substrate consumed. This is illustrated in Figure 7.11. In this figure, a_x is the amount in kmols of ATP required to synthesize one kg of cells and a_P amount in the kmols of ATP required to synthesize one kg of excreted complex biochemicals. Because of the uncertainty of the values of a_x and a_P (for example, we do not know exactly how many different complex compounds are formed and excreted from the cells), we use yield factors instead of yield coefficients.

In anaerobic yeast fermentation, the energy required for synthesis is provided by the ethanol-forming reaction. Since ethanol production can be measured easily, we can calculate how much energy in the form of ATP is generated.

We can calculate the cell and ethanol process yields as follows:

$Y_{x/S}$ = yield coefficient for cells on carbohydrate = 0.82
$Y_{P/S}$ = yield coefficient for ethanol on carbohydrate = 0.51

In those cases where we do not explicitly use the oxygen uptake rate, the substrate requirement to provide energy for maintenance is assumed to be first order with respect to cell concentration, i.e.:

$$r_{Sm} = m_S x_v \tag{7.37}$$

Thus the more common expression for volumetric rate of substrate utilisation is

$$r_S = r_{Sx} + r_{Sp} + r_{Sm} \tag{7.49}$$

$$r_{S} = \frac{r_{x}}{Y'_{x/S}} + \frac{r_{P}}{Y'_{P/S}} + r_{Sm} \qquad (7.50)$$

where $Y'_{x/S}$ and $Y'_{P/S}$ are yield factors, not stoichiometric yield coefficients.

7.3 Batch Culture of Microorganisms

7.3.1 Introduction

For a simple definition, a batch culture involves adding the nutrients, usually as aqueous solutions into a vessel, and then seeding the vessel with a collection of viable cells called the inoculum or seed culture, then closing the vessel and letting the bioactivity take its course with time. Since the concentrations change with time, a batch culture is an unsteady state operation. There is gas evolution mainly in the form of carbon dioxide as a consequence of microbial activity. Likewise since most industrially important bioprocesses are aerobic, and we need to supply oxygen continuously in the form of sparged air through the liquid medium, batch cultures are therefore, not true batches. Nevertheless, we usually concentrate on the liquid medium contents for modelling purposes and, except for the gaseous dissolved components, these operations can be treated as batch cultures.

A bioreactor is simply a vessel in which bioreactions take place (See Chapters 2 and 3). It can be a simple open container with very limited control or regulation such as the temperature, or a very sophisticated vessel with strict containment and computer monitoring and control, e.g. for bioreactions involving genetically modified organisms. Very often, bioreactors are referred to as fermenters and we tend to use the words fermenter and bioreactor synonymously. Strictly speaking only some bioreactors are fermenters, for instance, those that are anaerobic and used for alcohol production, but the term fermenter is commonly used to describe any vessel in which living organisms are being cultured. Likewise the bioreaction process is commonly referred to as a fermentation process.

A typical bioreaction/fermentation process might involve the following stages:

- medium preparation (the food source for the microorganism – usually a liquid) (See Chapter 5);
- sterilisation of medium, bioreactor, ancillaries (exclusion of unwanted organisms);
- development of a suitably sized inoculum (the pure culture 'seed' for the bioreactor);
- fermentation/bioreaction (growth of the desired microorganism and product formation);
- downstream processing (product extraction and purification);
- disposal of process effluents (bioprocesses rarely generate by-products).

Central to the whole process is the bioreactor, however, all stages of the process are important and influence the choice and design of the bioreactor. This is particularly true of batch processes, which make up the vast majority of bioprocesses.

The major groups of commercially important bioprocesses include:

- production of microbial cells (biomass) for food or fodder use (SCP);
- production of microbial enzymes;

- production of microbial metabolites (e.g. ethanol, citric acid, vitamins, antibiotics);
- production of recombinant products, i.e. resulting from genetic engineering to produce 'foreign' proteins, e.g. interferon, insulin, calf chymosin (rennet, for vegetarian cheese);
- production of monoclonal antibodies;
- biotransformations, in which a compound added to the fermentation is transformed into a more valuable product, the cells or enzymes being used as highly specific catalysts; biotransformations can also involve the degradation of toxic compounds;
- vaccine production;
- waste (water, soil) treatment.

Increasingly bioreactors are also used for the following:

- production of animal cells;
- artificial tissue and organ cultivation;
- production of plant cells for the production of pharmaceuticals, pigments and other speciality chemicals as well as synthetic seeds through somatic embryogenesis.

Except for the production of biomass and waste water treatment, most of these bioprocesses involve batch bioreactors. Although the productivities of continuous operations can be higher, in general batch operations are preferred to reduce the risk of contamination, reduce financial loses due to spoilt cultures, to provide better containment and to accumulate products so that inherently low product concentrations can be thus increased, which then help with downstream processing. These aspects are discussed more fully in Chapter 12. The operational characteristics and practicalities of differing modes of bioreactor operation are dealt with in Chapter 4, along with a basic subset of the equations used in this chapter, which anyone seeking to operate fermentations should be familiar with.

Assumptions

The general assumptions in the modelling of batch cultures in this section are:

- Microbial cultures are either live (living) or dead.
- Within these classes of live or dead cells, they are homogeneous in size, morphology, physiology and other physical–chemical–biological characteristics so that they can be treated as a single mass of live or dead cells. This assumption is demonstrably not appropriate to some fungal or Actinomycete cultures. In the modelling of such heterogeneous systems, we need to use structured models that account for the heterogeneous nature of the system.
- The bulk liquid is well mixed so that all the intensive properties, such as concentrations, temperature, pH are the uniform in the bulk liquid. This implies that the bulk liquid can be chosen as the control region.
- These assumptions also imply that cells are freely suspended. The modelling of biofilms, cell aggregates, and immobilised cells involve heterogeneous systems where the control region is normally infinitesimal in size. These will not be treated in this chapter.

- Bulk liquid volume is constant. This implies evaporation from the bulk liquid, which can be quite significant in long operations at raised temperatures especially when sparged air not saturated with respect to water vapor, is ignored.
- The control region is the bulk liquid in the bioreactor.

Other assumptions may be introduced as necessary in the following sections.

7.3.2   General Balance Equations for Batch Cultures

Balance equations are written for every extensive property of interest in each control region. For the batch culture, with the assumptions listed above, we shall take the bulk liquid as the control region.

- Extensive properties are those that are additive over the whole of a system. Thus mass or energy are extensive properties whereas concentration or temperature are not (they are intensive properties).
- Each balance equation should be linearly independent of the others. Linearly independent means that no balance equation can be formed by adding together any combination of the others.

A simple point but a common mistake made by those new to modelling is this: a mass balance should be made for one species/substance at a time in the same control region, and balances for two or more species/substances should not be mixed. It is like counting only the apples in a basket of mixed fruits. In this respect, using clear identification of the species/substances in the units should help avoid mistakes. It is for this reason that if we need to account for the maintenance energy, we should have either the maintenance energy, m_S, in the substrate balance or the endogenous energy (respiration), k_e, in the cell balance but not both in the model.

In batch cultures, concentrations change with time within the bioreactor bulk liquid. We therefore, have the accumulation or the depletion term, $d(\ )/dt$ in the mass balances. There is no flow of liquid medium into or out of the bioreactor and hence the input and output terms due to mass transfer are zero. We shall deal with the air sparging through the bulk liquid and the dissolved oxygen transfer into the bulk liquid of the batch culture in a later section. The only contributions to the input and output terms in the mass balance for a batch culture are those due to reactions. We treat generation or formation of a substance by a reaction in the control region as a positive input term. Contrarily, the consumption of a substance by a reaction in the control region is a negative output term. One should not get confused with the terminology of 'input' and 'output' because there are no flows in a batch culture. These terms simply imply positive or negative contributions in the 'balances'.

The sign convention especially, for the rate expressions as well as the time rate of change, d/dt terms, and slopes calculated from plots, is another concern in the material balances for modelling. The rate expressions should be treated as positive terms when being considered them on their own. Then let the sign convention associated with the category of terms in the mass balance, for example positive for input and negative for output terms, determine the sign. For example, the volumetric death rate, r_d, is a concept for the speed at which viable cells become dead. When r_d is included in the mass balance for viable cells, it is an output (consumption) term as far as the population of viable cells is concerned and, therefore, it will be a negative term. However, when r_d is included in the mass balance for the dead cells, it is a positive term as far as the population of dead cells is concerned. It contributes to the dead cell population, so it is an input, a positive term.

Using the volumetric rates, r, the general material balance of Equation (7.1) for a batch culture becomes:

$$\frac{dVy}{dt} = \sum Vr_{\text{gen}} - \sum Vr_{\text{cons}} \qquad (7.51)$$

where y is the general extensive property, r_{gen} is the volumetric rate of generation (production), r_{cons} is the volumetric rate of consumption, V is the volume of the control regions, which is the bulk liquid for a well-mixed bioreactor. In our case, y can be, for example, one of the following: live biomass concentration (x_v), dead biomass concentration (x_d), total biomass concentration (x_T), substrate concentration (S), product concentration (P), dissolved oxygen concentration (C_o), etc. The sign of $\sum$ indicates the summation of the similar terms; for example the summation of all the different volumetric consumption rates of the carbon substrate for biomass formation, for the formation of an excreted product and for the maintenance energy requirements.

For constant volume operation, V can be deleted from Equation (7.51) to yield

$$\frac{dy}{dt} = \sum r_{\text{gen}} - \sum r_{\text{cons}} \qquad (7.52)$$

7.3.3 Kinetic Models for Batch Cultures

Making an individual mass balance for each biological and chemical species of interest, we will have, for a microbial culture in a batch bioreactor the following mass balances:

Live biomass balance:

$$\frac{d(x_v V)}{dt} = +r_x V - r_d V \qquad (7.53)$$

Dead biomass balance:

$$\frac{d(x_d V)}{dt} = +r_d V \qquad (7.54)$$

Substrate balance:

$$\frac{d(SV)}{dt} = -r_s V \qquad (7.55)$$

Product balance:

$$\frac{d(PV)}{dt} = +r_P V \qquad (7.56)$$

where x_v = concentration of viable cells; x_d = concentration of dead cells; S = substrate concentration; P = product concentration; V = volume of bulk liquid (culture volume); r_x = volumetric rate of growth; r_S = volumetric rate of substrate consumption; r_P = volumetric rate of product formation; r_d = volumetric rate of cell death.

In a typical batch bioreactor, liquid volume is constant, since no liquid is added to or removed from the bioreactor. Therefore, V can be cancelled and the equations that are given above can be rewritten as:

Viable cells:

$$\frac{dx_v}{dt} = +r_x - r_d \tag{7.57}$$

Non-viable cells:

$$\frac{dx_d}{dt} = +r_d \tag{7.58}$$

Substrate:

$$\frac{dS}{dt} = -r_S = -(r_{Sx} + r_{Sm} + r_{SP}) \tag{7.59}$$

Product:

$$\frac{dP}{dt} = +r_P \tag{7.60}$$

The kinetic expressions for various volumetric rates can then be substituted from the previous sections. Assuming Monod kinetics for biomass growth, Luedeking–Piret type of product formation kinetics and maintenance energy concept, we use Equations (7.14), (7.15), (7.30), (7.34), (7.37), (7.49) and (7.50) in Equation (7.57–7.60) and obtain the following (for convenience, we drop using + sign for the first term after the equality sign as shown later in the chapter):

$$\frac{dx_v}{dt} = +\mu x_v - k_d x_v \tag{7.61}$$

$$\frac{dx_v}{dt} = +\mu_m \frac{S}{K_S + S} x_v - k_d x_v \tag{7.62}$$

$$\frac{dx_d}{dt} = +k_d x_v \tag{7.63}$$

$$\frac{dP}{dt} = +\alpha r_x + \beta x_v = \alpha \mu x_v + \beta x_v = \alpha \mu_m \frac{S}{K_S + S} x_v + \beta x_v \tag{7.64}$$

$$\frac{dS}{dt} = -\left(\frac{1}{Y'_{x/S}} r_x + \frac{1}{Y'_{P/S}} r_P + m_S x_v \right) \tag{7.65}$$

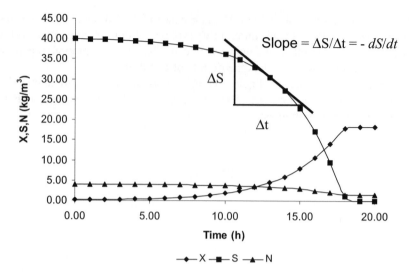

Figure 7.12   *Batch profile of biomass (x), carbon-and-energy substrate (S), and nitrogen source (N) concentration*

You should note that when maintenance energy has to be accounted for in the balances, it should appear as a consumption term in the limiting carbon and energy substrate balance. If on the other hand, the endogenous metabolism has to be accounted for instead of the maintenance energy, the endogenous respiration should appear as a consumption term in the viable biomass balance. The maintenance energy and endogenous respiration should not both be in the modelling of a biological system, only one of them. *Note*: The reader should try to write the balances above for the case when endogenous respiration has to be accounted for.

Figure 7.12 is a typical plot of biomass, carbon-and-energy substrate and nitrogen source concentrations against time in a batch culture, assuming that there is no death, no maintenance energy and no complex product formation that is excreted.

7.3.4   Growth Phases in Batch Culture

Now that we can construct the mass balances for viable cells, dead cells, substrates and products in a batch culture, we should know that d/dt, the time rate of change term cannot be automatically equated to a particular volumetric rate. For example, in Equation (7.61), dx_v/dt is not equal to the volumetric growth rate r_x if there is cell death, r_d. From a plot of experimental data of x_v against time, dx_v/dt can be calculated at any time t just by drawing a tangent to the viable cell curve and calculating the slope of this tangent. If there is no cell death, or if death can be ignored, then the numerical value of this slope is the numerical value of dx_v/dt which in turn is the numerical value of the volumetric growth rate, r_x. With this knowledge, we can now process the experimental data from batch cultures, calculate slope of tangents (Figure 7.12), and assign these numerical values to the appropriate volumetric rates or linear combination of these rates (e.g., Equation (7.65)).

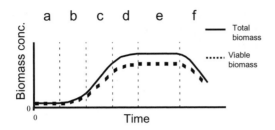

Figure 7.13   *Phases of a typical batch culture: (a) the lag phase, (b) the acceleration phase, (c) the logarithmic or exponential phase, (d) the deceleration phase, (e) the stationary phase, (f) the decline or death phase*

Figure 7.13 shows a typical growth curve for batch cultivation of bacterial cells. This curve includes six phases (a) to (f). In the lag phase (a), the cells adapt themselves from their previous to the new conditions, a period of metabolic adjustment. The lag period is usually prolonged if the organism is transferred from a rich medium to a poor one and vice versa. In phase (b), the acceleration phase, the cells begin to expand in volume and to divide; however, the culture is not synchronous. In the exponential phase, phase (c), cells grow at a maximum constant rate. However, the specific growth rate achieved is not an invariant property of a microorganism and it is markedly influenced by the environmental conditions, e.g. the complexity of the medium, the nature of the major carbon and energy source, pH and temperature. Duration of this phase depends partly on the initial concentration of growth-limiting substrate. The deceleration phase (d) is when the growth rate decreases. It is believed that limiting levels of oxygen and/or medium components or accumulation of toxic products could cause this decrease in growth rate. In the stationary phase (e), no net change in cell mass is observed, but this does not mean that all biosynthetic activity has ceased since a slow growth rate is balanced with cell death. The phase of decline or death phase (f), the last phase in growth curve, is when the rate of cell growth decreases to zero while the rate of cell death increases as a result of adverse physical conditions and hence the cell mass reduces.

As shown in Figure 7.13, the viable cell mass is usually slightly lower than the total cell mass. This is as a result of cell death occurring, usually simultaneously with growth. Cell death may occur as a result of many factors including mechanical stresses, lysis and mistakes in DNA biosynthesis.

Maximum Specific Growth Rate

The essential requirements for the growth of any microbial culture are as follows:

(i)   a viable inoculum;
(ii)   a carbon and energy source;
(iii) essential nutrients for biomass synthesis;
(iv)   suitable physicochemical conditions. This also includes the condition of absence of growth inhibitors or compounds/factors toxic to the cells.

The rate of increase in microbial biomass concentration when all the above requirements are provided and if cell death can be ignored, will be proportional to the viable microbial biomass concentration, i.e.:

$$r_x = \frac{dx}{dt} = \mu x_v \qquad (7.66)$$

where μ is described as the specific growth rate. When the specific growth rate is the maximum specific growth rate (i.e, $\mu = \mu_{max}$), which occurs during the exponential phase of microbial growth, the relationship between the variation biomass concentration and time may be deduced as follows:

$$\mu = \mu_{max} \qquad (7.67)$$

Inserting Equation (7.67) in to Equation (7.66) and rearranging:

$$\frac{dx_v}{x_v} = \mu_{max}t \qquad (7.68)$$

Taking the integral of Equation (7.68):

$$\int_{x_{v0}}^{x_v} \frac{dx_v}{x_v} = \int_{t=t_0}^{t} \mu_{max}t \qquad (7.69)$$

$$\ln \frac{x_v}{x_{v0}} = \mu_{max}(t - t_0) \qquad (7.70)$$

When there is no lag phase, then $t_0 = 0$. Inserting this in to Equation (7.70):

$$\ln x_v = \ln x_{v0} + m_{max}t \qquad (7.71)$$

or, using log instead of ln:

$$\log x_v = \frac{\mu_{max}t}{2.3} + \log x_{v0} \qquad (7.72)$$

where x_0 is the initial biomass concentration (inoculum concentration) introduced at zero time, t_0. A plot of $\ln x$ versus time can be used to obtain μ_{max} from the slope of the resulting line. It should be noted, however, that the equation above was derived under the assumption of constant specific growth rate (μ_{max}), which is usually true only in the exponential phase. A linear plot of $\ln x_v$ versus t cannot be expected for the whole batch period. In contrast to the specific growth rate μ, which is dependent on the environmental factors, μ_m is an inherent characteristic of an organism for a given set of physicochemical conditions.

Doubling Time

Doubling time, t_d, is defined as the time required for the biomass to be doubled and assuming a constant maximum value of specific growth rate this will be:

$$t_d = \frac{\ln 2}{\mu_{max}} = \frac{0.693}{\mu_{max}} \qquad (7.73)$$

Degree of multiplication

The degree of multiplication is given by x_v/x_{v0} and is equal to $e^{\mu t}$. Alternatively, if the biomass undergoes n doublings or generations, then:

$$\frac{x_v}{x_{v0}} = 2^n \tag{7.74}$$

Therefore,

$$n = 3.32 \log \left(\frac{x_v}{x_{v0}} \right) \tag{7.75}$$

Often, in culturing cells, it is useful to use an inoculum size which is 10% of the final biomass; then n will be 3.32.

7.3.5 Productivity

The productivity of a batch culture is given by the final concentration of whatever is being made divided by the complete time of the batch operation. This time period includes the running of the bioreactor as well as the emptying, cleaning, sterilizing and filling time of the bioreactor. The units of productivity, which is usually denoted by P (but in order not to confuse it with product concentration P, we can use the complete term) are, kg product m^{-3} h^{-1}.

7.3.6 Estimation of Parameters

As explained before, we can process the experimental data from batch cultures by first plotting concentrations against time to obtain the time profile (or time course) of the batch culture. Then we can calculate the slope of the tangents to experimental time course data at various time points, and assign these numerical values to the appropriate d/dt term (Figure 7.12). In turn, from the mass balances, the numerical values of d/dt terms are the numerical values of either individual volumetric rates (e.g., Equations 7.58 and 7.60) or linear combinations of these volumetric rates (e.g., Equations 7.57 and 7.59). In order to obtain the numerical value of a specific rate, we divide the experimental volumetric rate (calculated from the slope) by the biomass concentration at that time point.

Observed maximum specific growth rate can be determined if all the nutrients in the medium are in excess. Because under such conditions, growth is not limited and proceeds at its maximum, and the cell concentration increases exponentially. Therefore, a plot of $\ln x/x_o$ versus time can be used to determine μ_{max}.

The cell growth saturation coefficient, K_S as well as μ_{max} can be estimated by rearranging the Monod equation (Equation 7.15) in Lineweaver–Burk form, by taking the reciprocals of both sides of the equality sign, i.e.:

$$\frac{1}{\mu} = \frac{K_S}{\mu_{max} S} + \frac{1}{\mu_{max}} \tag{7.76}$$

According to Equation (7.76), the slope of a plot of $1/\mu$ versus $1/S$ will allow K_S to be evaluated, while the intercept is $1/\mu_{max}$.

The yield factor and maintenance coefficient can be estimated using the carbon substrate balance. If the substrate utilized for product formation is negligible, then r_P is zero and

$$r_S = \frac{r_x}{Y'_{x/S}} + m_S x_v \qquad (7.77)$$

Dividing both sides by x_v and substituting $r_x/x_v = \mu$ we get

$$\frac{r_S}{x_v} = q_S + \frac{\mu}{Y'_{x/S}} + m_S \qquad (7.78)$$

Therefore, a plot of specific substrate utilization rate, q_S, against specific growth rate, μ, will produce a straight line where the slope is equal to $1/Y'_{x/S}$ and intercept is m_S.

In order to obtain values for the product formation parameters α and β, the Luedeking–Piret equation (Equation 7.34) is rearranged dividing both sides by cell concentration, x_v, i.e.:

$$\frac{r_P}{x_v} = \alpha \frac{r_x}{x_v} + \beta \qquad (7.79)$$

Therefore, plotting r_P/x_v against μ should yield a straight line. α and β can then be obtained from the slope and y-axis intercept, respectively, as shown in Figure 7.14. Depending on the plot, one of the following may apply:

- growth associated product formation kinetics, $\alpha > 0$, $\beta = 0$ (line goes through the origin);
- mixed production kinetics, $\alpha > 0$, $\beta > 0$;
- nongrowth associated product formation kinetics, $\alpha = 0$, $\beta > 0$ (line is parallel to the x-axis).

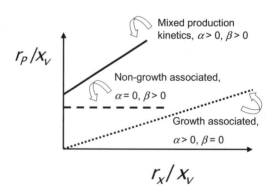

Figure 7.14   *Luedeking–Piret plot to determine the values of α and β for product formation rate expression*

7.4 Continuous Culture: The Chemostat

7.4.1 Definition and Assumptions

A continuous bioreactor is one that has a constant inflow and outflow of liquid, and usually at the same volumetric rate so that the volume in the bioreactor remains constant. Usually, the fresh medium is the inflow, and the bioreactor contents are assumed to be well mixed so that the concentrations are uniform throughout the bulk liquid. This means that the concentrations in the outflow from the bioreactor are the same as those in the bulk liquid in the bioreactor. A continuous culture starts as a batch culture first and then after the establishment of some growth, it is switched to continuous operation. After a transient state, a steady state is achieved when all the extensive properties of the system, such as the concentrations remain constant with time. This creates a physicochemical environment for the culture that is constant in time. For this reason, the continuous culture is often referred to as the chemostat. The chemostat is a very useful research tool (Chapter 12) and industrially it can be the most efficient mode of operation for biomass and growth associated product formation (Chapter 3). It does however, have an increased risk of contamination because of the flowing streams, and containment has to be carefully engineered.

7.4.2 The General Mass Balance

Figure 7.15 is a schematic diagram of a chemostat at steady state. The assumption that the bulk liquid is well mixed means that the output concentrations are the same as those in the bioreactor bulk liquid.

For a continuous culture, in addition to the kinetic rate terms, r_{gen} and r_{cons}, we have to consider the bulk flow terms, F_i and F_o in the mass balances. For the general extensive property y, the general mass balance of Equation (7.1) becomes:

$$\frac{d(Vy_o)}{dt} = \sum Vr_{gen} - \sum Vr_{cons} + F_i y_i - F_o y_o \qquad (7.80)$$

The input and output stream concentrations are indicated as subscripts.

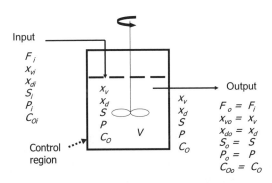

Figure 7.15 *A schematic diagram of a continuous culture, chemostat, at steady state*

For the chemostat, since the volumetric flow rates of input and output streams are equal, then

$$F_i - F_o = \frac{dV}{dt} = 0 \tag{7.81}$$

which means that the bulk liquid volume is constant. If we define γ as:

$$\gamma = \frac{F_o}{F_i} = 1 \tag{7.82}$$

then dividing Equation (7.80) by V gives:

$$\frac{dy_o}{dt} = \sum r_{gen} - \sum r_{cons} + D(y_i - \gamma y_o) = 0 \tag{7.83}$$

where $D = F/V$ is the dilution rate (h^{-1}). This is a key parameter, the reciprocal of which is the mean residence time of all the material flowing through the system.

In a chemostat operating at steady state, there is no accumulation or depletion of any extensive quantity, that is the concentrations remain constant with time so long as the dilution rate is kept constant. Thus, the specific material balance in every case can be obtained from the general equation by setting $dy_o/dt = 0$.

Equation (7.83) is the general mass balance equation that can be applied to individual species/compounds and the appropriate rate expressions can be inserted in order to obtain relationships that give steady state concentrations of biomass, substrates and products in terms of operational variables, kinetic and stoichiometric parameters.

7.4.3   Mass Balances for the Individual Compounds

Using Equation (7.83), we have the following mass balances for the individual compounds/species in the bulk liquid.
Viable cells:

$$0 = r_x - r_d + D(x_{vi} - x_{vo}) \tag{7.84}$$

Nonviable cells:

$$0 = r_d + D(x_{di} - x_{do}) \tag{7.85}$$

Substrate:

$$0 = (r_{Sx} + r_{Sm} + r_{SP}) + D(S_i - S_o) \tag{7.86}$$

Product:

$$0 = r_P + D(P_i - P_o) \tag{7.87}$$

We now need to substitute the appropriate volumetric rate expressions in order to obtain the equations that give the steady state concentrations for biomass, substrates and products.

Assumptions

Unless stated otherwise, for the following treatments we shall assume that:

- The culture is at steady state: the volumetric flow rates of the input and output streams are equal and remain constant, the volume of the bulk liquid remains constant (no evaporation losses).
- The bulk liquid is well mixed: the concentrations in the output stream are the same as those in the bulk liquid.
- The feed is sterile and hence does not contain any viable biomass: $x_{vi} = 0$.
- The feed does not contain any product: $P_i = 0$.
- The control region is the bulk liquid.

7.4.4 Kinetic Models for Chemostat

Assuming that the cells grow according to Monod kinetics, cell death, maintenance energy and product formation can be ignored, by inserting the volumetric rate expressions in to the steady state mass balances for cells and the carbon substrate we can obtain:

for viable cell balance:

$$0 = \mu x_{vo} - D x_{vo} \tag{7.88}$$

for substrate balance:

$$0 = -\frac{\mu x_{vo}}{Y'_{x/S}} + D(S_i - S_o) \tag{7.89}$$

From Equation (7.88), we get:

$$\mu = D \tag{7.90}$$

and this can be substituted into Equation (7.89) to give:

$$x_{vo} = Y'_{x/S}(S_i - S_o) \tag{7.91}$$

The substrate concentration S_o is obtained from the Monod expression:

$$\mu = \mu_{max} \frac{S_o}{K_S + S_o} \tag{7.15}$$

or by rearranging Equation (7.15) and substituting D for μ we get:

$$S_o = \frac{K_S D}{\mu_{max} - D} \tag{7.92}$$

It should be emphasized that Equation (7.90) holds only when a growth limiting medium is used in the chemostat. Otherwise, with the rich, nongrowth limiting medium cells grow at their maximum rate but their apparent growth can be anything, so Equation (7.90) becomes indeterminate. μ is only numerically equal to the dilution rate, D, in Equation (7.90) as the consequence of mathematics.

In order to solve the general model equations without the simplifying assumptions made to obtain Equations (7.91) and (7.92), we must select some suitable kinetic rate expressions. As we did before in the treatment of the batch bioreactor, we shall use Equations (7.14), (7.30), (7.34), (7.37), (7.49) and (7.50).

Inserting these rate expressions in to the mass balance equations (Equations 7.84–7.89, and rearranging the resulting equations we get the following expressions for the steady state concentrations:

From Equation (7.86):

$$x_{vo} = \frac{D(S_i - S_o)}{\dfrac{(D + k_d)}{Y'_{x/S}} + m_S + \dfrac{\alpha(D + k_d) + \beta}{Y'_{P/S}}} \tag{7.93}$$

From Equation (7.85):

$$x_{do} = \frac{k_d x_{vo}}{D} \tag{7.94}$$

From Equation (7.84):

$$S_o = \frac{K_S(D + k_d)}{\mu_{max} - (D + k_d)} \tag{7.95}$$

From Equation (7.87):

$$P_o = \frac{[\alpha(D + k_d) + \beta]x_{vo}}{D} \tag{7.96}$$

with the following constraints:

$$0 \le S_o \le S_i;\ 0 \le x_{vo};\ 0 \le x_{do} \le x_{vo};\ 0 \le P_o$$

7.4.5   Washout and Critical Dilution Rate

In a chemostat, the outlet limiting substrate concentration is independent of the input limiting substrate concentration. At a fixed dilution rate, called the critical dilution rate D_{crit}, the cell concentration drops to the constraint $x_{vo} = 0$ and the limiting substrate concentration reaches the upper constraint $S_o = S_i$. Beyond this critical dilution rate, cells are said to be washed out, since they are leaving the bioreactor at a higher rate than they are growing. Assuming that growth follows Monod kinetics, the critical dilution rate is found by using the onset of washout conditions, that is

$$S_o = S_i \quad \text{when} \quad D = D_{crit} \tag{7.97}$$

in the Monod expression (Equation 7.15)

$$D_{crit} = \frac{\mu_{max} S_i}{K_S + S_i} \tag{7.98}$$

Since S_i is usually very much greater then K_S, D_{crit} is approximately equal to μ_{max}.

7.4.6   Productivity

The productivity of a continuous bioreactor is given by multiplying the dilution rate, D, by the concentration of the product (which may be biomass) in the outlet stream. The cell productivity is Dx_v (kg biomass m^{-3} h^{-1}), and the product productivity is DP (kg product m^{-3} h^{-1}).

7.4.7 Estimation of Parameters

μ_{max} and K_S

Assuming that growth follows Monod kinetics and a growth limiting medium is used in the chemostat. Again, as for batch culture, the parameters μ_{max} and k_S are determined from a Lineweaver–Burk plot. Using the same equations as before we get a steady state chemostat version of the Lineweaver–Burk plot. By substituting the chemostat equation (Equation 7.90) into the Monod expression (Equation 15) and taking double reciprocals we have:

$$\frac{1}{D} = \frac{K_S}{\mu_{max}} \cdot \frac{1}{S_o} + \frac{1}{\mu_{max}} \tag{7.99}$$

A plot of $1/D$ against $1/S_o$ using the experimental steady state data, should yield a straight line if the assumptions are correct. The y-axis intercept is equal to $\frac{1}{\mu_{max}}$ and the slope is equal to $\frac{K_S}{\mu_{max}}$.

α and β

If the product formation follows a Luedeking–Piret model and there is no cell death, we have a model (Equation 7.96) that can be transformed to a linear relationship from which two parameters can be determined in a similar fashion to that shown in Figure 7.14. Normally, however, the number of kinetic parameters in any one expression is greater than two and this direct procedure must be extended. A typical example would be the steady state solution to the product formation equation when we cannot ignore cell death, Equation (7.96). By dividing by x_{vo} this equation can be transformed to:

$$\frac{P_o}{x_{vo}} = (\alpha k_d + \beta)\frac{1}{D} + \alpha \tag{7.100}$$

from which it is possible to obtain any two of α, β or k_d by plotting P/x_v against $1/D$ if any one of the kinetic parameters is known.

This example allows for the estimation of two parameters only if the third is known. Thus we can estimate α from the intercept and either β or k_d from the slope, given that the other is known. For example, we may already be satisfied that k_d can be ignored if cell death rate is low.

μ_{max}, K_S and k_d

From the mass balance we had Equation (7.95), which provides another example where there are three parameters, and thus by analogy, we would expect that we could only estimate two of the parameters given that the third was known. However this is an example of an equation where one of the parameters (k_d in this case) occurs only in a linear association with one of the variables (D). We can therefore use the observation that there will be only one value of this parameter for which we will be able to obtain a straight line plot. The transformed equation is:

$$\frac{1}{S_o} = \frac{\mu_{max}}{K_S} \cdot \frac{1}{(D+k_d)} - \frac{1}{K_S}$$
(7.101)

which will give a Lineweaver–Burk type of plot. Thus the procedure is to plot $1/S_o$ against $1/(D + k_d)$ for various values of the parameter k_d. For only one of these k_d values will we get a straight line and all others will give a curve. Hence this one value gives us the value of k_d that best fits the data, and subsequently μ_{max} and K_S can be calculated.

7.4.8   Chemostat with Recycle

In order to overcome washout problems and increase cell concentration within the continuous bioreactor, a portion of the cells from the outlet stream can be recycled back in to the bioreactor (Chapter 2).

For the purposes of mathematical analysis, recycle systems can be viewed as having an outflow stream with a cell concentration equal to the internal cell concentration multiplied by a constant, δ, which can be called the separation constant as shown in Figure 7.17.

$$0 \le \delta \le 1$$
(7.102)

As shown in Figure 7.17, including the cell recycle stream within the control region, and assuming that steady state is reached, simplifies the construction of mass balances. Please also note that, the recycle affects only the cells, that is only the cells are recycled. There is no change in the soluble component concentrations, substrates and products before and after the recycle point in Figure 7.17.

The steady state equations are obtained by modifying Equations (7.84 to 7.87) given for the chemostat for zero values of x_{vi}, x_{di} and P_i (no cells and no product in the input stream):

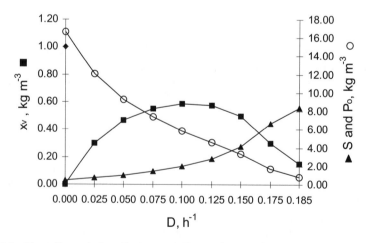

Figure 7.16 *The effect of the dilution rate D on the steady state concentrations of live biomass, carbon substrate and an excreted product*

Viable cells

$$0 = r_x - r_d - \delta D x_{vo} \tag{7.103}$$

Non-viable cells

$$0 = r_d - \delta D x_{do} \tag{7.104}$$

Substrate

$$0 = (r_{Sx} + r_{Sm} + r_{SP}) + D(S_i - S_o) \tag{7.105}$$

Product

$$0 = r_P - D P_o \tag{7.106}$$

Assuming growth follows Monod kinetics (Equation 7.15) and first order cell death rate kinetics, from the viable cell balance (Equation 7.103) we obtain:

$$S_o = \frac{K_S(\delta D + k_d)}{\mu_{max} - (\delta D + k_d)} \tag{7.107}$$

We can then insert Monod expression for growth, Luedeking–Piret expression for product formation, and first order death rate kinetics in to the balance equations Equations (7.103)–(7.106).

From the substrate balance (Equation 7.105), we get:

$$x_{vo} = \frac{D(S_i - S_o)}{\dfrac{(\delta D + k_d)}{Y'_{x/S}} + m_S + \dfrac{\alpha(\delta D + k_d) + \beta}{Y'_{P/S}}} \tag{7.108}$$

From Equation (7.104), the dead cell balance, we obtain:

$$x_d = \frac{k_d x_v}{\delta D} \tag{7.109}$$

From the product balance (Equation 7.106):

$$P_o = \frac{[\alpha(\delta D + k_d) + \beta] x_v}{D} \tag{7.110}$$

with the following constraints:

$$0 \le S_o \le S_i; \ 0 \le x_v; \ 0 \le x_d \le x_v; \ 0 \le P_o$$

As a simple check, note that these equations reduce to those for the simple chemostat when $\delta = 1$.

As before, 'washout' occurs when cells are being removed from the bioreactor at a rate that is just equal to the maximum rate at which they can grow in the bioreactor plus the rate at which they are recycled back in to the bioreactor (Figure 7.17). At the washout, as before, the value of x_{vo} becomes zero, and the outlet substrate concentration becomes equal to the inlet substrate concentration since it is not being consumed. The washout dilution rate for cell recycle is now found by inserting $S_o = S_i$ in the steady state equation for the substrate concentration (Equation 7.107):

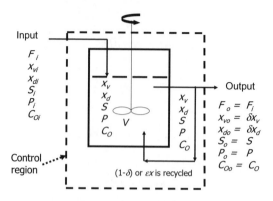

Input
F_i
X_{vi}
X_{di}
S_i
P_i
C_{oi}

Control region

$(1-\delta)$ or εx is recycled

X_v
X_d
S
P
C_o
V

X_v
X_d
S
P
C_o

Output

$F_o = F_i$
$X_{vo} = \delta X_v$
$X_{do} = \delta X_d$
$S_o = S$
$P_o = P$
$C_{oo} = C_o$

Figure 7.17   *Schematic representation of a chemostat with a cell recycle*

$$S_o = \frac{K_S(\delta D_{crit} + k_d)}{\mu_{max} - (\delta D_{crit} + k_d)} = S_i \qquad (7.111)$$

Rearranging this equation we get:

$$D_{crit} = \frac{S_i(\mu_{max} - k_d)}{\delta(K_S + S_i)} - \frac{K_S k_d}{\delta(K_S + S_i)} \qquad (7.112)$$

Since the numerical values of K_S and k_d are small compared with S_i, we can ignore the multiplication value of K_S with k_d and addition of K_S to S_i in this equation, giving:

$$D_{crit} \cong \frac{\mu_{max} - k_d}{\delta} \qquad (7.113)$$

The smaller is δ, the higher is the critical dilution rate.

7.5   Modelling Dissolved Oxygen Effects

7.5.1   Introduction

Most bioprocesses are aerobic, and aerobic bioprocesses require oxygen to be supplied to the bioreactor to allow growth of microorganisms. Oxygen is usually supplied in the form of air bubbled (sparged) through the bulk liquid, and this air must be sterilised, usually by filtration. Oxygen is only sparingly soluble in water, 6000-times less so than glucose. Typically, the saturation dissolved oxygen (DO) concentration in water, in equilibrium with air, is 7–10 ppm (= 7–10 mg/litre). This value depends on temperature, partial pressure of oxygen, and the concentration of other components, especially the ionic compounds in the aqueous phase. It is impossible to provide all the required oxygen at the start of the fermentation because of its low solubility. The dissolved oxygen must be replaced continuously throughout the fermentation, irrespective of the mode of operation, in order to prevent oxygen limitation slowing growth and affecting the metabolism.

Since oxygen is a sparingly soluble gas, in many aerobic microbial processes productivity is frequently limited by the transport of oxygen from the gas phase to the aqueous phase where the microbial culture grows. Oxygen transfer rate (OTR) is a physical phenomenon whereas the oxygen uptake rate (OUR) by the cells is a biological phenomenon. As long as the volumetric oxygen transfer rates to the culture medium exceed the volumetric rate of oxygen utilization by the cells, and no other nutrient is limiting, cell growth continues unimpeded. At some critical cell concentration, oxygen can no longer be supplied to the culture fast enough to meet the oxygen demand. Under these conditions oxygen becomes the limiting nutrient for cell growth, a factor causing both low cell densities and low product concentrations. In this section we shall model oxygen transfer and oxygen uptake along with other simple biological activity in continuous culture and investigate the transition from nongrowth-limiting to growth-limiting cases. We shall seek simple equations that can indicate the conditions under which such transitions will occur so that we can design and run experiments, equipment and bioprocesses under the desired regimes.

7.5.2 Oxygen Transfer Rate

Oxygen from the gas phase transfers to the bulk liquid through the gas–liquid interface. This interface is created when air (or other oxygen containing gas) is bubbled (sparged) through the bulk liquid using compressors and a sparger, which is often like a showerhead turned upside at the bottom of the bioreactor. The agitator of the bioreactor interferes with the rising air bubbles and creates smaller bubbles. Collectively, these small air bubbles create the large gas–liquid interfacial area that is necessary for satisfactory oxygen transfer.

Figure 7.18 is a schematic representation of the conditions at the gas–liquid interface. We describe the rate of mass transfer from a gas to a liquid phase using the two-film theory. On either side of the gas–liquid interface is a boundary layer or film of relatively stagnant fluid (liquid or gas). Molecules of oxygen must move from the bulk gas through

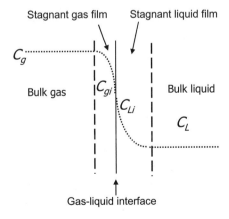

Figure 7.18 *Schematic representation of concentrations at either side of the gas–liquid interface*

the stagnant gas film, go into solution (become dissolved) at the interface, move through the stagnant liquid film, eventually reaching the bulk liquid. Mass transfer will only occur if there is a driving force for mass transfer in the form of a concentration difference.

A major assumption in modelling oxygen transfer is the treatment of all those hundreds of thousands of gas bubbles of different sizes as one, uniform, gas phase. We shall assume that the bulk phases are well mixed so that the bulk concentrations are the same every-where. So the concentration of oxygen in the bulk gas is greater than that on the gas side of the interface, providing a concentration driving force to transfer oxygen molecules to the interface. The concentration of oxygen on the liquid side of the interface is greater than that in the bulk liquid, providing a concentration driving force to transfer oxygen molecules into the liquid. At steady state, these two rates of mass transfer must be equal:

$$\text{Oxygen flux} = k_G \left(C_g - C_{gi} \right) = k_L \left(C_{Li} - C_L \right) \tag{7.114}$$

where k_G and k_L are the individual, gas and liquid, respectively, mass transfer coefficients with units of $(m)h^{-1}$.

At the interface, the oxygen concentrations on the gas and liquid side are in equilibrium described by Henry's Law, which is a thermodynamic relationship:

$$C_{gi} = HC_{Li} \tag{7.115}$$

However, we do not know and cannot measure the concentrations at the interface, C_{gi} and C_{Li}. So instead, we recognise that there would be a bulk liquid concentration in equilibrium with the bulk gas concentration, if we were to allow the system to reach equilibrium. It is the level away from equilibrium that provides the mass transfer driving force. We therefore, need to express Equation (7.115) in terms of a concentration we can measure.

C_g^* is defined as the fictitious bulk liquid concentration of the dissolved gas (in this case dissolved oxygen), which would be in equilibrium with the bulk gas concentration, C_g. We can then write:

$$C_g = HC_g^* \tag{7.116}$$

The difference between C_g^* and C_L provides the driving force for mass transfer. The mass transfer rate (expressed as a flux, i.e. rate per unit area) can be described in terms of this driving force and an overall mass transfer coefficient K_L as:

$$\text{Oxygen flux} = K_L \left(C_g^* - C_L \right) \tag{7.117}$$

The overall mass transfer coefficient K_L is the inverse of the sum of the individual resistances, k_L for the liquid film and k_G for the gas film:

$$\frac{1}{K_L} = \frac{1}{k_L} + \frac{1}{Hk_G} \tag{7.118}$$

It turns out that of these, $1/k_L$ is the largest, i.e. essentially all of the resistance to mass transfer of oxygen from gas to liquid phase is on the liquid-film side. So K_L can be replaced by k_L:

$$\text{Oxygen flux} = k_L \left(C_g^* - C_L \right) \tag{7.119}$$

Overall oxygen transfer rate per unit volume = (Flux) (Interfacial area)/
(Liquid Volume)   (7.120)

Then:

$$\text{OTR} = N_a = k_L \frac{A}{V}(C_g^* - C_L)t \qquad (7.121)$$

and

$$a = \frac{A}{V} \qquad (7.122)$$

where a is the specific interfacial area giving:

$$\text{OTR} = N_a = k_L a \left(C_g^* - C_L\right) \qquad (7.123)$$

Henry's law (Equation 7.160) is often written as:

$$C_g^* = \frac{\overline{P_O}}{H'} \qquad (7.124)$$

where $\overline{P_O}$ is the partial pressure of oxygen in the gas phase (N m^{-2}); H' is the Henry's law constant for this case [(N m^{-2})/(kg oxygen m^{-3})] C_g^* is the hypothetical saturation value of the dissolved oxygen in the liquid at the gas–liquid interface (kg O$_2$ m^{-3}).

For an ideal gas that is not pure and has more than one component (e.g., air), the partial pressure $\overline{P_O}$ is:

$$\overline{P_O} = Py \qquad (7.125)$$

where P is the total pressure of the gas and y is the mole fraction of oxygen in the gas phase.

From this point on, we shall use C_o instead of C_L to indicate the dissolved oxygen concentration in the liquid phase. The volumetric rate of mass transfer of oxygen from the gas to liquid phase, for constant a volume V of the bulk liquid, can be expressed mathematically as follows:

$$N_a = k_L a \left(C_g^* - C_O\right) = \text{OTR} \qquad (7.126)$$

where N_a = volumetric mass transfer rate (kg O$_2$ m^{-3} h^{-1} or mg O$_2$/Lh); $k_L a$ = volumetric oxygen transfer coefficient (h^{-1}); C_g^* = hypothetical saturation value of the dissolved oxygen in the liquid at the gas–liquid interface (kg O$_2$ m^{-3}); C_O = dissolved oxygen concentration in the bulk liquid (kg O$_2$ m^{-3}) OTR = oxygen transfer rate (kg O$_2$ m^{-3} h^{-1} or mg O$_2$/Lh).

The value of $k_L a$ obviously depends on the value of interfacial area, a, the thickness of boundary layer and the resistance to the diffusion of the oxygen through the boundary layer. These depend in a complex way on the hydrodynamics of the bioreactor and the physical properties of the medium and its constituents.

7.5.3   Microbial Oxygen Demand

The term specific oxygen uptake rate, q_O, shows the rate of oxygen consumption by the cells per unit amount of biomass:

$$q_O = \frac{r_O}{x_V} \qquad (7.127)$$

Oxygen requirement can also be expressed as growth and product yields on oxygen as defined in the following equations:

$$Y'_{x/O} = \frac{\Delta x}{\Delta C_O} \qquad (7.128)$$

$$Y'_{P/O} = \frac{\Delta P}{\Delta C_O} \qquad (7.129)$$

The dissolved oxygen taken up by cells can be used for the oxygen requirement in the biosynthesis of new cell material, excreted products and maintenance energy requirements:

$$r_O = r_{Ox} + r_{OP} + r_{Om} \qquad (7.130)$$

where r_{Ox} = rate of oxygen consumption for growth; r_{Om} = rate of oxygen consumption for maintenance energy; r_{OP} = rate of oxygen consumption for product formation.

$$r_{Ox} = \frac{r_x}{Y'_{x/O}} \qquad (7.131)$$

$$r_{Om} = \frac{r_{Sm}}{Y'_{S/O}} \qquad (7.132)$$

$$r_{OP} = \frac{r_P}{Y'_{P/O}} \qquad (7.133)$$

7.5.4   General Balance Equations for Chemostat Including Dissolved Oxygen

Ignoring cell death, we write the mass balances as follows:

For cells:

$$\frac{d(Vx_{vo})}{dt} = Vr_x + F(x_{vi} - x_{vo}) \qquad (7.134)$$

For soluble (carbon) substrate:

$$\frac{d(VS_o)}{dt} = -V(r_{Sx} + r_{SP} + r_{Sm}) + F(S_i - S_o) \qquad (7.135)$$

For dissolved oxygen:

$$\frac{d(VC_{Oo})}{dt} = -V(r_{Ox} + r_{OP} + r_{Om}) + VN_a + F(C_{Oi} - C_{Oo}) \qquad (7.136)$$

$$\frac{d(VC_{Oo})}{dt} = -V(r_{Ox} + r_{OP} + r_{Om}) + Vk_La(C_g^* - C_{Oo}) + F(C_{Oi} - C_{Oo}) \tag{7.137}$$

For product:

$$\frac{d(VP_o)}{dt} = Vr_P + F(P_i - P_o) \tag{7.138}$$

At steady state, the Equations (7.134) to (7.138) above will be equal to zero. V is also constant if we ignore evaporative losses, and hence V can be cancelled from both sides of these equations. As before, $F/V = D$, the dilution rate. Usually, the feed into the continuous bioreactor contains negligible amounts of dissolved oxygen and therefore, usually C_{Oi} is zero. In the most efficiently run bioreactors, the C_{Oo} will also be zero. We shall therefore use C_o in order to indicate the steady state dissolved oxygen concentration at the outlet and in the bioreactor in order to simplify the nomenclature.

As before, we can then substitute the appropriate volumetric rate expressions in Equations (7.134) to (7.138). Volumetric growth rate can still be assumed to follow Monod kinetics, but in this case, in addition to the carbon-and-energy substrate limiting the growth, we additionally have the dissolved oxygen, which may be limiting the growth. Double substrate limitation can, therefore, be applied for cell growth (both carbon substrate and dissolved oxygen limiting growth):

$$r_x = \mu x_v = \mu_{max} \left[\frac{S_o}{K_S + S_o} \right] \left[\frac{C_o}{K_o + C_o} \right] x_{vo} \tag{7.139}$$

We shall start with only one substrate limiting the growth, either the carbon substrate or the dissolved oxygen, and check the operating conditions when the other substrate starts becoming growth limiting, as this indicates the transition from one type of limitation to another. In between these two different limitations, there will be a region of operating conditions where both carbon and dissolved oxygen will be limiting the growth.

7.5.5   Carbon Substrate Limiting, Oxygen in Excess

For the sake of simplicity, we shall ignore cell death and product formation in the following analysis. Assuming growth follows Monod kinetics, and the growth limiting substrate is the carbon source, we get Equation (7.15). As before, the material balance equations give the solutions in Equation (7.92), and setting cell death and product formation related terms to zero in Equation (7.93), we obtain:

$$x_{vo} = \frac{DY'_{x/S}(S_i - S_o)}{D + m_S Y'_{x/S}} \tag{7.140}$$

Even though the carbon source is the growth limiting substrate, the cells do consume the dissolved oxygen and since its saturation concentration is low to start with, the steady state dissolved oxygen concentration becomes lower than the saturation value. We find the steady state dissolved oxygen concentration, for carbon-limited growth from the dissolved oxygen balance, Equation (7.137), by setting it equal to zero at steady state, ignoring the dissolved oxygen in the inlet stream, ignoring product formation and using Equations (7.15), (7.37), (7.90), (7.131) and (7.132) in Equation (7.137). We thus obtain:

$$C_o = C_g^* - x_{vo} \frac{D + m_S\left(\dfrac{Y'_{x/O}}{Y'_{S/O}}\right)}{Y'_{x/O}k_L a}$$

(7.141)

As $k_L a$ decreases, the value of C_o decreases until at some stage it will hit the constraint and become zero according to this model. Setting C_o to zero in Equation (7.141), we find:

$$C_g^* = x_{vo} \frac{D + m_S\left(\dfrac{Y'_{x/O}}{Y'_{S/O}}\right)}{Y'_{x/O}k_L a}$$

(7.142)

Since the numerical value of m_S is small, this is approximately equal to:

$$C_g^* \approx \frac{x_{vo}D}{Y'_{x/O}k_L a}$$

(7.143)

Equation (7.143) indicates the operational conditions for the dissolved oxygen concentration to be zero. This means that all the dissolved oxygen transferred from the gas phase is consumed. Carbon source in the medium is supplied at a level that is limiting growth, that is, it is not in excess. Furthermore, since the transfer of oxygen is costly, such an operating condition corresponds to an economically efficient operation because we are transferring oxygen at just the sufficient level. From Equation (7.143), for given/set values of other parameters, we can calculate the viable cell concentration that can be supported for such a case. Alternatively, we can calculate which dilution rate to use in order to achieve a particular biomass concentration for an economically efficient operation

7.5.6   Oxygen Limiting, Soluble Carbon Substrate in Excess

This time, we can write the Monod expression for growth using dissolved oxygen concentration instead of carbon substrate concentration as the growth limiting substrate. The mass balance on biomass gives:

$$D = \mu = \mu_{max} \frac{C_o}{K_o + C_o}$$

(7.144)

which then leads to:

$$C_o = \frac{K_o D}{\mu_{max} - D}$$

(7.145)

Setting the dissolved oxygen balance, Equation (7.137), equal to zero at steady state, assuming that the inlet stream contains no dissolved oxygen, and using Equations (7.15), (7.37), (7.90), (7.131) and (7.132) in Equation (7.137), we obtain:

$$x_{vo} = \frac{Y'_{x/o}\left[k_L a(C_g^* - C_o) - DC_o \right]}{D + m_S\left[\dfrac{Y'_{x/O}}{Y'_{S/O}}\right]}$$

(7.146)

Since numerically $k_L a \gg D$, and DC_o is a small number, Equation (7.146) reduces to:

$$x_{vo} = \frac{Y'_{x/o}k_La(C^*_g - C_o)}{D + m_S\left[\dfrac{Y'_{x/O}}{Y_{S/O}}\right]} \qquad (7.147)$$

The amount of oxygen required for maintenance, m_O is

$$m_O = \frac{m_S}{Y_{S/O}} \qquad (7.148)$$

Although the growth limiting substrate is dissolved oxygen, cells do consume the carbon substrate, and its steady state value can be obtained from the mass balance for carbon substrate, Equation (7.135) by setting it equal to zero at steady state and ignoring product formation. We then get:

$$S_o = S_i - x_{vo}\frac{D + m_S Y'_{x/S}}{DY'_{x/S}} \qquad (7.149)$$

Washout under conditions of oxygen limitation occurs when $C_o = C^*_g$ and from Equation (7.144) this is when:

$$D_{crit} = \mu_{max}\frac{C^*_g}{K_O + C^*_g} \cong \mu_{max} \qquad (7.150)$$

At low dilution rates the cell concentration is relatively high and carbon substrate concentration S_o drops to zero; that is, the lower constraint is encountered. The value of D at which this occurs can be obtained by setting $S_o = 0$ in Equation (7.149), which then gives:

$$S_i = x_{vo}\frac{D + m_S Y'_{x/S}}{DY'_{x/S}} \qquad (7.151)$$

Rearranging Equation (7.151) to give the biomass concentration x_{vo} and equating it to Equation (7.147) gives:

$$x_{vo} = \frac{S_i DY'_{x/S}}{D + m_S Y'_{x/S}} = \frac{Y'_{x/o}k_La(C^*_g - C_o)}{D + m_S\left[\dfrac{Y'_{x/O}}{Y_{S/O}}\right]} \qquad (7.152)$$

In Equation (7.152), we may ignore C_o and the terms multiplied with m_S since these are numerically small, which then gives us:

$$k_LaC^*_g = \frac{Y'_{x/S}}{Y'_{x/O}}DS_i \qquad (7.153)$$

which can be rearranged to give the value of the dilution rate at which the carbon substrate concentration S_o drops to zero (our starting supposition):

$$D = \frac{k_LaC^*_g Y'_{x/O}}{S_i Y'_{x/S}} \qquad (7.154)$$

This value of D decreases as either k_La or C^*_g decrease or S_i increases.

7.5.7   Double Substrate Limitation in Chemostat

Using Equation (7.184) for double substrate limitation (considering carbon substrate and oxygen as the limiting substrates), from the cell balance, we have:

$$r_x = Dx_{vo} = \mu x_{vo} = \mu_{max}\left[\frac{S_o}{K_S + S_o}\right]\left[\frac{C_o}{K_o + C_o}\right]x_{vo} \tag{7.155}$$

which has the non-trivial solution:

$$D = \mu = \mu_{max}\left[\frac{S_o}{K_S + S_o}\right]\left[\frac{C_o}{K_o + C_o}\right] \tag{7.156}$$

Unfortunately this cannot be solved directly for S_o or C_o as could be done in the previous cases with only one limiting substrate. Instead, we may adopt the following approach.

For carbon substrate limiting the growth we had Equation (7.140), and for the dissolved oxygen limiting the growth we had Equation (7.147). In the real world, the concentration of biomass cannot take on two values. Therefore, when both substrates are limiting the growth, x_{vo} will be equal to whichever is the lower value resulting from Equations (7.140) and (7.147).

If the numerical value of x_{vo} calculated from Equation (7.140) is lower than that calculated from Equation (7.147) then, we have the steady-state concentrations in the chemostat that were derived for carbon substrate limited growth:
for biomass, Equation (7.140). The corresponding value of the limiting substrate concentration is given by Equation (7.92) and the corresponding other substrate concentration is given by Equation (7.141).

If the numerical value of x_{vo} calculated from Equation (7.147) is lower than that calculated from Equation (7.140), then we have the following steady-state concentrations in the chemostat that were derived for dissolved oxygen limited growth:
For biomass, from Equation (7.146):

$$x_{vo} = \frac{Y'_{x/O}k_L a}{D + m_S\left[\dfrac{Y'_{x/O}}{Y_{S/O}}\right]}\left[C_g^* - \frac{K_O D}{\mu_{max} - D}\right] \tag{7.157}$$

The corresponding value of the limiting substrate concentration is given by Equation (7.145) and the corresponding carbon substrate concentration is given by Equation (7.149).

7.5.8   Oxygen Limitation in Batch Cultures

Similarly to continuous (chemostat) cultures, oxygen has to be continuously transferred from a sparged gas phase (normally air) in to the bulk liquid. The same oxygen mass transfer equations apply. If we assume that there is no cell death, and no excreted product formation, the mass balance equations for the batch culture are:

For biomass:

$$\frac{d(Vx_{vo})}{dt} = Vr_x \tag{7.158}$$

For the dissolved oxygen:

$$\frac{d(VC_o)}{dt} = -V(r_{Ox} + r_{Om}) + VN_a \tag{7.159}$$

Using the Monod rate equation for growth, with dissolved oxygen as the growth limiting substrate:

$$r_x = \mu x = \mu_{max} \frac{C_o}{K_O + C_o} x_v \tag{7.160}$$

Inserting the appropriate volumetric rate expressions and Equations (7.131), (7.132) and (7.148) as before, and making various substitutions in Equations (7.158) and (7.159), we get:

$$\frac{dx_v}{dt} = r_x = \mu_{max} \frac{C_o}{K_O + C_o} x_v \tag{7.161}$$

$$\frac{dC_o}{dt} = -\left[\frac{r_x}{Y'_{x/O}} + m_O x_v\right] + k_L a(C_g^* - C_o) \tag{7.162}$$

Equations (7.161) and (7.162) give the rate of change of biomass over time and dissolved oxygen concentration in a batch culture when the dissolved oxygen is limiting the growth.

When C_o drops to a very low value, the derivative dC_o/dt will be almost zero. For such a case, from the last equation above we get:

$$r_x = Y'_{x/O}(k_L a C_g^* - m_O x_v) \tag{7.163}$$

This equation gives approximately, the volumetric growth rate as a function of some operational and kinetic and stoichiometric parameters.

In closing this section, we should define the respiratory quotient that links carbon dioxide production as a consequence of aerobic respiration:

$$RQ = \text{Respiratory Quotient} = \frac{\text{Moles } CO_2 \text{ produced}}{\text{Moles } O_2 \text{ consumed}} \tag{7.164}$$

7.6   Conclusion

Mathematical models describe the behaviour of a system in response to changes in the system variables, parameters and environment. They are used in interpreting the experimental results, designing experiments, hypothesis creation and testing, and new knowledge generation. Industrial bioprocesses are designed using models from laboratory experiments. Industrial process and laboratory equipment are also designed using models.

The principles of modelling involve the following steps:

• choose the system;
• choose the control region;

- list the assumptions;
- decide on the type of model (for example, structured or unstructured; in this chapter we used a homogeneous culture system and therefore the models were unstructured);
- list the basic biological reaction rate expressions;
- list any thermodynamic equation to be used (we used Henry's law in the dissolved oxygen transfer section);
- construct the material balances for live biomass, dead biomass, substrate(s), dissolved oxygen, product(s);
- substitute kinetic and mass transfer rate expressions and thermodynamic equations into the mass balances;
- rearrange the mass balance equations for the rates of change of the amount or concentration of compounds (or species) with time for batch/unsteady state operation, or rearrange the mass balance equations for the steady state values of the concentrations of the compounds;
- estimate the values of model parameters using experimental data, and check these against the system constraints;
- test the model by comparing the simulation results from the model to the results from an independent set of experiments;
- develop, refine and improve the model in an iterative cyclic process by performing experiments, and comparing experimental and model results;
- use the developed model: design and plan experiments, design equipment, design bioprocess, create and test hypothesis, create new knowledge.

For those who are new to modelling, it is advisable to start with simple systems or simplified systems using appropriate assumptions. Once the principles of modelling are mastered, more complex systems can be attempted. Simple or complex, a valid mathematical model is a very useful and effective tool in fermentation science.

Further Reading

There are several books that can be suggested for detailed studies of modelling. Some of these are out of print, but can be found in the libraries, and some of the new ones specialise in more advanced modeling, which a beginner may find difficult to follow. Below is a list of suggested reading material which should be easy to follow:

Bioprocess Engineering Principles, Pauline M. Doran, Academic Press, London, 1995, Chapters 2, 3, 4, 11.
Basic Biotechnology, Colin Ratledge and Bjorn Kristiansen (Eds), second edition, Cambridge University Press, Cambridge, 2001, Chapters 3, 6, 7.
Biochemical Engineering, Harvey W. Blanch and Douglas S. Clark, Marcel Dekker, Inc., New York, 1997, Chapters 1, 3 and 4.
Biochemical Engineering Fundamentals, James E. Bailey and David F. Ollis, second edition, McGraw-Hill, Singapore, 1986, Chapters 3, 4, 5, 7, 8 and 9.

Nomenclature

a gas-liquid interfacial area; $m^2 m^{-3}$
a (with subscript) stoichiometric coefficient; $kg\,kg^{-1}$

B Contois coefficient; (kg substrate) (kg cells)$^{-1}$

C_g^* oxygen concentration in liquid medium in equilibrium with gas phase; (kg oxygen) m^{-3}

C_o O_2 concentration in bulk liquid medium; (kg oxygen) m^{-3}

D dilution rate; m^3 m^{-3} h^{-1} [that is, h^{-1}]

DF   driving force for mass transfer, see $k_L a$

E total amount of enzyme; kg

E activation energy; kJ mol^{-1}

F liquid volumetric flow rate; m^3 h^{-1}

h customary unit of time (hour)

I inhibitor concentration; kg m^{-3}

k kinetic rate constant; kg (or mol) kg^{-1} h^{-1}

K_S Monod constant, saturation constant; kg m^{-3}

kg   unit of mass (kilogram)

k_L oxygen transfer coefficient (on unit area basis); (kg O_2) m^{-2} h^{-1} (DF unit)$^{-1}$ (DF = driving force; if DF unit is (kg O_2) m^{-3}, then k_L unit becomes m h^{-1}

$k_L a$ Oxygen transfer coefficient (on unit volume basis); (kg O_2) m^{-3} h^{-1} (DF unit)$^{-1}$ (DF = driving force; if DF unit is (kg O_2) m^{-3}, then $k_L a$ unit becomes h^{-1}

m_S maintenance rate constant; kg(or mols) (kg cells)$^{-1}$ h^{-1}

m   unit of length (metre)

N O_2 transfer rate per unit liquid volume; (kg O_2) m^{-3} h^{-1}

P product concentration; (kg product) m^{-3}

r volumetric rate (of generation, consumption, production); kg m^{-3} h^{-1}

S (limiting) substrate concentration; (kg substrate) m^{-3}

t time; h (customary unit)

x cell concentration; (kg cells) m^{-3}

y general extensive property, concentration; (kg y) m^{-3}

$Y_{a/b}$ Stoichiometric yield coefficient; the order of subscripts is important, e.g. $Y_{a/b}$; (kg a) (kg b)$^{-1}$

$Y'_{a/b}$ yield factor (or yield); the order of subscripts is important, e.g. $Y'_{a/b}$; (kg a) (kg b)$^{-1}$

Greek

α growth-related product formation coefficient; (kg product) (kg cells)$^{-1}$

β nongrowth-related product formation coefficient; (kg product) (kg cells)$^{-1}$ h^{-1}

γ ratio of outlet flow rate to inlet flow rate

δ separation constant; ratio of outflow cell concentration to cell concentration in fermenter

μ specific growth rate; (kg cells) (kg cells)$^{-1}$ h^{-1} (that is, h^{-1})

Subscripts and Superscripts

ATP   adenosine triphosphate

c CO_2 from oxidative phosphorylation

crit   critical value (of dilution rate)

cons   consumption

d   non-viable (dead)

e   endogenous
$*_g$ in equilibrium with gas phase
gen   generation
i   inlet
i   inhibition
l   lysis
L   liquid phase
max   maximum
m   maintenance
N   nitrogen
o   outlet
O   oxygen
P   product
S   substrate
S   saturation
v   viable
x cells
x_d non-viable cells
x_v viable cells
$'$ (prime) superscript for yield factor

Appendix: Problems and Solutions

Batch Culture

What can You Do with the Experimental Data?

The experimental data obtained in a microbial batch culture are given in Table A.1. N refers to the nitrogen source and S indicates the carbon and energy source.

Find the volumetric rate expressions of growth, substrate consumption and product formation and estimate the numerical values of the relevant kinetic parameters. In doing so,

(a) Estimate the values of the maximum specific growth rate, μ_{max} and the Monod coefficient, K_S from the Lineweaver–Burk plot.
(b) Estimate the values of the maintenance constant, m_S and the yield for biomass on carbon substrate, $\dfrac{1}{Y'_{x/S}}$.
(c) Estimate the value of the yield for biomass on nitrogen substrate, $\dfrac{1}{Y'_{x/N}}$.

Solution

First plot all the concentrations against time and try to plot smooth curves (or lines, if it looks linear) through the experimental points either by hand or using a suitable graphics programme such as Excel. At this stage we do not know the mathematical expression (or the model) that would give us how the concentrations change with time. We shall use

Table A.1 *Experimental data obtained in a microbial batch culture*

t (h)	x_v (kg/m³)	S (kg/m³)	N (kg/m³)
0	0.100	40.00	4.00
1	0.134	39.93	4.00
2	0.180	39.83	3.99
3	0.241	39.20	3.98
4	0.323	39.50	3.97
5	0.433	39.30	3.96
6	0.581	39.10	3.94
7	0.778	38.50	3.92
8	1.040	37.80	3.88
9	1.400	37.20	3.84
10	1.870	35.40	3.78
11	2.500	34.80	3.70
12	3.350	32.90	3.59
13	4.490	29.50	3.44
14	6.000	27.20	3.24
15	8.000	21.80	2.97
16	10.700	17.10	2.57
17	14.100	9.60	1.89
18	17.900	1.11	1.50
19	18.300	0.00	1.49
20	18.300	0.00	1.48

these smooth curves through the data points in the calculation of time derivatives, d(...)/dt.

We assume that growth follows Monod kinetics and the growth limiting nutrient is the carbon and energy source, S. Since there is no decrease in the values of the biomass concentration at the end of the batch, we can also assume that microbial death can be neglected.

(a) Estimation of μ_{max} and K_S from the Lineweaver–Burk plot. The Lineweaver–Burk plot is given by Equation (7.76):

We need to plot $1/\mu$ versus $1/S$ and evaluate K_S from the slope, and $1/\mu_{max}$ from the y-axis intercept. We therefore need to obtain the values of the specific growth rate μ which changes with the carbon substrate concentration S, which in turn changes with time, t. From the mass balance on biomass in a batch culture, using Equations (7.57), (7.61) and (7.62), and assuming that there is no death, we have:

$$\frac{dx_v}{dt} = r_x = \mu \cdot x_v \qquad \text{so that} \qquad \frac{1}{\mu} = \frac{1}{x_v} \cdot \frac{dx_v}{dt}$$

According to this, $1/\mu$ value at a time t can be obtained from the time derivative of cell concentration, dx_v/dt. For this, use the plot of experimental cell concentration against time, as shown in Figure A.1.

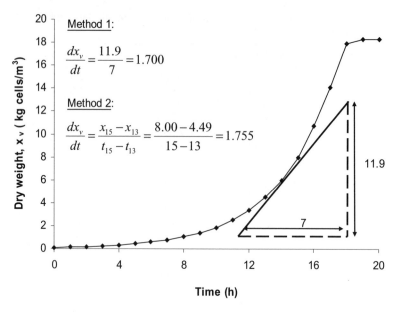

Figure A.1 *Biomass concentration against time plot and the slope-of-the-tangent method to obtain μ values at different times*

How to obtain the time derivatives, d(. . .)/dt:

Method 1: Hand-drawn tangent: Hand draw a tangent to the smooth curve drawn through the experimental data points and measure its slope as shown in Figure A.1 for $t = 14\,$h.

Method 2: Numerical differentiation: Take the difference between the values equally spaced on either side of the data point being analysed ($t = 14\,$h) and divide by the appropriate time interval as shown in Figure A.1.

Method 3: Curve fitting: You may have already used a software to plot a smooth curve through your experimental data points. Some software gives you the equation used for curve fitting and from the equation you can find the value of the time derivative.

Some values of $1/\mu$ obtained from the experimental data are given in Table A.2.
Use the data in Table A.2 to obtain the Lineweaver–Burk plot of Figure A.2

From the *y*-axis intercept on can calculate:

$$\mu_{max} = \frac{1}{3.4} = 0.294\,\text{h}^{-1}$$

From the *x*-axis intercept, one can calculate:

$$K_S = \frac{1}{0.6} = 1.67\,\text{kg m}^{-3}\ \text{glucose.}$$

These values can be checked from the slope of the line, $\dfrac{K_S}{\mu_{max}} = 5.67\,\text{kg glucose h m}^{-3}$ as shown in Figure A.2.

Table A.2   *Calculation of 1/μ and 1/S for the Lineweaver–Burk plot*

t (h)	x_v (kg/m³)	$r_x = dx_v/dt$	$\mu = r_x/x_v$	$1/\mu$	S	$1/S$
0	0.100				40.00	0.0250
1	0.134	0.040	0.299	3.350	39.93	0.0250
2	0.180	0.054	0.297	3.364	39.83	0.0251
3	0.241	0.072	0.297	3.371	39.70	0.0252
. . .	. . .	. . .	. . .	. . .	. . .	. . .
12	3.350	0.995	0.297	3.367	32.90	0.0304
13	4.490	1.325	0.295	3.389	30.50	0.0328
14	6.000	1.755	0.293	3.419	27.20	0.0368
15	8.000	2.350	0.294	3.404	22.80	0.0439
16	10.700	3.050	0.285	3.508	17.10	0.0585
17	14.100	3.600	0.255	3.917	9.60	0.1042
18	17.900			8.547	1.11	0.9009

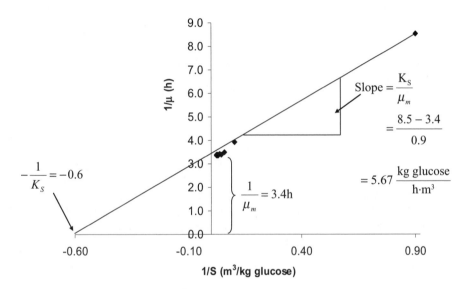

Figure A.2   *Lineweaver–Burk plot*

(b) Estimation of the maintenance constant m_S and biomass yield on carbon source, $Y'_{x/S}$. Inserting Equations (7.6) and (7.37) into Equation (7.50) (or from Equation (7.65) and assuming that there is no product formation other than the cells, we have:

$$r_S = \frac{r_x}{Y'_{x/S}} + m_S x_v = \frac{\mu x_v}{Y'_{x/S}} + m_S x_v$$

Dividing both sides with x_v we get:

$$\frac{r_S}{x_v} = \frac{1}{Y'_{x/S}} \cdot \frac{r_x}{x_v} + m_S$$

Table A.3   *Estimated values of the volumetric rates for growth (r$_x$), carbon (r$_S$) and nitrogen (r$_N$) substrate consumption from the time derivatives of biomass, carbon and nitrogen substrate concentration against time plots*

t (h)	x$_v$ (kg/m³)	S (kg/m³)	N (kg/m³)	r$_x$	r$_S$	r$_N$	r$_S$/x$_v$	r$_x$/x$_v$
0	0.100	40.00	4.00					
1	0.134	39.93	4.00	0.040	0.085	0.005	0.634	0.299
2	0.180	39.83	3.99	0.054	0.115	0.010	0.639	0.297
3	0.241	39.70	3.98	0.072	0.165	0.010	0.685	0.297
...	...	...	...	...	...	...	...	...
12	3.350	32.90	3.59	0.995	2.150	0.130	0.642	0.297
13	4.490	30.50	3.44	1.325	2.850	0.175	0.635	0.295
14	6.000	27.20	3.24	1.755	3.850	0.235	0.642	0.293
15	8.000	22.80	2.97	2.350	5.050	0.335	0.631	0.294
18	17.900	1.11	1.50	2.100	4.800	0.300	0.268	0.117
19	18.300	0.00	1.49					
20	18.300	0.00	1.48					

This indicates that a plot of r_S/x_v values against corresponding (at the same time point) r_x/x_v values should yield a linear plot with a slope of $1/Y'_{x/S}$ and an y-axis intercept of m_S value. For such a plot, we need to estimate the values of the volumetric carbon substrate uptake rate, r_S at various points in time.

From the mass balance on the carbon substrate, in batch culture, we have:

$$\frac{dS}{dt} = -r_S$$

Similar to the calculations in (*a*) above, the time derivative of S and hence the value of r_S can be obtained from the curve of experimental S against time (Table A.3).

Data from Table A.3 are used to obtain the plot in Figure A.3. As shown in Figure A.3, we estimate the following values:

$m_S = 0.06$ (kg glucose) (kg cells)$^{-1}$ h^{-1} and $Y'_{x/S} = 0.51$ (kg cells) (kg glucose)$^{-1}$

(c) Estimation of the biomass yield on nitrogen substrate.   Nitrogen source balance in this batch culture is:

$$\frac{dN}{dt} = -r_N = -\frac{r_x}{Y'_{x/N}} = -\frac{1}{Y'_{x/N}} \cdot \mu x_v$$

According to this equation, the volumetric rate of nitrogen substrate uptake at different points in time can be obtained from the time derivative of the plot of nitrogen substrate concentration against time using the experimental data.

A plot of the volumetric rate of nitrogen uptake, r_N against the volumetric growth rate, r_x should yield a straight line going through the origin since we assume that the nitrogen source is used only for growth and not for maintenance energy requirements. This plot is given in Figure A.4. From the slope of the line we have:

$$Y'_{x/N} = 6.84 \text{ (kg cells) (kg nitrogen substrate)}^{-1}$$

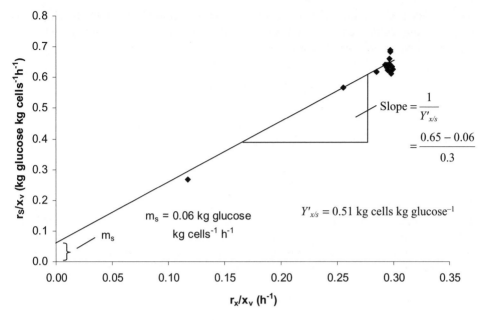

Figure A.3 *Estimation of the values of the maintenance constant and yield of biomass on carbon substrate*

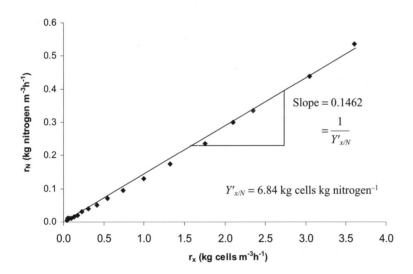

Figure A.4 *Estimation of the biomass yield on nitrogen substrate*

Inserting all these parameters into the kinetic expressions used above, the batch performance of this culture can be expressed with the following equations for growth, and carbon and nitrogen substrate consumption:

$$\frac{dx_v}{dt} = r_x = 0.294 \frac{S}{1.67 + S} x_v$$

$$\frac{dS}{dt} = -r_S = -\frac{1}{0.51} \cdot \left(0.294 \frac{S}{1.67 + S} x_v \right) - 0.06 x_v$$

$$\frac{dN}{dt} = -r_N = -\frac{1}{6.84} r_x = -\frac{1}{6.84} \left(0.294 \frac{S}{1.67 + S} x_v \right)$$

These three differential equations describe how the concentrations of biomass, carbon and nitrogen substrate change with time during the time course of this particular batch culture. These can be integrated numerically with software such as MathCad, using the concentrations at zero time as the initial conditions, in order to obtain the model-predicted concentration curves against time. Such a plot is shown in Figure A.5.

Death and product formation in a batch culture

In a completely mixed batch bioreactor a microbial culture grows at its maximum specific growth rate, $\mu_m = 0.425\,h^{-1}$, and simultaneously becomes nonviable at a volumetric rate

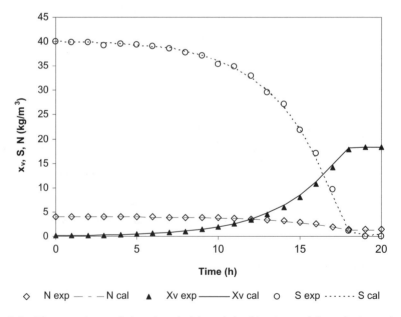

Figure A.5 *The experimental data (symbols) and the kinetic model predictions, drawn as curves*

that follows first order kinetics with respect to viable cell concentration with a death rate constant, $k_d = 0.06\,h^{-1}$. The inoculum concentration is $2\,kg$ biomass m^{-3} and it is 100% viable. A product is formed by the viable cells. Product formation kinetics can be expressed by the Luedeking–Piret model and is both growth and nongrowth associated (mixed kinetics), with $\alpha = 0.2\,kg$ product (kg biomass)$^{-1}$. Assume that nonviable cells remain intact and do not autolyse:

(a) If there is no lag phase, what is the viable cell concentration at the 8th hour of the batch culture?
(b) If the initial product concentration is zero, and the product concentration at the 8th hour of the batch culture is $12\,kg$ product m^{-3}, what is the value of β in the Luedeking–Piret model?
(c) If growth stops when an important substrate, with an initial concentration of $88\,kg\,m^{-3}$ is completely consumed and growth yield on this substrate is $Y'_{X/S} = 0.5\,kg$ biomass (kg substrate)$^{-1}$, what is the final total biomass concentration?
(d) What is the doubling time for this culture?

Solution:

The following expressions can be written based on the description of the system:

Growth: $r_{x_v} = \mu_m x_v$ $(\mu = \mu_m = \text{constant})$
Death: $r_d = k_d x_v$
Product formation: $r_p = \alpha r_x + \beta x_v$
                      at $t = 0$, $x_v = x_{vo} = 2\,kg\,m^{-3}$ (inoculum concentration)

(a) Mass balance on live biomass (x_v) gives:

$$\frac{dx_v}{dt} = r_{x_v} - r_d = (\mu_m - k_d)x_v$$

$$\frac{dx_v}{x_v} = (\mu_m - k_d)dt$$

Taking the integral:

$$\int_{x_{vo}}^{x_{v_t}} \frac{dx_v}{x_v} = (\mu_m - k_d)\int_0^t dt, \qquad (\mu_m - k_d) \text{ is a constant}$$

$$\ln x_v \Big|_{x_{vo}}^{x_{v_t}} = (\mu_m - k_d)t \Big|_0^t$$

$$\ln x_{v_t} - \ln x_{v_t} = (\mu_m - k_d)t$$

$$x_{v_t} = x_{v_t} \cdot e^{(\mu_m - k_d)t}$$

Using the given values of $\mu_m = 0.425$, $k_d = 0.06$ and $t = 8$ (the units should be written):

$$x_{v_t} = (2)e^{(0.425-0.06).8} = 37.08\,\frac{kg}{m^3}$$

(b) Mass balance on P gives:

$$\frac{dP}{dt} = r_P = \alpha r_x + \beta x_v = \alpha \mu_m x_v + \beta x_v$$

at $t = 0$, $P_o = 0$ (no product in the medium at time zero)

$$\int_0^P dP = (\alpha \mu_m + \beta) \int_0^t x_v dt, \qquad (\alpha \mu_m + \beta) \text{ is a constant}$$

substituting:

$$x_{v_t} = x_{v_o} \cdot e^{(\mu_m - k_d)t}$$

$$\int_0^P dP = (\alpha \mu_m + \beta) \int_0^t x_{v_o} \cdot e^{(\mu_m - k_d)t} dt$$

$$P = (\alpha \mu_m + \beta) \left[x_{v_o} \cdot \frac{1}{(\mu_m - k_d)} \cdot (e^{(\mu_m - k_d)t} - 1) \right]$$

Substituting the known values of parameter,

$$\beta = 0.04 \frac{\text{kg product}}{\text{kg cells h}}$$

(c) $Y'_{x/S} = \dfrac{\Delta x}{\Delta S}$

$x_T = x_v + x_d$
at $t = 0$, $x_T = x_o$ (inoculum is 100% viable)
$\Delta S = S_o - 0 = S_o$ and $\Delta x = x_f - x_o$ ($x_f = x_T$ at $t = t_f$)

$x_f = S_o \cdot Y'_{x/S} + x_o$, therefore $x_f = 46 \dfrac{\text{kg cells}}{\text{m}^3}$

(d) $t_d = \dfrac{\ln 2}{\mu_{max}} = \dfrac{0.693}{0.425} = 1.631 \text{h}$

Oxygen Limitation In a Batch Culture

(a) Develop the equations for the rate of change of cell and dissolved oxygen concentrations with time in an oxygen-limited batch culture. Soluble carbon substrate is in excess. Assume Monod kinetics for cell growth with the dissolved oxygen as the growth limiting substrate. Ignore maintenance energy requirements, cell death and product formation other than the biomass itself.
(b) In the production of a secondary metabolite in a 'batch' culture, nutrients are fed continuously in the solid form to the 'batch' in order to keep the cells growing at a constant specific growth rate equal to 10% of their maximum specific growth rate. At the same time, in order to achieve a desired product formation, the dissolved oxygen concentration must not fall below 10% of the saturation value and when it does the 'batch' is terminated. Using the data given below, calculate how long the 'batch' could theoretically run and what the final biomass concentration is?

Assumptions:

There is no lag phase in growth.
There is no maintenance energy requirement.
There is no cell death.
Inoculum is 100% viable.
There is no change in the volume of the batch despite the addition of solids.
Cell growth can be described by Monod kinetics and the growth limiting substrate is the dissolved oxygen.
Oxygen transfer rate is equal to oxygen uptake rate without any accumulation/depletion terms in the mass balance for the dissolved oxygen.
Oxygen consumed is used for biomass formation only.

Data:

Inoculum concentration $= 0.02 \, \mathrm{kg \, m^{-3}}$
Maximum specific growth rate of cells $= 0.2 \, \mathrm{h^{-1}}$
Saturation (equilibrium) dissolved oxygen concentration $= 0.007 \, \mathrm{kg \, m^{-3}}$
The maximum achievable volumetric oxygen mass transfer coefficient $= 80 \, \mathrm{h^{-1}}$
Yield factor for cells on oxygen $= 1.5 \, \mathrm{kg \, cells \, kg \, oxygen^{-1}}$

Solution

(a) Balance equations give:

$$\text{Cells:} \quad \frac{d(Vx)}{dt} = V \cdot r_x$$

$$\text{Oxygen:} \quad \frac{d(VC_o)}{dt} = -Vr_{Ox} + VN_a$$

where $r \equiv$ volumetric rates $(\mathrm{kg \, m^{-3}})$; $r_{Ox} =$ oxygen uptake for biomass production; $V =$ liquid volume in the bioreactor; $N_a =$ volumetric oxygen transfer rate; Rate equations give:

$$r_x = \mu \cdot x_v = \mu_m \frac{C_o}{K_o + C_o} \cdot x_v \qquad \text{(Monod kinetics)}$$

$$r_{ox} = \frac{1}{Y'_{x/O}} \cdot r_x$$

$$N_a = k_L a \, (C_g^* - C_o)$$

Making the various substitutions into the material balances gives:

$$\frac{dx}{dt} = r_x = \mu_m \frac{C_o}{K_o + C_o} x$$

$$\frac{dC_o}{dt} = -\frac{r_x}{Y'_{x/O}} + k_L a (C_g^* - C_o)$$

(b) Given:

$$x_o = 0.02\,\text{kg}\,\text{m}^{-3}, \; \mu_{\text{max}} = 0.2\,\text{h}^{-1}, \; C_g^* = 0.007\,\text{kg}\,\text{m}^{-3},$$

$$Y'_{x/O} = 1.5\,\text{kg}\,\text{kg}^{-1}, \; k_L a = 80\,\text{h}^{-1}, \; \mu = 0.1\mu_{\text{max}}$$

Since oxygen is continuously supplied in order to keep it at 10% of the saturation value (of course it can be higher than this but we are calculating for the worst or minimum allowed scenario):

$$C_o = 0.10C_g^* \quad \text{and} \quad \frac{dC_o}{dt} = 0$$

Since the maximum $k_L a$ is fixed at $80\,\text{h}^{-1}$, a point in time will be reached when the oxygen transferred just meets the biological demand of a culture whose cell concentration keeps increasing with time and DO concentration still kept at 10% of saturation. At this point in time, the batch will be terminated.

Therefore,

$$\frac{dx_v}{dt} = r_x = \mu \cdot x_v = (0.1)\mu_{\text{max}}x_v$$

Inserting this r_x value in to the DO balance equation below:

$$\frac{dC_o}{dt} = 0 = -\frac{r_x}{Y'_{x/O}} + k_L a(C_g^* - C_o)$$

$$\frac{(0.1)(0.2)x_v}{1.5} = (80)(1 - 0.1)(0.007)$$

$$x_{\text{final}} = 37.8\,\text{kg}\,\text{m}^{-3}$$

Now that we know the final and the inoculum cell concentrations, and the constant growth rate, we can calculate the time (no lag phase):

$$\frac{dx}{dt} = (0.1)(0.2)x \quad \text{which can be inegrated:} \quad \int_{x_o}^{x} \frac{dx}{x} = (0.02)\int_0^t dt$$

$$\ln\frac{x}{x_o} = (0.02)t \quad \text{and} \quad \ln\frac{37.8}{0.02} = (0.02)t$$

$$t = 377\,\text{h}$$

Continuous Culture

Two-stage Continuous Bioreactors in Series

A two-stage steady state continuous (chemostat) system (two bioreactors in series) is used for the production of a secondary metabolite. There is cell growth but no product formation in the first bioreactor and there is product formation but no growth in the second. The volume of each bioreactor is $0.5\,\text{m}^3$ and the volumetric flow rate of the feed is $0.05\,\text{m}^3\,\text{h}^{-1}$. The concentration of the growth limiting substrate in the feed is $10\,\text{kg}\,\text{m}^{-3}$. Growth is assumed to follow Monod kinetics. Product formation follows nongrowth-associated kinetics and is first order with respect to the viable cell concentration. Assume that there is no death of cells in the system. The feed to the first bioreactor is sterile. Kinetic and yield parameters for the microorganism are given below:

Biomass yield factor on substrate: 0.5 (kg cells) (kg substrate)$^{-1}$
Monod saturation constant: 1.0 (kg substrate) m^{-3}
Maximum specific growth rate: 0.12 (kg cells) (kg cells)$^{-1}$ h^{-1}
Maintenance energy constant: 0.025 (kg substrate) (kg cells)$^{-1}$ h^{-1}
Specific product formation rate: 0.16 (kg product) (kg cells)$^{-1}$ h^{-1}
Product yield factor on substrate: 0.85 (kg product) (kg substrate)$^{-1}$

(a) Sketch the arrangement of the two stages as a flowchart (Figure A.6) and label the system parameters on the streams and the two bioreactors.
(b) Write the steady state mass balance equations for cells, substrate and product for each stage. List your assumptions.
(c) Determine the cell and substrate concentrations entering the second reactor.
(d) What is the concentration of the product leaving the system?
(e) What is the overall substrate conversion in the system?

Solution

(a) The sketch of the two continuous stirred tank bioreactors, at steady-state is shown in Figure A.6.
(b) First stage, at steady state:

Cell balance:

$$Vr_x - Fx_1 = 0, \qquad \text{and hence} \qquad V\mu_{max}\frac{S_1}{K_S+S}x_1 = Fx_1$$

Substrate balance:

$$-V(r_{Sx} + r_{Sm}) + F(S_0 - S_1) = 0$$

$$F(S_0 - S_1) = V\left[\frac{1}{Y'_{x/S}}\mu_{max}\frac{S_1}{K_S+S_1}x_1 + m_S x_1\right]$$

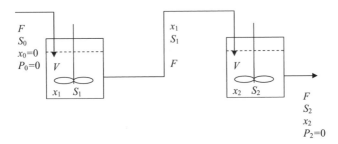

Figure A.6   *Flowchart indicating system parameters*

There is no product entering and being produced in the first stage bioreactor.
Second stage, at steady state:
Cells do not grow nor do they die in this stage. Substrate is consumed for product formation and cell maintenance.
Substrate balance:

$$-V(r_{SP} + r_{Sm}) + F(S_1 - S_2) = 0$$

$$F(S_1 - S_2) = V\left[\frac{1}{Y_{P/S}}\beta x_1 + m_S x_1\right]$$

Product balance:

$$Vr_P + F(0 - P_2) = 0$$

$$F(S_1 - S_2) = V\left[\frac{1}{Y'_{P/S}}\beta x_1 + m_S x_1\right]$$

(c) From the cell balance in stage one we get:

$$(0.5)(0.12)\frac{S_1}{1.0+S_1}\cdot x_1 = (0.05)x_1 \quad \text{which gives } (x_1 \text{ cancels}) \quad S_1 = 5\,\text{kg m}^{-3}.$$

From the first stage substrate balance,

$$(0.05)(10-5) = (0.5)\left[\frac{1}{0.5}(0.12)\frac{5}{1.0+5}x_1 + 0.025x_1\right]$$

which gives $x_1 = 2.22\,\text{kg m}^{-3}$

(d) From the product balance in the second stage:

$$(0.5)(0.16)(2.22) = (0.05)P_2$$

which gives $P_2 = 3.55\,\text{kg m}^{-3}$

From the substrate balance in stage two:

$$(0.05)(5-S_2) = (0.5)\left[\frac{1}{0.85}(0.16)(2.22)+(0.025)(2.22)\right]$$

which gives:

$$S_2 = 0.266\,\text{kg m}^{-3}$$

(e) Overall substrate conversion in the system is:

$$\left(\frac{S_0 - S_2}{S_0}\right)\times 100 = 97.3\%$$

8

Scale Up and Scale Down of Fermentation Processes

Frances Burke

8.1   Introduction – Why Scale Up/Down?

The detection or induction of a novel product, metabolite or protein is the first stage in the development and study of the molecule of interest. However, once an interesting metabolite has been detected, invariably it is needed in larger volumes for characterization, trials, structural elucidation and potentially for commercialization.

The process of generating product in increased quantities can take the form of setting many replicates of the cultivation system originally used for detection or induction. Conversely it could require establishing an efficient process in a larger volume vessel optimized for cultivation in larger volumes, e.g. a stirred tank reactor. In contrast, once a product has been commercialized, there may be a need to investigate changes observed or that require introduction at the large scale, and for economic, strategic or environmental constraints unable to be studied or investigated at this scale. A scaled-down model of the production process is an extremely cost effective, environmentally aware, option for exploring potential change options in a production operation.

The intermediary activities translating a successful bench product to production scale, or alternatively a system for translating a production process back to bench scale, is the concept of scale up/scale down, tending to involve microtitre plates, shake flasks, laboratory, pilot and production scale equipment. Typically, large scale fermentations have raw material and utilities costs of the magnitude of £10 000–£1 000 000, and depreciation on large scale plant can be significant. Carrying out a large scale experimental fermentation

Practical Fermentation Technology Edited by Brian McNeil and Linda M. Harvey
© 2008 John Wiley & Sons, Ltd

is therefore costly, potentially renders fermenters unavailable to generate product, and difficult and/or costly to dispose of. The nature of a production environment and necessary constraints to ensure manufacture takes place under GMP can also mean that some experimentation is impossible to do at the large scale. It makes sense therefore to develop small scale venues where experimentation can take place.

Most scale up/down projects involve the exploitation of a biological opportunity; typically, the potential for fermentation systems to make natural products, the potential to make a recombinant product, to understand the physiology of prokaryotes or eukaryotes in a model system, or to study molecular genetics and genome function under closely regulated conditions. All these approaches can benefit from a knowledge of the types of change that occur as we move from small fermentation systems such as microtitre plates, through shake flask to small laboratory fermenters, and beyond. In order to illustrate some of the challenges in scale translation of a process, characteristics that may change or are important in scaling processes will be briefly discussed in turn; then approaches to scaling will be discussed, finally description and comments of issues that typically arise during movement through each scale will be reviewed. For theoretical data explored with case examples and from first principles the reader is referred to key reference articles. From this it should be clear that as process scale changes, consequential changes in other variables occur, some can be predicted, others will have to be studied and it is for this reason that scale translation is a marriage of art, experience and science.

The author's experience has been largely spent in scaling natural product fermentations involving *Streptomyces*, *Bacillus* and *Fungi* with some experience of working with recombinant systems and in strain improvement. These will be discussed in detail, with additional comment made to issues that may can be relevant if commercializing novel biotechnology products or scaling expression systems for prokaryotic and eukaryotic organisms for product characterization.

8.2 Variables to be Considered when Changing Fermentation Scale

8.2.1 Aeration and Agitation

Aeration provides two related functions in an aerobic fermentation process: (i) to provide mixing, and (ii) to supply oxygen.

In almost any fermentation system, oxygen is supplied by diffusion across a gas–liquid interface. This interface may be present on the surface of the liquid or at the surface of a bubble suspended in liquid. Both systems behave differently and have benefits and consequences for mixing, aeration, shear and biological responses.

Microtitre Plates and Shake Flasks

In these cultivation systems oxygen is present in the gaseous phase above the liquid surface, and transfer is by diffusion through the gaseous phase, through the gas/liquid interface and into solution. The demand for oxygen by respiring organisms provides the continuous diffusion gradient maintaining supply by this method, and agitating the surface helps to prevent stationary boundaries developing and promotes diffusion and mixing in both gaseous and liquid phases. This agitation and aeration system is a relatively low shear mixing system, and it is necessary to limit the liquid volume/flask volume to a maximum of 20% to maintain useful gas transfer rates. Aeration can be increased or decreased by raising or lowering the agitation rate, by changing the liquid volume, by reducing or increasing liquid viscosity, and by changing temperature to influence the solubility of oxygen in the liquid phase.

Stirred Tank Reactors

In stirred tank reactors, supply of air is optimized to provide maximum gas transfer by the use of shear, power and turbulence, to maximize the gaseous surface area to liquid volume ratio and gas velocity.

There can be up to eight transfer steps for gaseous oxygen in suspended bubbles to transit through to become available intracellularly for energetic transformations. It can be helpful to understand these in order to understand the complex interrelated contribution made by tank design, power input and air flow rate on oxygen transfer. It is also useful to understand some of the theories around gas transfer for working with Process Engineers on scale-up projects, particularly for commercialization of novel products in a new facility or to refine scale-down models to match more closely the production scale.

Firstly there needs to be convection and diffusion of oxygen to the surface of a bubble, then transfer across the interface of the bubble where there may be films on either side of the interface. Oxygen must then dissolve in the fermentation fluid and be transferred in solution by bulk mixing. Finally, oxygen must diffuse across the interfacial films surrounding the cell boundary, either into or between cells to be available for reaction with cytochromes or be directly incorporated into product molecules.

It is possible therefore to define an overall transfer coefficient for use in comparing the efficiency of oxygen transfer in different cultivation systems (K_{La} – which is used in determining the Oxygen Transfer Rate – OTR for a cultivation system).

$$\text{Flux} = K_L \, x \text{ (the driving force or concentration gradient}$$
$$\text{of } O_2 \text{ from gaseous phase to intracellular availability)}$$

where K_L is the overall transfer coefficient.
   Therefore

$$\text{Flux} = K_L \, (C^* - C) \text{ per unit area of bubble} \qquad (8.1)$$

where C^* = dissolved oxygen concentration, which would be in equilibrium with the bubble (mmol/mL) and C = dissolved oxygen in liquid phase in mmol/mL.

However, as small bubbles have less buoyancy than large bubbles, they tend to rise more slowly having a negative impact on gas transfer by reducing gas velocity, but also a positive impact on gas transfer as surface area/volume ratio is increased. Hence, the overall flux incorporates a bubble area term, to provide a normalized point for comparison. This gives:

$$\text{Overall flux of oxygen} \qquad = K_L \, a \, (C^* - C)$$
$$\text{(the oxygen transfer rate or OTR)} \qquad (8.2)$$

where a = surface area of bubbles per unit of liquid volume (mm^2/mL)

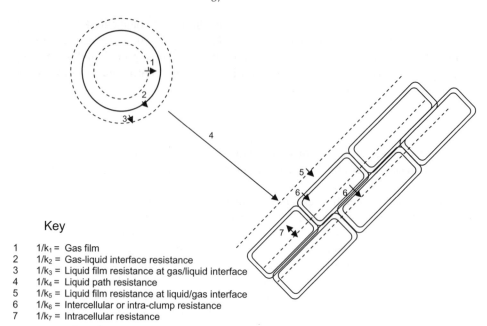

Key

1	$1/k_1$ =	Gas film
2	$1/k_2$ =	Gas-liquid interface resistance
3	$1/k_3$ =	Liquid film resistance at gas/liquid interface
4	$1/k_4$ =	Liquid path resistance
5	$1/k_5$ =	Liquid film resistance at liquid/gas interface
6	$1/k_6$ =	Intercellular or intra-clump resistance
7	$1/k_7$ =	Intracellular resistance

Figure 8.1 *Diagrammatic representation of the different resistances to gaseous transfer through liquid for use in intracellular reactions*

In general, for a continuously agitated optimized fermentation vessel $(C^* - C)$ is considered to be constant, and hence $K_L\,a$ is often used as the term to describe the overall OTR. In effect, the oxygen transfer rate is a mean value for a particular cultivation system, but it is also a useful comparison parameter particularly across different scales of operation where aeration and mixing may be significantly different.

This explanation of gaseous transfer is illustrated diagrammatically as a series of resistances in Figure 8.1. The resistances of highest value are those where engineering intervention to optimize transport is most useful. In spite of transfer being considered a sum of resistances, transfer across each boundary reaches a steady state and can be observed and considered to be one step – hence the value in having a summation described as the OTR. It is worthwhile noting that transfer responses could vary across a fermentation time course and in response to changing bubble size, broth viscosity, changes in solutes in solution, stagnant areas, and biological changes (e.g. morphology changes, see Figure 8.2, or changes in oxidative state due to more or fewer cytochromes being operational). However the OTR concept is useful for comparing different fermentation regimes. Needless to say, measurement of these resistances is problematic and resides in the annals of fermentation history.

It is typical for large fermenters to adhere to a common design philosophy which summarises the available knowledge for fermenter design for optimal gas transfer into a 'generalized standard tank design', see Figure 8.3. This design is considered to constrain the major design parameters in useful ranges for efficient gas transfer and mixing specifically for fermentation processes. For further detail on fermenter design, the reader is referred to the engineering texts listed under References for further information.

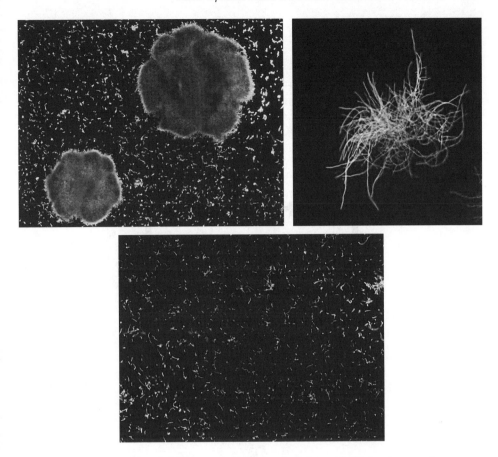

Figure 8.2 *Photographs of different submerged colonial morphologies that can impact gas transfer to hyphal interior. (Photo: Michelle Lea, John Moores University, Liverpool)*

At the smaller scales, there is likely to be less stringent adherence to the optimal geometric relationships as power input to a small scale system is not typically a restrictive feature. However an inefficient tank design or lack of appropriate correlation between airflow and agitation rate can easily move a fermentation into an inefficient mixing and gas transfer mode, and limit potential growth rates unnecessarily. Impeller flooding (where airflow rate is typically beyond 1 vvm and tends to cause bubble coalescence to occur around the impeller, reducing bubble interfacial area/volume ratio) in particular is something to be avoided.

The 'standard tank design' (STD) philosophy is one that permeates production scale manufacture. Typically an organization would use its own variation of the STD (often optimized by its own process engineering group specifically for use on its major molecules) and use it typically to design fermenters at different scales across the corporation. The concept is that if vessels are geometrically equivalent, then flow velocities, gas transfer rate turbulence and shear can be maintained as consistently as possible across scales. If facilities are built around this concept, then transfer of similar processes between

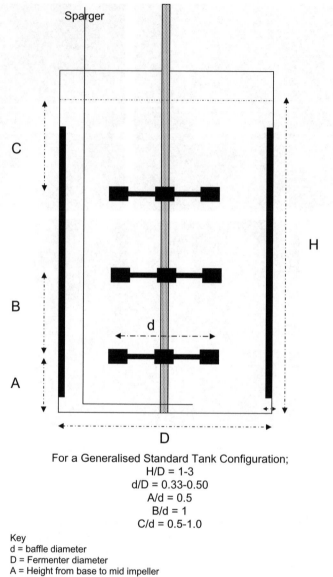

For a Generalised Standard Tank Configuration;
H/D = 1-3
d/D = 0.33-0.50
A/d = 0.5
B/d = 1
C/d = 0.5-1.0

Key
d = baffle diameter
D = Fermenter diameter
A = Height from base to mid impeller
B = Distance between impeller
C = Distance between top impeller and liquid level
H = Fermenter height to liquid level

Figure 8.3   *Generalised standard tank design. Traditionally used as the basis to scale stirred tank reactors while maintaining geometric similarity based on optimal ratios of design variables*

facilities helps to make process transfers more straightforward. This concept is valuable for traditional high capacity natural product fermentation manufacture, but is less valuable for low volume, high value products, e.g. recombinant protein manufacture or mammalian cell operations, where specific design for the product of interest may be more valuable as capacities are generally much lower. In addition, as new technologies to study mixing become available, it can be valuable and worthwhile commissioning work to optimize mixing criteria for novel products where facilities are newly built and commissioned. Time lapse photography of scaled glass vessels is specifically useful for understanding bubble coalescence, which can be a problem when optimizing in low shear environments. For further detail, please see the engineering texts in the reference list as starting points.

STD also does not tend to hold for products used in facilities that have been acquired by purchase, and often installation of new pilot scale vessels entails the use of 'off the shelf' fermenter designs.

Once air has been supplied to the cultivation vessel in a useful range and sheared by the impeller, it then needs to be transferred by turbulence both vertically and horizontally. Typically a volumetric air flow rate between 0.5–1 vvm is recommended for optimum gas transfer. A series of impellers from fermenter base to close to liquid surface is the typical method for achieving this. Combinations of bladed impellers, or bladed impellers combined with propellers, can be used efficiently to cause vertical turbulent transfer, with baffles helping to promote horizontal turbulent transfer and providing shear, thus promoting efficient bulk mixing.

Draught Tube/Bubble Colum Fermenters/Other Designs

Some production operations employ bubble columns at the production scale. These fermenters are extremely energy efficient, low shear vessels, but are typically not directly scaleable from stirred tank reactors, as shear is significantly reduced and superficial gas velocity plays a much greater role in gaseous transfer processes. In spite of some scaling limitations, however, it is typical to use a standard scaling route using stirred tank reactors at the laboratory and pilot scale, merely adjusting to equipment constraints when the process reaches production scale.

There are also other instances of large scale novel fermenter designs (Figure 8.4). At Eli Lilly Clinton Labs Manufacturing site, where natural product fermentation vessels (greater than 200 000 litres in volume) are sited horizontally rather than using the traditional vertical design. This design optimizes utilities consumption for both agitation and cooling systems and reduces the extent of hydrostatic pressure experienced at the fermenter base. They scale reasonably well from traditional stirred tank reactors.

Nutrient transfer and product transfer would be expected to follow similar principles to oxygen delivery; therefore, if oxygen transfer is used as a focus for optimizing equipment, it would be expected that other transfer steps could be similarly modelled on the same principles.

8.2.2 Sterilization of Fermentation Media and System

Batch thermal sterilization processes tend to be used for liquid and equipment in fermentation systems, as these tend to be reliable and cost-effective options for both small and large scale systems, and there is a degree of confidence in being able to assure sterility in a batch operation. Continuous sterilization of medium can be used for fermentation systems, but it tends to be the result of specific process benefit, e.g. reduction in utilities or specific nutrient characteristics, which are provided by continuously sterilised media,

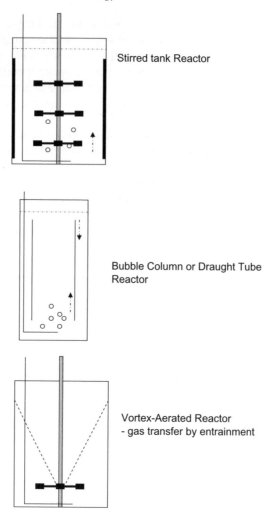

Stirred tank Reactor

Bubble Column or Draught Tube Reactor

Vortex-Aerated Reactor
- gas transfer by entrainment

Figure 8.4 *Diagrammatic representation of reactors with different gas transfer designs*

e.g. continuous sterilization has been advantageous for fermentations that are phosphate sensitive, although practical considerations can often dictate that batch sterilization is preferred. The sterilization methodology at different scales and in different items of equipment may be a major source of variation due to chemical degradation of medium components and should be considered a key variable and one that is valuable to monitor closely in the scaling process.

Small Scale Fermentation Systems

Typically sterilization methods for small scale cultivation systems (equipment and/or media) rely on the use of autoclaves, where the autoclave cycle time is dictated by the use of a thermocouple in a test volume of medium reaching a temperature of above 121 °C for a minimum cycle time. In using an autoclave to sterilize medium there may be a requirement to test that the sterilization cycle is suitable for the specific medium being used, e.g. if the medium contains particulate material or material that may change viscosity during heating (e.g., starches), or to evaluate an optimum quantity of antifoam to use to prevent volume loss during sterilization.

Chemical degradation can be limited by sterilizing vulnerable components separately (e.g., glucose sterilized separately from amino nitrogen sources can limit degradation and the generation of Maillard compounds – to which some fermentation processes can be extremely sensitive). Chemical degradation during sterilization may lead to significant pH changes, which may require adjustment after sterilization to bring the pH into a more useful range. Interestingly, this seems to be a phenomenon that shows large differences between different scales, even when parameters are closely controlled for scaling purposes.

High value or sensitive products (e.g., antibiotics required for retention of a plasmid) may usefully be filter sterilized rather than exposing them to heat and potential degradation.

Pilot and Production Scale Fermentation Processes

As scale increases, so autoclaves become impractical for sterilizing liquid medium, and typically fermenters greater than 5 litres in volume are sterilized *in situ* using live steam injection. Often, as processes move from laboratory autoclaves to pilot vessels, differences in process performance are observed, due to chemical changes affecting medium components during the sterilization process. If the differences are found to be significant, then often the best way of offsetting these changes is to record each sterilization process and adjust to longer or shorter cycles (if possible) and observe the impact.

For scaling sterilization processes taking place by live steam injection, generally reasonably consistent sterilization characteristics can be generated at the different scales using controlled heat up/cool down (potentially decelerated at the small scale when required to mimic larger scale operations), and a similar or the same sterilization time. There are also useful scaling parameters that can be directly applied if computer data logging is available to generate sterilization curves; a particularly useful set of parameters to use for scaling sterilization is the concept of F_o/R_o. F_o is the integral of sterilization from 90 °C where it is considered that significant heat degradation of components starts to occur, and R_o is the integral of sterilization from 120 °C, where it is considered that significant biological 'kill' starts to take place, For a process susceptible to heat degradation of media, a low F_o/R_o would be preferred, whereas for a process that benefits from some heat degradation of media, a longer F_o/R_o would be required.

If medium components are relatively heat insensitive, there is a tendency to sterilize for longer than the minimum time period or at a temperature greater than 121 °C in order to provide the optimum assurance of sterility for the entire vessel. On the large scale, fermentation vessels may typically require a minimum sterilization time of 20–30 minutes, but times of 45 minutes or 1 hour may be used to provide an absolute assurance of sterility with a large degree of confidence. On the large scale, the heat up and cool down times tend to be much longer than in smaller scales and may become considerably significant in terms of their contribution to nutrient degradation. It is typical to monitor sterilization cycles not just in terms of time at a specified temperature, but also including the chemical influence of heat up and cool down.

The impact of sterilization on media can be significant to the fermentation; however, it is important to note that sterilization temperature is the key sterilization parameter only when the following prerequisites are in place:

(i)   The sterilization time relates to the slowest-to-heat part of the fermenter.
(ii) Complete evacuation of air takes place. enabling 100% steam heat penetration to all parts of the vessel.
(iii) No breach of sterile barrier occurs during heat up or cool down (so vacuum creation at high temperatures is prevented by efficient air management, or equipment to permit vacuum release without breaching the sterile barrier).
(iv) Efficient systems for sterilizing air filters exist, which permits them to dry at high temperatures prior to vessel cooldown.
(v)   No holes or cracks in vessel interior or exterior are present that could permit foreign growth ingress.
(vi) There are no pockets inaccessible to cleaning processes that could harbour solid residue that would prevent steam heat penetration.

Hence, the design of a large scale vessel may predetermine an appropriate sterilization time and temperature and, in this case, it is important that the sterilization cycle is scaled to the most constrained scale.

8.2.3   Inoculum Development and Culture Expansion

It is essential to have some understanding of the organism under study and how the changing scale influences it, as unexplained results may occur due to biological change in the population occurring but remaining unnoticed. Minimizing variation in culture expansion phases can help reduce this risk (covered in this section), and identifying when it does occur can be critical to reducing population change on process output (covered in Section 8.2.8).

The starting point for any fermentation system is a pure culture single cell isolate, or as close to this starting point as possible. Assuming a pure culture isolate has been achieved, it is necessary to prepare a stock of this biological material and preserve it. Best practice would then require working cultures be generated in a consistent manner from the stock culture for use in further fermentation process development work; see Chapter 6 for preservation techniques.

Should the starter unit be the product of a strain selection programme, a clone from a library, or a recombinant organism displaying a preferred phenotype, it may be that the starter culture is maintained on agar plates pending the results of the screening programme; in which case, once the isolate is considered worthy of further study, it should be prepared as a stock culture, followed by working cultures. It is important that the stock cultures are not used as working cultures or subjected to freeze–thaw cycles that will stress and induce novel selection pressures on the cell bank.

In inoculum generation it may be relevant to explore any of the variables listed below for optimizing biomass yield and generating biomass suitable for producing product. Vegetative stage methods can be species specific, but in general, they tend to be less sensitive to slight media differences than are production stages, and vegetative phases tend to have evolved from media developed as general purpose media. In general, a good starting point is to test a variety of vegetative general purpose media and then take results from these evaluations and explore further. Typically these evaluations are carried out at shake flask scale, with the mature inoculum being transferred to another shake flask (containing medium supporting production of molecule of choice) to test whether productivity has been affected.

Variables that may be valuable to explore in developing vegetative inoculum processes include:

- Inoculum type (e.g., vegetative cells, spores, pregerminated spores).
  Extremely different results can be generated depending on type and number of inoculum points; e.g. for *Streptomyces* cultures, 'pregermination' of spores can sometimes be helpful for promoting synchronous growth. (Hopwood *et al.* 1985)
- Cultivation temperature.
- Agitation/mixing rates.
- Medium development. It is advisable to consider traditional medium development activities of vegetative media, but also to consider whether or not catabolite induction and repression processes need to be accounted for, ensuring that in the inoculum condition, the organism has the potential to develop efficient methods for using the carbon and nitrogen sources that are to be supplied during the production phase. This is exemplified by the use of lactose, an inducible carbon source for *E. coli*, and in this case it may be necessary to ensure that the organism has the opportunity to 'adapt' to efficient growth on lactose source during the inoculum phases. Conversely, a carbon source may be required as an inducer for a natural biosynthetic conversion process at the production scale, exemplified by a source of slowly assimilated acetate being an inducer for biosynthesis, and it would need to be excluded from the inoculum process in order to promote fast rapid growth, without any carbon and energy drain to metabolite production during the culture expansion phases.
- Transfer criteria for vegetative phase. It is helpful to have a specific criterion that describes inocula as suitable for transfer to the next stage, as consistency is a key factor in contributing to reducing variation. This could be dependent on an estimate of cell density such as optical density for *E. coli*, for filamentous organisms, dry weight, respiratory profile if off-gas measurement is available, and if no specific measure is available, then time can be as useful a transfer parameter as any.
- Growth form. Although generally outside the control of the experimenter, it can be useful to observe whether there are any gross differences apparent when varying vegetative culture conditions; e.g. for filamentous organisms, is the growth form largely pelletted or in the form of mycelial mats? For *S. capreolus*, it is known that pelletted growth is the preferred inoculum condition for high yields, whereas for other *Streptomyces* species, mycelial growth is the preferred inoculum condition.

Once an inoculum process permitting productivity has been developed, there may be a need to expand it – introducing another transfer step to a culture expansion or 'seed' vessel, with the aim of increasing biomass in the vegetative phase to sufficiently larger volumes to use a sufficient volume of mature inoculum usefully to inoculate a production vessel.

Traditionally it has been recommended that attempts be made to get as close as possible to transfer of 10% by volume of inoculum into a final stage production scale, so maximizing equipment utilization at the largest scale of operation. This can be considered to be a scale-up operation in its own right, with the aim being the maximum yield of consistent (vegetative) biomass that retain potential for optimal production of product in the final volume. Typically a culture expansion phase will not yield any secondary metabolite product as culture conditions are optimized for rapid, synchronous vegetative growth.

It is useful to aim at keeping medium components as similar as possible across all the culture expansion steps, so that the culture does not go through any transition phases where either biosynthetic systems become induced or repressed or where oxygen becomes depleted significantly. Any stress is likely to lead to slight suboptimal performance, which will take biological time to reverse or alleviate. However, a culture expansion operation merely has to be fit for purpose, and it can be useful to move as quickly as possible to optimizing the fermentation steps, where there is much more potential for process improvement.

For small scale cultivation systems, it may be sufficient to inoculate shake flasks from stocks maintained in glycerol at −70 °C, then to transfer several millilitres of, hopefully consistent, synchronous inocula to the small fermenter or large volume shake flasks for production of product. For large scale cultivation systems, it may be typical for frozen inoculum to be inoculated into multiple shake flasks or a small fermenter vessel to generate 2–4 litres of inoculum. This inoculum is then used in entirety in a production seed (or culture expansion) vessel, representing as close to 10% volume of a production vessel as possible, or in smaller scaled volumes in the laboratory or pilot stirred tank vessel. By developing the culture expansion phase as a part of scaling processes and maintaining it as a consistent operation, hopefully there will be no significant variation between scales caused by variation in the generation of vegetative inocula.

8.2.4   Raw Materials and Nutrient Availability

Media invariably change during scaling. The constraints of microtitre and shake flask cultivation systems introduce compromises that can change once the process is operating in stirred tanks. For example, particulate containing media can be used and pH control becomes possible in stirred tanks. Conversely, the constraints of the production environment may limit the type of materials that can be used at large scale and this information may need to be introduced into the pilot scaling stage. In addition, in making changes of scale there may be changes due to chemical change during sterilization and the opportunity to intensify the process by feeding concentrated nutrients to sustain phases of productivity.

A medium that is useful for producing product in small vessels may not be viable from a cost basis at the large scale, e.g. bacteriological grade yeast extract is a useful general purpose medium component (see Chapter 5 on medium formulation). However, it is likely to be prohibitively costly to use in large-scale fermentation operations, and cheaper raw material options supplying a similar mixture of vitamins, minerals and cofactors may require evaluation at pilot tank scale; e.g. testing corn steep liquor, dried yeast, or mixed bulk protein sources as a replacement.

It becomes possible to start feeding nutrients to a stirred tank reactor and prolong and intensify the productivity of the process, as final yields in excess of twice that achievable typically from batched materials becomes possible. This is one of the huge benefits of scale up: that productivity increases can be achieved by focusing on the biochemistry and induction kinetics of the organism of interest. As a recovery operation can cost the equivalent or more of a fermenter operation, it can be much more effective to increase titre per unit volume than to increase the fermenter vessel volume to achieve the same final yield per fermenter. For intensifying and prolonging a phase of production, there needs to be sufficient headspace available in the fermenter, and ideally the feed is highly concentrated, so the starting volume can be relatively unaffected by the requirement to feed additional volume, supplied through a system that will permit as close to ideal mixing as possible. Often feeds are relatively low cost, extremely concentrated raw materials, e.g. ammonium hydroxide, oils and protein isolates, providing concentrated readily available nitrogen and carbon sources. Often identifying a suitable timing and rate can be critical for achieving increases in yield, and it is usual to identify a biochemical trigger, change or process marker, which may indicate that feeding additional nutrient could help relieve a nutrient limitation.

The general approach to medium design in scaling is similar to any medium development project.

8.2.5   pH

pH tends to be a sensitive fermentation output variable and potentially a controlling parameter, so it is therefore valuable to track pH at all stages of a scaling operation as it can be a route for optimization. There are many historical instances of pH effects not being identified as the root cause of variation in a dependent variable; and it is recommended that a typical operational pH range be developed for each scale for application at the next scale wherever possible.

At shake flask scale stages, it may be relevant, to consider using buffers to ensure pH does not drift to an extreme. Consideration should be given as to whether an inorganic buffer (e.g., phosphate) could be useful, or whether it may interact with an inducible process, particularly relevant for secondary metabolite processes that may rely on nitrogen or phosphate derepression for biosynthesis, e.g. phosphate derepression with *S. fradiae*. The biological buffers, although costly are a useful tool to use to understand if pH variation is impacting the experiment without intentionally impacting nutritional variables in the experiment. At a minimum it is recommended always to check the pH at the end of a shake flask cultivation programme, perhaps just one control flask, to give assurance that pH has not drifted into a non-useful zone.

The pH of a fermentation tends to be both a dependent and independent variable, in that at extreme pH values, growth can be impacted. However pH changes can be the result of degradation or uptake processes, e.g. proteolytic breakdown can result in release of ammonia, causing pH increases (Figure 8.5).

Carbon source can also be a key component that can impact the pH response, and it has been documented that carbon supplied in excess in some fermentations can lead to the conversion of the carbon source into organic acids using the TCA cycle, with the

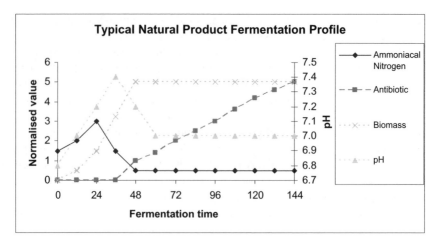

Figure 8.5 *Typical pH profile for chlortetracycline production, where pH naturally rises and falls as a result of ammoniacal nitrogen release and consumption*

organic acids being excreted into the growth medium. When carbon sources become limiting, the organic acids are then consumed, altering pH (Holms *et al.* 1990). As nutrient availability tends to change with changing scale, then pH changes can also alter and the changes may have positive or negative impacts on the required outcome. It is advisable to monitor, track, and attempt to interpret pH changes, and recommend a course of action, which may require the use of pH controlling actions, based on observation of the individual cultivation system.

8.2.6 Shear

Shear tends to be low in shake flask cultivation systems, increasing dramatically when a process is scaled to stirred tank reactors, as these are designed to be high shear mixing systems for optimal gas transfer.

Shear can be increased artificially in shake flasks by introducing baffles, springs and glass beads, but the inclusion of these shear-increasing elements cannot be scaled. In a stirred tank reactor, shear is increased by changing agitator geometry, the inclusion of baffles and by increasing agitation speed.

Shear rates can also change during a fermentation, if the organism produces a polymeric product such as, for example, *Xanthamonas campestris*, producing xanthan gum, which shows pseudoplastic tendencies, or if it produces enzymes which can degrade viscous substrates e.g. pectinases from some filamentous fungi such as *Cochliobolus sativus*, or the organism goes through changes in submerged morphology, e.g. moving from a pelletted growth to a more filamentous growth form.

For unicells such as *E. coli*, fermentation processes tend to be described as shear insensitive, whereas shear sensitivity may be experienced in cultivation of eukaryotes and filamentous organisms. In the worst case, cells are friable and lyse in a high shear environment. Responses to shear may be culture specific even within the same *genus* – e.g. *S. albus* fragments readily achievea a filamentous growth form and rates increased growth in flasks supplied with springs, whereas *S. coelicolor* or *S. lividans* merely show increased tendency to generate extremely dense oxygen limited pellets with asynchronous secondary metabolite production. *S. capreolus* is an interesting organism in that it responds morphologically by evoking a pelletted growth form, but this corresponds with elevated levels of capreomycin output.

Shear requirements and responses can therefore generally be considered culture and cultivation system specific.

8.2.7   Temperature Maintenance

At a small scale venue, biomass quantities are relatively low and heating can be supplied relatively easily to an incubator or stirred tank reactor. However, as reactor volume increases, so significant heat is generated during aeration and agitation. Processes are intensified, often biomass increases are achieved, and although the specific heat output may stay the same, the overall cooling requirement would increase due to an increase in biomass density.

Cooling capacity can be delivered to a large fermenter by using a jacketed vessel supplied with cooling or chilled water, or the vessel may have internal cooling coils (Figure 8.6), or the vessels can be sited in the open air and be sprayed with water to cause cooling by evaporative loss. Mixing characteristics are considerably different in vessels where cooling capacity is supplied by the use of internal cooling cools, and this may interfere with aeration characteristics, creating a partial draught tube.

In addition, if cooling capacity is limited, then it is valuable to introduce temperature as a variable in a scale-up process or to get some estimations of cooling requirements for changes introduced at the small scale. It has been proposed that the use of thermotolerant organisms for production processes could alleviate this constraint (Stowell and Bateson

Figure 8.6 *Interior of a large scale production vessel illustrating the use of internal cooling coils for providing cooling capability (Eli Lilly and Company Limited)*

1983). The intensification of a process that includes a significant increase in biomass concentration may not be realizable at the large scale if cooling is a constraint or if it is not ameliorated during scale up.

8.2.8   Partial Pressures

In larger fermenters, back pressure tends to be used to help protect the sterile envelope of the fermenter. In addition, in production scale venues of 100 000 litres plus, hydrostatic pressure at the base of the fermenter can be significant. Both of these pressures would tend to influence gaseous partial pressures in the liquid medium, potentially changing gaseous gradients. For oxygen, typically slight increases in vessel overpressure would not be expected to have a significant effect on metabolism, but if the organism is sensitive to a gaseous product that can dissolve in broth, then unexpected events can be observed. Typically carbon dioxide is the molecule most often observed as having an adverse impact on processes with increasing scale as pCO_2 increases – noted with industrial fermentations but not published. However, there can also be a positive impact in increasing partial pressures, and for *S. rimosus* and *S. aurofaciens* a doubling in productivity was observed for a 6.2-fold increase in oxygen partial pressure for wild type strains achieved by increasing the total pressure (Liefke *et al.* 1990). Probes exist for monitoring pCO_2 for instances where pCO_2 is thought relevant to the process.

8.2.9   Genetic Changes in the Population

As every cultivation volume starts from a single cell or unit of inoculum, then with increasing volume, a population will transition through many generations of itself (with errors in DNA replication occurring at roughly 10^6 base pairs – rule of thumb for *Streptomyces* Hopwood *et al.* 1985) then natural variants (mutants) will build up in the population over time. The population will change if there is a selective advantage over another mutation, e.g. if loss of antibiotic production gives an increased growth rate, but the majority have no impact on the process. It is an unusual event for production strains to display the characteristics of culture degeneration, as they have been developed to be robust and have demonstrated stability over time.

Culture degeneration can be minimized by ensuring that a master culture stock is used to generate a working cell bank and to ensure that each inoculum unit is taken from the working cell bank and not from serial subculture (see Chapter 6). Ensuring consistent culture propagation is the best method to help ensure low variability between batches as well as ensuring that the culture does not go through an unusual number of generations and therefore risk of population change at the genetic level. For novel processes and organisms, it is valuable to closely monitor the process development phases for any potential incidence of genetic instability in the producing strain, in addition to designing the system so that there is a rigid maintenance of selection pressure for the genetic construct. It can be helpful to ensure that a minimum of three strains is proposed for stirred tank work for an entirely novel process, with one of the first evaluations being to establish that the organism does not show any tendency to genetic instability during initial experimental runs. A protein gel or DNA-level characterization of the starter organism is useful where the organism has been genetically manipulated, to that verify the fingerprint of the final organism has not varied from starter culture.

Finally, in general, if unexplained results are demonstrated during scaling, it can be valuable to check the culture microscopically and by plating out to establish that the single colony phenotype has not altered during the scaling operation.

8.3   Implementing a Scaling Activity

8.3.1   Introduction

There are traditionally two major approaches for deciding the number of stages in a scale-up programme, but common practice now is to employ something between the two extremes:

(i)   To have a scale-up plan with a relatively large number of scale changes, ideally each scale change being 10 times larger than the former volume, to avoid large increases of volume for each stage of scale up, and to minimize the risk of volume increase providing novel variables during the scaling exercise. This approach focuses on supply-side requirements. Production operations were relatively typically set up with vessels for four or five scale transitions in the 1960s to 1980s.

(ii)   To employ the minimum number of stages, so enabling maximum focus of resource (manpower particularly) for investigating the biological responses at each stage. This approach focuses more on demand-side requirements than does the first approach. New production facilities are generally set up with one, or at most two, scale transitions.

The first approach relies on following a largely process-engineering-focused scaling exercise based around maintaining a standard geometric tank design from small to large scale and designing transport phenomena through fluid to be equivalent at each stage. The process would be intensively optimized at the small scale, then evaluated at each subsequent scale with minimal adjustment and rapid transit time through each stage. It is based on a concept that predicts that supply side requirements are of critical importance to scale up and can be maintained relatively consistently from scale to scale.

In general it it is recommended to scale up on the basis of constant power consumption for most fermentation processes. For filamentous (*Streptomyces*/fungal) fermentations, impeller tip speed is often the most useful correlation; for nonstandard fermenter designs, scale up on the basis of constant volumetric transfer coefficient is to be recommended, and for specific conditions where rapid mixing at the molecular level is important, scale up on the basis of constant mixing time is best. For comparison of the benefits and disadvantages of these major engineering correlations for scale up across the different scale stages for stirred tank reactors, see Table 8.1.

If supply-side requirements are consistently scaled, then the prediction is that biological performance should be consistent. However, it relies on no or minimal change in demand-side (biological) requirements during the scaling operation. This may be a relevant assumption for growth-associated products but can be confusing for secondary metabolite processes, where nutrient supply, repression systems or changes in submerged morphology may have a significant contribution to make to biological performance, and may change as the culture changes and responds to vessel configuration and nutrient supply strategy.

Table 8.1 Advantages and disadvantages of the four major scale-up correlations

Scale-up method	Primary reference	Engineering correlation	Benefits	Disadvantages	Assumptions
Constant power consumption/unit volume	Cooper et al. (1944)	Constant K_La directly proportional to gassed power consumption/unit volume	Relatively straightforward correlation, easy to scale factor.	Shear increases with scale increase.	Turbulent flow, K_La is limiting factor, tends to assume geometric similarity between vessels, assumes bubbles would coalesce without adequate bulk turbulence. Only strictly applicable to small fermenters having a single impeller. Actually shown that Pg/V decreases with increasing scale – dropping to roughly 0.5 for production scale plant (Bartholemew, 1960) due to power draw on upper impellers without proportional impact on aeration efficiency.
Constant mixing time	Norwood and Metzner (1960)	$N_2 = N_1(D_2/D_1)^{1/4}$ Where N = impeller tip speed D = impeller diameter	Time required for a liquid droplet to be completely and uniformly dispersed in the bulk fluid in an agitated vessel, constant across scales; the mixing should be at the molecular level. Particularly useful for low shear systems with rapid reaction kinetics, often where microbial growth is of secondary importance.	If this scale up method is used, power consumption/unit volume will increase significantly with scale. It is therefore not advisable to use this route unless necessary for mixing-specific issues.	Extremely difficult to scale as delivery system may have a large impact and delivery system may not be capable of being maintained across scales.

Criterion	Reference	Relationship	Description		
Constant impeller tip speed	Steel and Maxon (1962)	$N_2 = N_1(V_1/V_2)^{1/3}$ Where N = impeller tip speed D = impeller diameter V = liquid volume	As the maximum shear experienced by the medium is at the tip, then it has been found advantageous for organisms susceptible to shear or mechanical damage, e.g. protozoa or shear-responsive filamentous organisms.	Power consumption/unit volume will decrease.	Tends to assume geometric similarity between vessels, found that gas bubbles in nonnewtonian systems do not readily coalesce, and this is probably the reason the correlation is useful. Typically used for fermentations of filamentous organisms.
Constant volumetric transfer coefficient	Aiba et al. (1973)	Constant K_La used for bubble aeration, i.e. nonmechanically agitated systems	Can be useful for non-standard agitation systems e.g draught tube or bubble columns. Once an optimal aeration efficiency has been demonstrated at small scale, conditions are then found by experiment on the large scale to support the same aeration efficiency. Equations are complex, involving estimations of bubble diameter.	Need a method for monitoring K_La; sulfite oxidation, gassing out or exhaust gas oxygen balance are typical. Difficult with live fermentations.	Assumes the optimal aeration efficiency at the small scale can be determined.

However, without adequate gaseous and liquid transport rates, it is difficult to optimize processes where oxygen supply and mixing are important, and so focusing and planning a scaling activity based on optimizing transport phenomena at each scale is an entirely practical starting point and one to be recommended.

The second approach relies on optimizing transport phenomena together with gathering a reasonably detailed understanding of the biological responses shown by the process. An array of analytical tools helps considerably and it is the improvements in analytical technology over the past few decades that have led to this being a realistic approach to scaling. There is less need and less opportunity to take the process through a large number of scale changes because of the increased time taken in detailed observation at the first stage. The data is used to optimize the cultivation system at the final scale directly, taking into account both supply-side and demand-side requirements on the final scale. In this scaling system, data may be gathered at 10–20 litre scales and used to develop process changes for scales in excess of 100000s litres. Technology to deliver appropriate optimization techniques (e.g., feedback control of nutrient feeds based on RQ or respiration characteristics or methods for numerically correlating mixing phenomena at each stage) are typical examples. In this instance, it can be more effective to use resources to focus intensively on understanding the biological process and nutritional responses at each stage, and to have as few stages for scaling as possible. It does mean that the scaling arena needs to be equipped with similar or scaleable instrumentation to that available on the target scale; typically this can be difficult to achieve but can be immensely valuable.

For improvement of existing processes through new strain introductions, where typically there is a large body of existing data available from the target scale for similar processes, scaling through a single pilot venue can be straightforward and rapid.

An understanding of the biochemistry of the pathway for the metabolite of interest (or knowledge of the structure of the product, enabling some extrapolation into potential biosynthetic routes) is extremely valuable for pilot studies, e.g. if the molecule of interest is a polyketide, derived from acetate units, then the feeding of oil and acetate precursors could be usefully incorporated into the experimental approach. If the molecule of interest is derived from amino acid units, then protein and nitrogen source evaluation can usefully be studied. For growth-associated product, attention to providing easy to transport and rapidly utilized carbon sources, and training or inducing the inoculum to use this carbon source will be important. Finally, if the molecule of interest is to be induced, then provision of active inducer (or maintenance of its activity) needs to be considered, e.g. beta lactamase activity in a host organism is less than ideal in a system where penicillin derivatives, e.g. ampicillin, are used for maintenance of a plasmid, and strategies for minimizing its impact or designing it out of the host could be useful.

In practice, both routes tend to be used interchangeably, often dependent on the producing organism product being scaled and equipment availability.

8.3.2   Scaling Approaches

Traditional scale up was recommended to take place over a minimum of three and up to five or more stirred tank venues, relying heavily on maintaining a specific, chosen, engineering parameter consistently across scales.

Since the 1980s, microtitre (or miniwell) plate to shake flask is a stage that tends to feature prominently in scale-up routes, particularly when preliminary screening for strains

Table 8.2 *Potential scaling routes for different groups of products*

Scale route for structural elucidation of recombinant protein	Scale route for production of material for clinical trials	Scale route for natural product fermentation
Agar plates	Microtitre plates	Microtitre plates
Shake flasks	Minifermenters	Shake flasks
Laboratory fermenters (2–5 litres volume)	Laboratory fermenters/small pilot vessels (1–5 litres)	Pilot vessels
	Small production vessels (20–100 litres volume)	Production vessels. (>100 000 L)

of novel products is the experimental lead for the scaling programme. In the 2000s, new 'mini-fermenter' technology has been introduced and although it is early in the use of this technology, it is gaining acceptance as a useful investigational and potentially scaling venue. In addition, equipment, instrumentation and understanding have become more sophisticated and have increased the potential capability of output from smaller scale venues.

Currently it can be common for a scaling route to follow one of the schemes shown in Table 8.2.

8.3.3   Planning a Scaling Activity

Each new scaling stage tends to differ from the previous in the type of constraint it applies to the cultivation system (see Table 8.3). It is therefore helpful to devise a scale-up strategy at the start to enable appropriate focus on the end result to be applied at each stage. It is valuable to take the constraints of subsequent stages into account and design systems to work with or reduce the impact of the constraint. For example scaling strains exposed to mutagenic agents from microtitre plates to a large-scale antibiotic production operation would require a different set of targets at each scale, to those required when scaling a genetically modified nematode shake flask system expressing a protein to laboratory fermenters. Also, nutrient source availability, utilities constraints, material storage, fermenter broth make-up capability, and sterilization and effluent constraints, may all be usefully incorporated into a scaling strategy.

The recommended approach to a scaling exercise is to:

(i)   Identify the most appropriate equipment to use based on preliminary analysis of available information about process, potential impact of scaling on the process variables and equipment availability.

(ii) Identify minimum success criteria for each scaling stage – criteria that have to be met before it is worth transferring the process to the next scale stage. For example a target for a microtitre stage may be relatively crude, such as the identification of a condition that would generate a result 10% statistically different from control. Criteria for a transferring a process to the next stage for a pilot result are likely to be much more similar to the target criteria, such as achieving a product quality meeting a minimum specified limit, and yield sufficient to enable generation of material in quantities for clinical trials.

Table 8.3   Table to illustrate process constraints (benefits and disadvantages) of each scaling stage

Scale stage	Benefits	Disadvantages
Microtitre plates	Potential for automation Equipment for 96 well plates available Can attempt many variables and many replicates of variables in a single experiment Relatively inexpensive Low labour requirement/variable Laboratory space only required	Small volumes only Media restricted; soluble media Limited opportunity to feed nutrients Mixing and oxygen supply limited to rotary shaking, although some systems can have mini-fleas to promote agitated cultivation. Evaporation can be problematic – cultivation times can be limited due to evaporative loss Limited sample volume for detecting product; end point determinations only.
Shake flasks	Greater volume over microtitre plates; 10–50mL typically Improved aeration characteristics; high K_La through surface transfer Potential for introducing different shear regimes; glass beads, springs, baffles (although not a scaleable shear function) Potential for running longer fermentations without total evaporation of liquid volume Capacity per run limited by number of shaker spaces Potential to take limited numbers of samples or add nutrients (μl amounts), e.g. To check pH on paper or assay for nutrients using colorimetric assays. DOE strategies can be employed productively; relatively easy to replicate as not experimental numbers are not severely equipment constrained. Tend to be of most use for giving fermentation end-point data.	Shear can only be increased or decreased; it isn't a scaleable parameter from shake flasks Continuous nutrient feeds not possible; so basal nutrients must provide sufficient for the entire fermentation. Carbon catabolite repression or other repression effects may dominate responses and this may differ from response in final production venue pH can vary to extremes; alternatively medium can be buffered using mineral or biological buffers As a result of the large numbers of replicates per experiment, assay requirements to support a shake flask programme can be significant. May not be an ideal venue for scale down due to inability to supply feedstocks continuously May not be an ideal venue for scale down if culture morphology changes between different aeration regimes. Any sampling significantly breaches the sterile envelope

Autoclavable stirred tank reactors	Straightforward to set up a preliminary growth curve experiment by setting multiple flasks and harvesting entire flasks – often the first step in any scaling project.	Sterilization In an autoclave relates more to shake flask scale than stirred tank reactor scale where live steam injection is used.
	Because of increased volume, can take samples on a fermentation time course.	Operations run at ambient pressure without any opportunity to use steam to help maintain axenicity.
	Potential to feed nutrients; e.g. ammonium hydroxide, glucose, protein hydrolysates, oils, precursors	Evaporation may have a more significant impact on process than that impacting processes with greater volume.
	Potential for pH control	Because limited turbulence during sterilization, it may restrict medium components to soluble materials or very fine particles
	Potential for volume maintenance by feeding water an minimizing evaporation	Can be used flexibly; location can be variable; just bench space and access to an autoclave.
	Potential for on-line monitoring; pH, nutrient levels, off gas	
	Sterilization process similar for an entire batch of fermenters autoclaved together, so reducing variabilitly	
	Agitation, aeration and shear regimes can be crudely scaled to a production mimic	
	Shear regimes can relate better to scaleable shear parameters than in shake flask.	
	Opportunity ffor electronic data collection.	
Steam sterilizable stirred tank reactors	As above	Sterilization characteristics for each fermenter mahy be slightly different.
	Sterilization using *in-situ* steam more similar to large scale production environment.	Typically run using daytime operations crew, so sometimes timings cannot be the equivalent of the larger scale.
	Opportunity for maintenance of axenicity using sophisticated seals, steam traps and backpressure.	Feeding systems tend to be by peristaltic pump; this differs from the feeding strategies typically employed for large tanks and can be a source of difference.

Table 8.3 *Continued*

Scale stage	Benefits	Disadvantages
	Media components can be exactly the same as employed on large scale as sterilization operation can be run in a scaled mimic e.g. particles are agitated during sterilization therefore soygrits, or particulate material can be effectively sterilized.	Safety constraints can be more significant than those for autoclavable fermenters due to the use of on-line steam and pressure.
	Sterilization can be scaled using F_o/R_o values.	Location is fixed due to need for piped utilities.
	Agitation, aeration and shear regimes can be more precisely scaled than for autoclavable fermenters.	
	Culture expansion phases can mimic production scale.	
	Assays for production scale can be developed in this venue, or equivalent samples to those from production can be taken.	
Production fermenter	Ideal final test venue.	Limited scope for variation; generally demonstrating final proposed process. Must take into account production constraints that may include GMP, safety, regulatory and effluent requirements.
	Excellent opportunity for training operations staff on a changed process	Raw materials signficant cost for the experiment
	Requirement for process validation data from final scale venue in production-like conditions.	Losses significant; either due to failure to execute the trial as planned, and due to effluent costs and risks.
	Ideal venue for generation of regulatory submission data.	

From available information and experience, a crude estimation can be useful to assess resourcing, and timing requirements; see Table 8.4. The available data is obviously extremely limited at the start of the exercise, but will develop as the exercise progresses. A crude resource utilization value can be used to factor in holiday time, equipment failure, foreign growth, e.g. 80% utilization can be typical value to use in the absence of more precise data. The biological probability will be dependent on knowledge of the system, e.g. whether input to a microtitre screen has a large degree or small degree of variation based on mutagen used or source of isolates. In the absence of this information, an estimate of 1–10% probability rate for optimization may be a useful starting point (i.e., identifying an improvement in 1 in 10 or 1 in 100 runs). Even if this proves to be far from true, it will help considerably to focus resource on the bottleneck steps. It tends to be extremely important in the commercial environment, especially as timescales may need to be set for each stage in order to achieve time deadlines that may be outside the control of the scaling project, e.g. timing to deliver product for clinical or farm trials to fit in with submission plans for new licences. This type of planning activity can help to identify where bottlenecks may occur, other options explored or additional resource support be made available to achieve targets. In the academic environment it can help in identifying resource requirements and possibly help in accommodating plans of other colleagues, meeting schedules and project deadlines. It will need to be renegotiated at key points as biological information becomes available, but is extremely useful as a management and communication tool:

(iii) Transition to the next stage: start by trying to replicate the most productive output of the previous stage, then explore a response surface to key variables and focus on those results that shift the process closer to target.

(iv) Transfer to the next scale of operation once at least the minimum success criteria have been achieved. If time and resource are available, continue to optimize at the smaller volume stage until time runs out. Invariably it is more cost effective to optimize in pilot scale vessels than in the production scale. It is also extremely helpful to have licence submissions of at least partially optimized processes, with practical ranges relating to product quality attributes. There are many instances where production scale optimization is unnecessarily hampered by historical submission data specifically due to original submission data not relating to product quality impacting variables.

8.3.4 Executing a Scaling Activity

Microtitre/Minidish Scale Process Stage (10–1000 μL)

Microtitre dishes are first stage venues for liquid cultivation in high throughput screening and are typically in arrays of 8×12 wells in both deep- and shallow-well plate format (with working volume of 100–400 μL working volume). They can be used in manual and automated modes and can be incubated in static and shaking conditions (Figure 8.7). These types of array permit a rapid assessment of a large number of variables in a high-throughput laboratory.

Table 8.4 Example model to illustrate resourcing, timing and equipment issues relating to scale up of a new cultivar or novel product

Scale up	Microtitre plates	Shake flasks	Small STR scale	Preparative/production scale
Planned input	100 000 isolates	100 isolates	Five isolates	One variable
Equipment required (assuming 80% utilization)	Microtitre laboratory	Microbiology laboratory plus two shaking incubators of 25 stations	5–20 fermenters	Preparative/production scale equipment
Manpower required	Three people (two micro, one analytical)	Three people (two Micro, one analytical)	Four people (three operations, two analytical)	Defined by operations
Analytical requirements	Assay: 1000 per week	Potency assay; 50 per week Simple nutrient assay; dry weight, pH More sophisticated analysis; e.g. plasmid retention	Potency assay; five × timecourses per week Yield assessment Nutrient assays/off gas analysis Potential requirement to assess downstream processing efficiency at bench scale Tests intended for preparative/production trials, e.g. potential fold purification, product quality assay, plasmid retention assessment.	Potency assay Toxicity studies Yield Preparative efficiency
Constraints	5 day incubation. Lab only operates weekdays	7 day incubation, lab only operates weekdays	No sampling/transfers during night hours. FG rate of 5% or less.	
Maximum timescale	3 months	3 months analysis of all isolates (100 experiments/month run in duplicate) with 1 month detailed interrogating the five hits	3–6 months	1 month

Timescale assumptions	Single isolate/per well repeated in triplicate	100 shake flasks/month run in duplicate	7–14 day fermentation, 1–5 day inoculum development. No weekend working.	2 weeks from start of inoculum stages to harvest (6 weeks if triplicate fermentations are run in same vessel) One evaluation run in triplicate.	
Target	Five hits per 1000	Five hits	Two optimized hits		
Planned numbers of experiments	10000 isolates	100 × 3 plus 100 (more detailed investigation on the five hits)	10 × 5 – 20 experiments		
Typical strategy	Hit; 10% difference from control (above or below)	Use statistical design of experiments; (DOE) investigating the response surface to key variables; five variables for each isolate in duplicate. Look for yield statistically different from control. Take the above isolates with yield above control and interrogate further nutrient requirements based on developing response surface analysis experiments;	Establish performance compared to control. Then improve based on magnitude of response to variable seen in statistical experiments in shake flask. Use DOE for basal medium manipulation. Take output from above and use DOE with different nutrient feed strategies. Take most promising and develop to maximum within timescale available.	Establish performance reproducibility and verification of data from pilot work.	Data for qualification and GMP verification of process. Data to qualify product meets forward process criteria and performs as expected in any downstream recovery evaluations. Qualification that equipment and operations can deliver the required process control.

Table 8.4 *Continued*

Scale up	Microtitre plates	Shake flasks	Small STR scale	Preparative/production scale
Information output		Basal medium responses for the isolates Potential yield probabilities	Central cell bank; inoculum plans and requirements Data package plans for clinical submissions/registration updates Preliminary information to develop validation master plan for permanent introduction of new process – explored in production trials For new process; final data on equipment requirements. Cost estimates for full scale production.	
Probability; time	Unknown			
Probability; resource	90%			
Planned output	5 × 10	Two		

Figure 8.7 *Microtitre plate and microtitre incubator*

The fermentations are constrained by:

(i)   medium must be particulate free, undergo efficient sterilisation (either heat or filter sterilized) and permit rapid, accurate dispensing;
(ii) the mixing/aeration regime is system specific and nonscaleable;
(iii) the inoculum train is system specific;
(iv) fermentation time is limited by evaporation rate;
(v)   fermentations are batch, end-point, determinations.

However the benefits far outweigh the constraints providing effective media are available for the organism and product of choice. The system will be suboptimal for production and so data generated must rely heavily on low variability, high replication and consistent assay for product, as the criteria for the 'successful result' tends to be relatively crude, often just a performance significantly different from control. It is most suited for evaluating novel isolates for improved yield or detecting novel molecules to input into a scale-up programme. It has little use as a scale-down venue.

Shake Flask Scale Process Stage (50–500 mL) and Minifermenter

Shake flasks are useful screening systems for starting to evaluate the output of a high throughput screening programme or evaluate a range of clones, to explore potential for expression of a specific product, evaluating raw material options, or raw material variability. They tend to be the most accessible culture venue for liquid cultivation systems, so will be the first investigative stage for many projects (Figure 8.8).

The shake flask scale is ideally suited to evaluating:

(i)   Temperature ranges (temperature of incubator)
(ii) Responses to high and low agitation conditions (speed of shaking)
(iii) pH
(iv) Range of nutritional responses on a media using either inorganic, organic or mixed sources of carbon and nitrogen.
(v)   Inoculum preparation

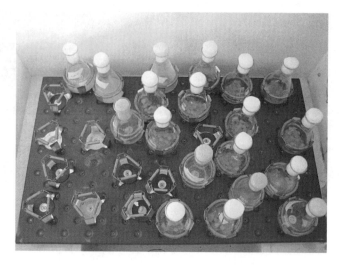

Figure 8.8 *Single stage flat bed shaking incubator (Photo: Michelle Lea, John Moores University, Liverpool)*

For scale-up projects, this may be the first opportunity to test out the effect of the listed variables, and the evaluations can provide data to support input into further evaluation in stirred tanks.

For scale-down operations, ideally the shake flask screen contains raw materials that are as similar as possible to the materials that would be used at the production scale, e.g. mixed nitrogen sources if both protein and ammonia are used at the production scale. Shake flasks have proven their worth as venues for screening raw material options in well established production processes where large numbers of replicates can be used to minimize variation and enable factorial designed studies to be used prior to further evaluations in stirred tanks.

If the aim of the scale up activity is to generate sufficient biomass to give suitable quantities of material or product for purification and isolation, it is possible that setting many replicates in large-volume flasks may give sufficient volume for preparation of product – in which case, the scaling activity stops at this point.

It can be helpful to look at the shake-flask scale as one where the aim is to look for potential value as a route for optimization in an experimental variable. The shake flask is unlikely to be the ideal venue for a product destined for agitated submerged culture; however, it can be a useful for exploring ideas and reducing them to key items that impact the product generated, enabling improved focus in the next scaling stage. For a novel process, where constraints at the stirred stage are have not materialized, the shake flask stage could screen out variants that could have better physical performance in stirred tank reactors and so it is valuable to maintain several high-yielding options as the output from the shake flask programme, investigating each option either simultaneously, or sequentially in the next scaling stage.

There are instances where there is good correlation between shake-flask experimentation and stirred tank results, and it is possible that some parameters can be scaled directly

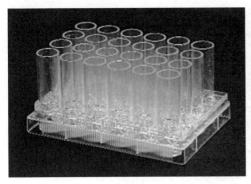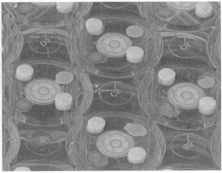

Figure 8.9   *Photograph of the minifermentation system (Application Biotechnology BV)*

from shake-flask results. However these are few and far between and tend to be relatively simple changes, e.g. straightforward replacement of a raw material ingredient.

Minifermenters.   This is a technology untested by the author, but showing promise as an investigational and scaling venue. It has some of the benefits of microtitre plates, shake flasks and laboratory fermenters combined into an array of 5×10 minifermentation units of 10 ml volume capable of running with stirred agitation, independent temperature and pH control (Figure 8.9). It is showing promise as a venue for exploring a range of variables rapidly, and with minimal equipment and medium components prior to using stirred tanks for further evaluation.

Stirred Tank Reactor Process Stage (1 litre – 500 + litre)

Laboratory fermenters Typically these are of the order of 500 ml–5 litre working volume equipped with agitators, temperature control and are sparged with air (Figure 8.10). Laboratory fermenters are a more sophisticated version of the shake flask system with the following benefits:

(i)   Increased K_{La} over shake flask; typically being in the range 20–100 times increased volume over shake flask, permitting sampling.
(ii)   The option of controlling pH using acid and base.
(iii)   The option of feeding nutrients, typically carbon or nitrogen sources, using continuous or shot-fed/ramped feed designs.
(iv)   The option of using off-gas monitoring.
(v)   The option of computer control and datalogging.

This still differs from a typical production scale system in that medium is sterilized by autoclave rather than the more usual *in-situ* live steam injection used on production scales.

Pilot-scale fermenters (in-situ sterilizable).   Pilot-scale fermenters usually tend to have working volumes of 20, 100, and 1000 litres. They may be of in-house design, tending to be scaleable versions of the production operation, or may follow the design intent of

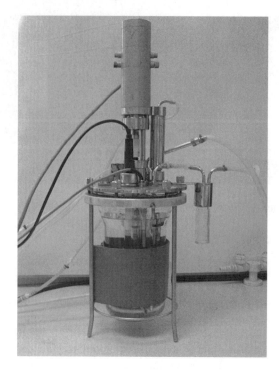

Figure 8.10 *Photograph of laboratory fermenter (Photo: Michelle Lea, John Moores University, Liverpool)*

the manufacturer. As a result of the increased volume over laboratory fermenters, it is normal to use *in-situ* sterilization either by live steam injection and/or jacket heating, which enables the scaling of sterilization characteristics to those of production scale operations. It is possible to use similar computer control systems as used in production scale, similar feeding technologies, sampling and on-line monitoring. The benefit of having improved control and data capture is extremely valuable prior to production-scale work, and although costly to install, equip and resource, the cost is more than offset by the savings achieved in a successful scale up, or successful production support over a period of time.

A pilot scale up stage can be valuable for establishing which, if any, medium components are critical for the fermentation, for identifying, defining and exploring critical process parameters, which will be required for GMP documentation, and to investigate any physical or business risk areas prior to introduction to the larger scale. Typical physical and business risk areas can include dimensions of cooling capacity constraints, foaming considerations for optimized media, product chromatographic species distributions for product, and conformation considerations for expressed proteins. Risk factors for high value products or biotech products may differ from those of natural products and it could be valuable to share information describing potential process options relatively early in

Figure 8.11   *Photograph of scaled down geometric fermentation vessels used for fermenting* Streptomyces spp *(Eli Lilly and Company, Inc)*

the pilot evaluation, in order to gather suggestions and commitment to problem solving or developing loss prevention strategies. See Figures 8.11 and 8.12.

Suggested variables to explore at this scale stage are:

(i)   Establish and define culture expansion processes. Establish seed transfer criteria for successful production of product in the production fermentation step.

(ii)   Define appropriate sterilization conditions for maintaining medium integrity. Use continuous sterilization equipment if this is to be used at final scale.

(iii)   Establish aeration and agitation conditions suitable for the process.

(iv)   Establish pH regimes for the process, and evaluate whether there is any value in using pH control.

(v)   Confirm temperature regimes for the process and consider taking specific heat output estimations, normalized to a biomass equivalent term, which could be used to estimate cooling capacity requirements for the fully scaled process.

(vi)   Explore basal medium constituents. Consider identifying any components that are risk items, e.g. unusual raw material or unusual supply chain. Consider evaluating options for reducing the degree of constraint for a risk item.

(vii)   Explore options for feeding nutrients for process intensification.

(viii) Establish that product meets accepted purity, species and quality requirements with consideration given to cross checking toxicological submission data or providing new data.

(ix)   Using process monitoring, develop the fermentation process description suitable for inclusion in regulatory, development and GMP report documentation.

(x)   Consider checking for phenotypic change and, where appropriate, genotypic change, particularly if unusual events are observed.

(a)

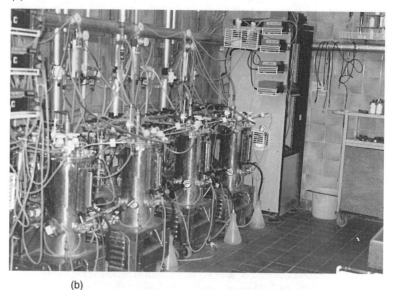

(b)

Figure 8.12 *(a) Photograph of supplier-designed first stage pilot tanks illustrating fermentation of Bacillus spp using a pH-stat feeding regime (Cyanamid of Great Britain Limited). (b) Photograph of supplier designed second stage pilot tanks illustrating a traditional multi-stage ten fold volume increase approach to scaling. For both laboratory and pilot stirred-tank work the types of recommended evaluation are the same (Cyanamid of Great Britain Limited)*

(xi) Consider evaluating processing options for recovery of product using bench- or pilot-scale models. Establish that the new fermentation process conforms to the requirements of downstream processing steps.

(xii) Develop the development data package and clinical trial documentation if the output from the pilot stage is to be used for clinical trials. Ensure that the production operation will not be constrained due to paucity of data in the development data pack, and ensure that practical operational ranges are proposed that specifically relate to parameters influencing product quality.

(xiii) Consider evaluating and comparing results from several different working cell banks to establish that consistency will be achieved in full scale process.

(xiv) Implement the final recommended process to generate data and product for trials (phase 1/2 clinical trials, farm trials, new product claims), submissions, reports and process description for production scale operation and harvest criteria suitable for appropriate processing downstream.

The benefit of evaluating variables at stirred tank scale is that it provides the opportunity not just to execute the experimental variable, but to optimize both inoculum and production phases, using more sophisticated control methods and data logging than can be achieved at earlier scales. The tanks are also the most similar to those used in production scale (high shear agitation and aeration) with potential for sampling during the fermentation and nutrient feeding, and hence the results are more valuable as they have higher likelihood of effective translation to the ultimate scale.

Pilot operation can be used iteratively, with data from one set of experiments leading to evaluation of a further set of variables, if data from earlier stages provides experimental leads. It can also be useful for identifying the response surface to a variable using factorial design techniques if clear experimental leads are not available at the start of the exercise. Generally, pilot areas are equipped with multiple vessels and the ability to sterilize and run experiments simultaneously with common raw materials, often inoculated from a common seed phase. The reduction in variation that this can provide can enhance the power of any statistical evaluations of the data generated.

Stirred Tank Reactor Process Stage – Full Scale Commercialization

A novel product in a new facility is more likely to be a biotech product and it is likely that the fermenter will have been specifically sized, designed and equipped for the output of the process, with a view to optimizing batch size for cost effectiveness, optimizing cleanability and optimizing data capture (see Figure 8.13). This differs from Classical Fermentations – natural product fermentations – where the production vessel is much more likely to be one that is routinely operational and where the production trials will be introduced into an operation familiar with the currently operating process (see Figure 8.14). However, in general, the considerations for implementing work at the final scale are likely to be similar.

The production venue is ideally the final verification venue for the new process, with the opportunity of generating at-scale data in the actual production venue, and for revising or rewriting an existing process flow document. It is highly likely that output from the trial will be subjected to additional GMP review and potentially the product will be put on hold pending receipt of test data. In view of the requirement for the data to be final

Figure 8.13   *Photograph of production fermentation facility used for expressing recombinant protein in an E. coli expression system*

test data, it is important that all the critical variables are explored at the pilot scale to ensure there is no undue risk to production, or unnecessary volumes of effluent generated by an unnecessary trial failure. For operation under GMP it is also typical to consider whether the process change may provide any additional cleaning requirements or constraints, and this can only be observed and evaluated fully in the production venue.

It is usual to carry out a three-lot evaluation of a new process or major process change, possibly forming part of a validation package or for appropriate consideration and comment in any requirement to validate the active pharmaceutical ingredient (typically the output of a downstream step). It provides an opportunity to establish that the scaled process can run repeatedly and consistently in the new venue, and to train operations staff in the new process and any changed biological responses, to establish that equipment and cleaning operations are likely to be adequate for the new process, and to generate GMP data to support any process changes. The product can generally be forward-processed down the relevant product recovery route and product release data generated, verifying that the output meets all processing and analytical requirements.

It can be useful to consider running the process within some boundary criteria during the 'process qualification' stage of implementation. This becomes even more critical for novel biotechnology products where development data regulatory submissions of process ranges suitable for defining the production process are required.

Figure 8.14 *Photograph of production fermentation facility used for natural product fermentations*

8.4   Summary

In this chapter an attempt has been made to provide strategies and advice to aid a researcher embarking on a scaling exercise. Features of major importance to be considered during scale changes are gaseous and liquid transfer processes, provision of nutrients from batched and fed materials, inoculum generation and methods for expanding inocula, impact of pressure changes during scaling, and any biochemical, genetic or morphological changes that may occur during the changes in scale. If the fermentation process under evaluation involves the creation of a growth-associated product, then there may be a linear relationship between the scale parameter changes and impact on the process. However if the product is a nongrowth-associated product, the impact is often not predictable without further study. At all stages in scale transfer, due consideration should be given to the inherent variability within fermentation processes, and this is discussed in Chapter 11.

Ideally the result of scale up would be an intensification of the biological process, which would mean a higher yield of product with a less than linear increase in cost, culture volume and energy requirements. The result of scale down would be to achieve the equivalent yield, potency and productivity as that achieved at the large scale.

The scaling environment is one where analysis, scientific knowledge, microbiological, biochemical, genetic and engineering skills and, most importantly, creativity, are required. If the scaling operation is intended to scale to a production environment, there is a need to understand the constraints of the production environment to create the effective focus for the scaling operation. If the scaling operation is intended to generate more product for purification or characterization, the approach will need to focus on the specific objective, for instance protein product requiring purification may impose restrictions in the range of medium options, so that medium protein does not adversely impact chromatographic purification steps. Invariably scale up/scale down is about uncovering information and data which help to feed into an iterative strategy for the next step at whatever scale or target is planned.

It is beneficial for scientists, molecular specialists and engineers to be involved in sharing and discussing data during the scaling process, as it is an organizational attempt to predict an optimized biological cultivation system at the target scale. There is no right or wrong way to do it; scaling is an approach for uncovering information, rather than 'creating' it. What is critical to success is to have a flexible strategy aimed at the goal with practical execution generating sufficient information to enable tighter and tighter focus and adjustment to be given to the system in order to achieve the end result.

Invariably, over time a practitioner gains experience by repeating the activity with different processes or products and sharing his or her learning with others. Much scale up/scale down expertise therefore resides within groups and organizations, and the ideal situation is to work alongside an experienced scientist or biochemical engineer. However, in the absence of practical experience, there is a number of key items that can systematically be evaluated and with keen observation, reproducible technique and appropriate measurement tools, the same path for successful scale up/down can be trodden by a relatively inexperienced researcher.

References and Further Reading

Aiba, S., Yamada, T. (1961) Oxygen absorption in bubble aeration. Part 1. *J. Gen. Appl. Microbiol.* **7**, 100–106.

Aiba, S., Hisamoto, F. (1990) Some comments on respiratory quotient (RQ) determination from the analysis of exit gas from a fermenter. *Biotechnol. Bioeng.* **36**, 534–538.

Aiba, S., Humphrey, A.E., Millis, N.F. (1973) *Biochemical Engineering*, second edition, Academic Press, Inc., New York.

Atkinson, B., Mavituna, F. (1991) *Biochemical Engineering and Biotechnology Handbook.* Stockton Press, New York.

Bailey, J.E., Ollis, D.F. (1986) *Biochemical Engineering Fundamentals*, McGraw Hill, New York.

Banks, G.T. (1979) Scale Up of Fermentation Processes, in *Topics in Enzyme and Fermentation Biotechnology*, A. Wiseman (ed.), Vol. 1 Chapter 3, John Wiley & Sons, New York, pp. 170–266.

Banks, G.T. (1977) Aeration of Mould and Streptomycete Culture Fluids, in *Topics in Enzyme and Fermentation Biotechnology*, A. Wiseman (ed.), Vol. 1, John Wiley & Sons, New York, pp. 72–110.

Bartholemew, W.H. (1960) Scale up of submerged fermentations. *Adv. Appl.Microbiol.* **2**, 289–300.

Charles, M. (1985) Fermenter Design and Scale-up, in *Compehensive Biotechnology*, C. Cooney, A. Humphrey (eds), Pergamon Press, Oxford.

Cooper, C.M. *et al.* (1944) Performance of agitated gas-liquid contacters. *Ind. Engng Chem.* **36**, 504.

Gravius, B., Bexmalinovic, T., Hranueli, D., Cullum, J. (1993) Genetic instability and strain degeneration in *Streptomyces rimosus. Appl. Environ. Mircrobiol.* **59**, 2220–2228.

Hamilton, B., Sybert, E., Ross, R. (1999) Pilot Plant, Chapter 27 in *Manual of Industrial Microbiology and Biotechnology*, second edition, A.L. Demain, J. Davies, R. Atlas, G. Cohen, C. Hershberter, W. Hu, D. Sherman, R. Wilson, J. Wu (eds), A.S.M., Washington.

Heijnen, S. (1994) Thermodynamics of microbial growth and its implications for process design. *Trends Biotechnol.* **12**, 483–492.

Holms, W.H., Hamilton, I.D., Mousedale, D. (1990) Improvements to microbial productivity by analysis of metabolic fluxes. *J. Chem. Tech. Biotechnol.* 0268-2575, p 138–141.

Hopwood, D.A., Bibb, M.J., Chater, K.F., Rieser, T., Bruton, C.J., Kieser, H.M., Lycliate, D.J., Smith, C.P., Ward, J.M., Schrempt, H. (1985) *Genetic manipulations of Streptomyces: A laboratory manual*. John Innes Foundation, Norwich.

Hosobuchi, M., Yoshikawa, H. (1999) Scale-up of microbial processes, Chapter 19 in *Manual of Industrial Microbiology and Biotechnology*, second edition, A.L. Demain, J. Davies, R. Atlas, G. Cohen, C. Hershberter, W. Hu, D. Sherman, R. Wilson, J. Wu (eds), A.S.M., Washington.

Hunt, G.R., Stieber, R.W. (1988) Inoculum development, Chapter 3 in *Manual of Industrial Microbiology and Biotechnology*, first edition, A. Demain, N. Solomon (eds), A.S.M., Washington.

Lea, M. (2007) A physiological study of *Streptomyces capreolus* and factors governing growth and capreomycin biosynthesis. PhD thesis, Department of Biomolecular Sciences, John Moores University, Liverpool.

Liefke, E., Kaiser, D., Onken, U. (1990) Growth and product form of actinomycetes cultivated at increased total pressure and oxygen partial pressure. *App. Microbiol Biotech.* **12**, 674–679.

Madden, T., Ward, J.M., Ison, A. (1999) Organic acid excretion by *Streptomyces lividans* TK24 during growth on defined carbon and nitrogen sources. *Microbiology* **142**, 3181–3185.

Monaghan, R.L., Gagliardi, M.M., Streicher, S.L. (1999) Culture preservation and inoculum development, Chapter 3 in *Manual of Industrial Microbiology and Biotechnology*, second edition, A.L. Demain, J. Davies, R. Atlas, G. Cohen, C. Hershberter, W. Hu, D. Sherman, R. Wilson, J. Wu (eds), A.S.M., Washington, p 29.

Moo-Young, M., Christi, Y. (1999) Considerations for designing bioreactors for shear-sensitive culture. *Biotechnology* 1291–1296.

Norwood, K.W., Metzner, A.B. (1960) *A.I.Chem. E.J.* **6**, 432.

Pandza, K., Pfalzer, G., Cullum, J., Hranueli, D. (1997) Physical mapping shows that the unstable oxytetracycline gene cluster of *Streptomyces rimosus* lies close to one end of the linear chromosome. *Microbiology* **143**(Part 5): 1493–1501.

Reisman, H. (1999) Economics, Chapter 23 in *Manual of Industrial Microbiology and Biotechnology*, second edition, A.L. Demain, J. Davies, R. Atlas, G. Cohen, C. Hershberter, W. Hu, D. Sherman, R. Wilson, J. Wu (eds), A.S.M., Washington, p 273.

Reusser, F. (1963) Stability and degeneration of microbial cultures on repeated transfer. *Adv. Appl. Microbiol.* **5**, 189.

Steel, R., Maxon, W.D. (1962) Some effects of turbine size on novobiocin fermentations. *Biotech. Bioeng.* **4**, 231.

Stowell, J.D., Bateson, J.D. (1983) Economic aspects of industrial fermentation, in *Bioactive Microbial Products. 11. Development and Production*, L.J. Nisbet, D.J. Winstanley (eds), Academic Press, London.

Tough, A.J., Prosser, J.I. (1996) Experimental verification of a mathematical model for pelletted growth of *Streptomyces coelicolor* A3(2) in submerged batch culture. *Microbiology* **142**, 639–648.

9

On-line, *In-situ*, Measurements within Fermenters

Andrew Hayward

9.1 Introduction

9.1.1 Why Monitor the Process?

Monitoring of the fermentation process provides data, but also allows the process to be automatically controlled. Temperature, gas flows, aeration and pH can all be controlled by measuring the parameter, providing a set point and then acting on the difference between the measured variable and the set point. In the same way that a thermostat controls temperature, the pH of the process can be controlled by the automatic addition of acid and base, or the dissolved oxygen level can be controlled by the speed of the agitator or the volume of oxygen in the gas sparge. In order to achieve this it is necessary to be able to make continuous, accurate, reliable measurements of each of the parameters.

9.1.2 What can be Monitored?

The number of continuous on-line measurements that can be practically measured within the fermenter is surprisingly small. Generally these are limited to pressure, level, flow (reagents, feed, etc.), temperature, pH (acidity/alkalinity), dissolved oxygen, dissolved CO_2 and off-gas analysis. The two most common analytical measurements in fermentation are pH and dissolved oxygen. An understanding of the sensor design and operation will enable the user to have a greater insight into the potential problems and pitfalls of these measurements.

Practical Fermentation Technology Edited by Brian McNeil and Linda M. Harvey
© 2008 John Wiley & Sons, Ltd

9.2   pH Measurement

9.2.1   Definition of pH

The variable pH is the negative logarithm of the hydrogen ion activity. For the purposes of this definition activity can be considered as the equivalent to concentration. The small p designates the mathematical relationship as power (log) and the H designates the ion as hydrogen.

9.2.2   Basic Sensor Design

Conventional glass pH sensors are made from two electrodes; a glass measuring electrode (Figure 9.1) and a reference electrode (Figure 9.2).

The measuring electrode consists of a pH sensitive glass membrane; a pH buffered filling solution, and a silver/silver chloride (Ag/AgCl) element, to form a galvanic half cell.

The reference electrode also utilises a silver/silver chloride (Ag/AgCl) element, submersed in a solution of potassium chloride (KCl) saturated with silver chloride. The inner liquid junction (porous ceramic frit) allows electrical continuity from the reference element to the KCl electrolyte salt bridge chamber. The outer liquid junction completes the electrical circuit to the process solution. With two liquid junctions, this type of reference is referred to as a double junction.

For convenience the measurement electrode and reference half cell are combined into a single unit called a combination electrode (Figure 9.3).

The glass measuring half cell is contained within an inner glass tube, whilst the reference electrode surrounds it. The half cells are completely isolated from each other, both electrically and chemically.

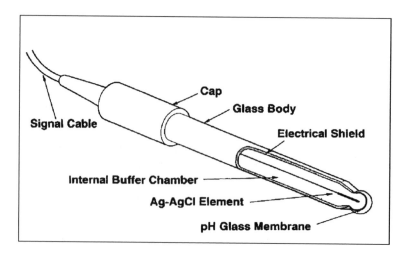

Figure 9.1 *Glass measuring electrode (Reproduced by permission of Broadley-James Corporation)*

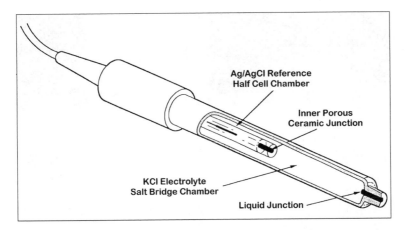

Figure 9.2 *Double junction reference electrode (Reproduced by permission of Broadley-James Corporation)*

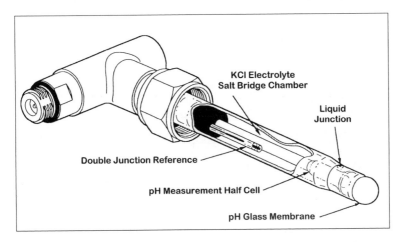

Figure 9.3 *Combination pH electrode (Reproduced by permission of Broadley-James Corporation)*

9.2.3 Principle of Operation

The glass used for the pH sensitive membrane is formulated from silica and doped with rare earths that make it pH sensitive. This allows hydrogen ions in solution to attach to the glass surface and create a potential across the glass membrane (Figure 9.4). This potential is proportional to the activity of the hydrogen ion (H^+) and therefore to the pH of the solution (see Section 9.2.1 above).

The reference electrode forms a stable potential and is connected to the process via two porous ceramic junctions and a solution of 3.8 molar potassium chloride (KCl). In this type of reference half cell, the filling solution is saturated with silver chloride (AgCl). This could react with the process solution blocking the ceramic frit and cause premature

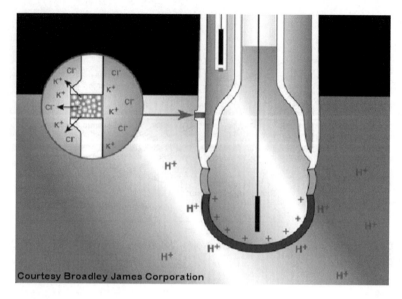

Courtesy Broadley James Corporation

Figure 9.4 *pH electrode construction (Reproduced by permission of Broadley-James Corporation)*

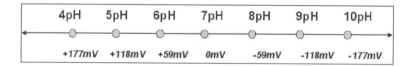

4pH	5pH	6pH	7pH	8pH	9pH	10pH
+177mV	+118mV	+59mV	0mV	-59mV	-118mV	-177mV

Figure 9.5 *pH electrode output (Reproduced by permission of Broadley-James Corporation)*

failure of the sensor. To prevent this, the two porous ceramic junctions isolate the silver chloride from the process, preventing any reaction taking place.

In sealed electrodes of this type, the 3.8-M KCl is thickened to reduce the migration of electrolyte out of the liquid junction whilst maintaining a good connection to the process. Whilst the reference half cell electrolyte is thickened, the measuring half cell solution can contain air bubbles. To ensure that the solution is in full contact with the inside of the glass bulb it is essential that the electrode is always mounted at least 15° above the horizontal. If the internal filling solution is not in contact with the inner of the glass bulb, the measurement half cell will give an incorrect and unstable output.

The combination pH electrode generates a mV output (Figure 9.5). In an ideal electrode the output is 0 mV at 7 pH increasing at 59.16 mV per pH unit at 25 °C.

9.2.4 Measurement Precision (What is Practically Achievable)

A number of factors contribute to errors in pH measurement. Each of these factors combines to give a practical measurement precision of approximately 0.05 pH.

Temperature

The output from a pH electrode varies with temperature at a rate of 0.03 pH per 10 °C per pH unit. So if the pH varies by less than ±1 pH (between 6–8 pH) and the temperature varies by less than 3 °C, the maximum error from the pH electrode's temperature sensitivity will be ±0.01 pH. This temperature effect is predictable and can be compensated for by the measuring instrument.

A second temperature effect is often ignored or misunderstood. The pH of the solution being measured will change with temperature. It is very difficult to predict what this change will be, because it depends upon the constituents in the process solution, but can be of the order of 0.01 to 0.03 pH per °C. This is of particular importance when comparing two solutions (e.g., when taking samples from the fermenter): if the solutions are not at the same temperature, you should not expect their pH to be the same. So if a sample was taken from a fermenter at 37 °C and allowed to cool to room temperature at 25 °C, then the difference in pH due to temperature change could be as much as 0.36 pH. This temperature effect is not corrected by temperature compensation in the measuring instrument.

Temperature Compensation

Temperature compensation can be either manual or automatic. Manual temperature compensation is achieved by entering the temperature value to the instrument. Automatic temperature compensation is achieved by using a temperature measurement sensor connected to the instrument and the instrument calculating the correcting factor to the measurement value.

9.2.5   Calibration

The output from a perfect pH electrode is 0 mV at 7 pH increasing to 59.16 mV per pH unit at 25 °C; however manufacturing tolerances and degradation of the electrode through use (autoclaving/steam sterilising) move the electrode output from the ideal (see Section 9.2.8). To overcome this, electrodes are standardised in solutions of a known, stable, pH; these calibration solutions are called pH buffer solutions. Buffer solutions are commercially available to traceable national standards in almost every pH value. The commonly used buffer values are 4, 7 and 10 pH.

Calibration procedures vary from instrument to instrument and the manufacturer's operating instructions should be followed. A recommended good practice would contain the following steps.

(i)   Calibrate the pH electrode in two buffer solutions prior to sterilisation.
(ii)   Normally one of these buffer solutions would be 7.00 pH, this will determine the mV offset of the pH electrode output at its zero point. The electrode offset should be displayed by the instrument in mV during or after the calibration procedure.
(iii) The second calibration point should be at least 3 pH units from the first, this will determine the slope of the pH electrode output, how many mV/pH unit it is generating. The electrode slope should be displayed by the instrument during or after the calibration procedure.

(iv) The instrument temperature compensation would normally be set to the fermentation running temperature (typically 37 °C). During calibration it must be set to the buffer solution temperature (20–25 °C). Do not forget to reset back to the fermentation temperature after the calibration is completed (see section on Temperature above).

(v) The following data should be logged at each calibration:
   (a) date of calibration;
   (b) operator ID;
   (c) vessel ID;
   (d) electrode ID;
   (e) batch ID;
   (f) electrode offset and slope;
   (g) speed of response.

This data will be valuable in determining the performance of the pH electrode and its suitability for use. Interpretation of this data is covered later under Troubleshooting, Section 9.2.8.

Good laboratory practice must be observed when handling and using buffer solutions. An operating procedure (SOP) should be completed:

(i) Care must be taken not to cross-contaminate the buffers when moving the electrode from one buffer to another. A good rinse in deionised water is essential; carefully blotting the electrode dry will avoid carry over.

(ii) The buffers will be date coded and will degrade with exposure to atmosphere. In particular the high value buffers (pH 9 and above) are susceptible to CO_2 absorption.

(iii) Buffer pH values vary with temperature (as do all solutions); ensure that the value standardised at is the *actual* value of the buffer solution. This data will be provided with the solution.

9.2.6   Comparing pH Data

After calibrating the pH electrode the fermenter will need to be sterilised. This subjects the electrode to a high temperature and cooling cycle and will effect the calibration of the electrode. For an electrode in good condition this will be of the order of 0.01 to 0.05 pH. It is common practice to compensate for this change by taking an off-line measurement and adjusting the on-line reading to match. There are a number of opportunities here to introduce errors into the calibration procedure. The following steps are recommended:

(i) Take the largest sample that is practical, as this should minimise contamination. Ensure that the pH electrode used with the off-line instrument is washed in deionised water and blotted dry before placing in the sample.

(ii) When the sample is taken from the fermenter, read off and note the on-line pH reading (e.g., 7.05). This will be the value that the off-line reading is subtracted from to give the calibration offset.

(iii) Take the sample immediately to the off-line pH meter. As the sample cools its pH will change, and it will be absorbing CO_2 from the atmosphere. If the off-line pH meter is temperature compensated it will only compensate for variations in the pH electrode with temperature and *not* the solution changes.

(iv) Note the off-line value (e.g., 7.09) and subtract it from the on-line value (7.05 − 7.09 = −0.04), this is the change in the electrode calibration that you will need to adjust the on-line reading by.

(v) The on-line reading should be adjusted by half the difference (−0.02). Note the current on-line value and add the half the difference to it (e.g., new on-line value 7.06 + (−0.02) = 7.04). The reason for adding only half the difference is the measurement uncertainty of both the on-line and off-line measurements. Experience has shown that adding in the whole difference tends to overcompensate for the calibration shift.

9.2.7 Maintenance (Cleaning, Storage)

To obtain the best life and performance from a pH electrode it will need to be maintained and stored correctly. The combination pH electrode does have two measurement elements: the glass membrane and the liquid junction. It is essential that both are treated appropriately.

The pH electrode needs to make intimate contact with the process solution to make reliable, accurate measurements; contamination of the glass surface or liquid junction will degrade the performance of the electrode.

The wetted materials on a pH electrode are glass, ceramic, and a sealing polymer (o-rings, etc.). These have a very good chemical resistance, so it is possible to use a number of different cleaning agents depending upon the contamination on the measurement surfaces. Warm soapy water and a toothbrush will remove the majority of deposits from a fermentation processes. More difficult deposits can be removed with 2% v/v NaOH (sodium hydroxide) or for hard scale deposits a 2% v/v solution of HCl (hydrochloric acid) can be used. **Normal laboratory safety procedures must be observed when handling these chemicals as they can cause severe burns.** The electrode should be thoroughly rinsed with deionised water after cleaning and placed in a storage solution for at least 2 hours prior to use.

Do not use anything abrasive on the glass electrode as the smallest scratch on the glass membrane will permanently damage the electrode.

The pH electrode must be kept wet at all times. Follow the manufacturer's recommendations on storage solutions. Never leave a pH electrode in deionised or distilled water for extended periods of time (hours) as this will dilute the electrolyte in the porous liquid junction, leading to unstable readings. If the pH sensitive membrane dries out it will need to be rehydrated by soaking in a saturated KCl solution (or manufacturer's recommendation) for several hours. If the liquid junction dries out it may not be possible to rehydrate and the electrode will need to be replaced. A dried out pH electrode will drift and not respond to buffer calibration.

Do not rest a pH electrode on the bottom of a beaker as this can scratch the glass membrane, leading to premature failure. Always support it in a retort stand or something similar.

Cables and connections are a critical component of the pH measurement loop. The pH electrode requires a very high electrical insulation to be maintained, any dampness or contamination on any of the connections will severely degrade the measurement integrity. Regularly check the connectors for signs of corrosion and/or build up of contamination. The cable should be neatly terminated into the back of the connector, if the insulation is damaged the cable should be replaced. Most connectors are crimped to the cable so it is not possible to service them.

Most manufacturers supply a cap for the pH electrode connector. This must be fitted when the electrode is autoclaved to prevent moisture entering the connector. Also check that any o-rings are fitted where they should be and that they are in good condition.

9.2.8   Troubleshooting (Sensor Diagnostics)

All pH electrodes degrade (wear out) with use. The single biggest contributor to this is the autoclave or steam cycle. Rapid heating and cooling of the electrode stresses the components, pressurises and depressurises the reference electrode via the liquid junction, dissolves the elements from glass measurement surface increasing the glass electrical resistance and reducing its sensitivity to measure pH. The number of sterilisation cycles that a pH electrode can withstand will depend upon a number of variables:

(i)   the manufacturer;
(ii)   how quickly the temperature is increased and decreased;
(iii) the maximum sterilisation temperature;
(iv)   how long the electrode is held at the sterilisation temperature;
(v)   what performance is acceptable to the process.

With all these variables it is very difficult to estimate how many cycles an electrode will be good for, so some other quantifiable method has to be used to determine the suitability of the electrode for the next run.

Interpretation of the data derived from the calibration log will provide invaluable information on the condition of the pH electrode and its suitability for use. It is possible to calibrate a pH electrode that is well beyond its useful life, but if then used in a fermentation it will give unreliable results and could jeopardise the run.

A typical pH electrode log could look like that in Table 9.1.

The third entry shows an electrode that would calibrate but which exhibits a degraded offset and slow response time. Clearly this electrode is nearing the end of its useful life and should be replaced. Limits can be placed on the acceptable values for offset, slope and response time. These values will vary according to how critical pH is to the process and the manufacturer of the pH electrode.

- Offset is the sensor's millivolt (mV) output in a pH 7 buffer. Theoretical offset is zero millivolts.
- Offset is measured at pH 7 and can be expressed in mV or pH units. A typical limit could be ±20 mV or ±0.35 pH.

Table 9.1   *pH electrode log (Reproduced by permission of Broadley-James Corporation.)*

Date	Operator	Vessel	Batch	Serial No.	Offset (mV)	Slope (%)	Response (sec)
30/04/2006	JWR	1	Ferm20	T34567	2.5	98	10
15/05/2006	JWR	1	Ferm21	T34567	3.8	95	15
22/05/2006	JWR	1	Ferm22	T34567	15.5	95	45

- The span is another useful diagnostic tool. A pH sensor should generate 177.5 mV in a pH 4 buffer at 25 °C.
- Span is calculated as the actual mV generated per pH unit over theoretical. It is usually expressed as % efficiency or as a decimal equivalent e.g. 98% or 0.98.
- Take the mV value at 7 pH and subtract it from the mV value at 4 pH, then divide by 177.5 (at 25 °C). A typical lower limit for slope could be 0.92 or 92%.
- The speed at which the sensor responds is important. A 'good' electrode should settle at the new pH value within about 20 seconds of submersion in a given buffer.

It is possible to determine other types of problems by observing the pH electrode response:

- Coating of the pH sensitive glass surface may result in sluggish measurements (slow response). Try to remove any coating in warm, soapy water. Do not abrade glass bulb.
- Media, antifoaming agents and biomass on the liquid junction may result in offsets or drifting signals. Try using warm, soapy water and a fine bristled toothbrush to remove contaminants.
- Loose, wet or dirty connections can produce pH signal problems, typically shorting out the pH signal and displaying a straight line measurement, close to 7 pH. Ensure that all connections are clean, dry and secure.

9.2.9 Summary

- When comparing data be sure it is like with like.
- Keep a log of the electrode calibration data.
- Keep the electrodes wetted at all times. Follow the manufacturers recommendations.
- Keep connections and cables in good condition.
- Use the sensor diagnostics to determine suitability for use.
- All electrodes will fail one day.

9.3 Dissolved Oxygen Measurement

Dissolved oxygen is an important variable in a fermentation process. Oxygen is consumed in large amounts in most fermentations, yet is sparingly soluble in the culture media. Replacing dissolved oxygen as it is consumed can be a major limiting factor in large-scale culture growth (see Chapter 7 on oxygen transfer, and Chapter 8 on scale-up).

9.3.1 Basic Sensor Design

Galvanic versus Polarographic

Two types of sensor, galvanic and polarographic, have been used to make dissolved oxygen measurements in fermentation processes. Both have two metal electrodes, a cathode and an anode. Both produce an electrical current at the cathode surface that is proportional to the amount of oxygen in the solution. The anode completes the electrical circuit to the dissolved oxygen transmitter, which converts the sensor output signal to the

Table 9.2 Output[a] of oxygen electrodes in air and air saturated water as a function of flow (Krebs and Haddad, 1972)

Electrode	Output		
	Air	Stagnant water	Stirred water
Polarographic	100	98.5	100
Galvanic	100	50.0	99.0

[a] The outputs for each electrode are referred to the output in room air for that electrode.
The deviation from 100 (room air) is in effect the percentage error.

unit of measurement of interest, e.g. % saturation. The difference between galvanic and polarographic sensor designs is the source of the mV potential (bias voltage) within the sensor.

Galvanic sensors derive their own bias voltage, which is not constant, from the internal reaction between the dissimilar metals chosen for the anode and cathode. However, since most galvanic sensors are designed to produce an output of sufficient magnitude to drive an ammeter directly, without electronic amplification, they require relatively large electrodes. This in turn means that they consume more oxygen as part of their operation. Without a constant update of the sample in front of the measuring electrode, a layer depleted of oxygen forms giving rise to a lower output signal.

Dissolved oxygen measurements are susceptible to motion of the sample. Data reported by Krebs and Haddad (1972), indicates that the galvanic sensor exhibits the greatest flow sensitivity (see Table 9.2). For this reason polarographic dissolved oxygen sensors have become the first choice for the majority of fermentation applications.

9.3.2   Principal of Operation of a Polarographic Sensor

Polarographic dissolved oxygen sensors consist of a silver anode (Ag) and platinum cathode (Pt), surrounded by electrolyte and separated from the process by a gas permeable polymer membrane. A polarising voltage is applied across the anode and cathode so that the cathode is sufficiently cathodic (negative with respect to the anode). A 675 mV polarization voltage is optimum for oxygen analysis.

Oxygen diffuses into the sensor through a gas-permeable polymer membrane (Figure 9.6). It undergoes a reduction reaction at the sensor's cathode surface that in turn produces a nanoamp (nA) current. This current is proportional to the partial pressure of the oxygen present in the solution, and is therefore a measure of the ratio of oxygen present in the sample if the temperature and pressure are known.

The majority of DO sensors designed for fermentation are manufactured from 316 L stainless steel and are electropolished to provide a surface that can be easily cleaned. A connector allows the cable to be removed from the sensor for convenience so that the sensor can be autoclaved whilst mounted in the vessel. There are different types of fitting to mount the sensor into the vessel, but the Pg13.5 has recently been established as an industry standard (Figures 9.7 and 9.8).

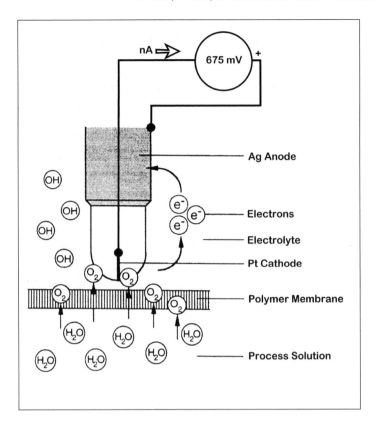

Figure 9.6 *DO sensor operation (Reproduced by permission of Broadley-James Corporation)*

9.3.3 Sensor Polarisation

All polarographic dissolved oxygen sensors have to be polarised for a set time prior to use. This can take up to 6 hours, but refer to the manufacturer's recommendations on the time taken for full polarisation.

An unpolarised sensor will give an output even under zero oxygen conditions, leading to a zero offset. This zero offset reduces as the sensor becomes fully polarised, so that the output from the sensor is only due to the oxygen that permeates through the membrane.

The sensor is polarised by applying a voltage (normally 675 mV) between the anode and cathode, with the cathode being negative with respect to the anode. The polarising voltage is derived from either the instrument the sensor is connected to, or from a battery-powered polariser (Figure 9.9) that is connected to the sensor. It is useful to store the sensor with a polariser so that it is ready for use.

9.3.4 Temperature Effects

The sensor is very temperature dependent and the output from the sensor increases as the temperature increases; this effect can be as much as 3% per °C and is caused by a change

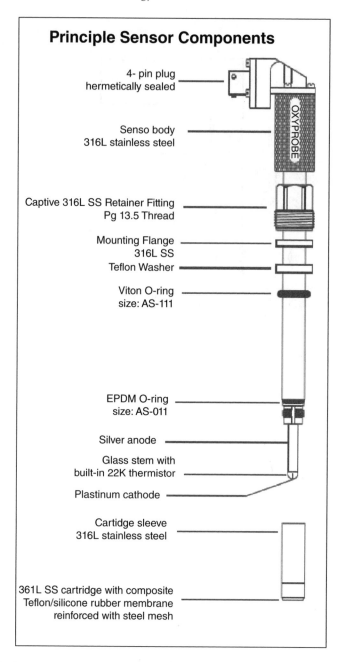

Principle Sensor Components

4- pin plug hermetically sealed

Senso body 316L stainless steel

Captive 316L SS Retainer Fitting Pg 13.5 Thread

Mounting Flange 316L SS

Teflon Washer

Viton O-ring size: AS-111

EPDM O-ring size: AS-011

Silver anode

Glass stem with built-in 22K thermistor

Plastinum cathode

Cartidge sleeve 316L stainless steel

361L SS cartridge with composite Teflon/silicone rubber membrane reinforced with steel mesh

OXYPROBE

Figure 9.7 *Sensor components (Reproduced by permission of Broadley-James Corporation)*

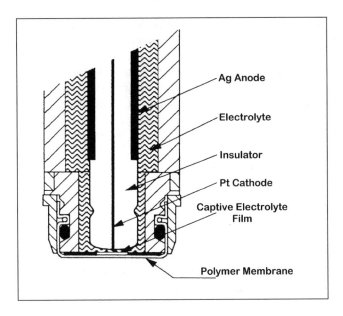

Figure 9.8 *DO sensor principal components (Reproduced by permission of Broadley-James Corporation)*

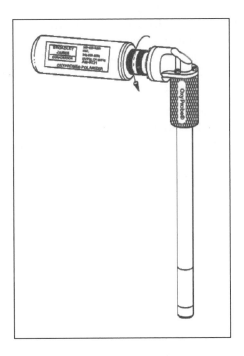

Figure 9.9 *DO sensor with battery polariser (Reproduced by permission of Broadley-James Corporation)*

in membrane permeability with temperature. As the temperature increases the membrane becomes more 'transparent' to the dissolved oxygen in solution.

Any dissolved oxygen measuring system expected to operate in a process with variable temperatures, or calibrated at a temperature different from the actual process condition, should incorporate a temperature compensation capability. To achieve this temperature compensation the sensor incorporates a temperature sensor. This measurement is taken by the dissolved oxygen measuring instrument and an adjustment is made to the displayed reading, to compensate for the membrane diffusion variability.

9.3.5   Autoclaving/Steam Sterilisation Effects

The high temperature cycle that the sensor is subjected to during either autoclaving or steam sterilizing expands and contracts the sensor. In particular, the membrane stretches and its tension across the cathode changes. This changes the relationship between the cathode and the membrane and so the sensor calibration is altered. The signal output will most likely shift upward; for example, a sensor can read 103% saturation after a steam cycle. It is therefore better to calibrate the sensor after sterilisation. Fortunately this is possible as the vessel can be run under set sparge, agitation, pressure and temperature conditions prior to inoculation.

9.3.6   Calibration

Calibration of the DO sensor is necessary for two reasons: (i) there is significant variability in the output from different dissolved oxygen sensors; and (ii) the output of the sensor is affected by the sterilisation cycle.

Calibration adjusts the display on the instrument to a set value (nearly always 100%) corresponding to the output from the sensor under operating conditions. The following would be a typical calibration routine:

(i)   steam or autoclave sensor in the media, cool slowly;
(ii)   saturate the media with filtered air;
(iii)   set the vessel to operating pressure, temperature and agitation;
(iv)   set the instrument display to 100%;
(v)   log the following data during calibration:
      (a) date of calibration;
      (b) operator ID;
      (c) vessel ID;
      (d) sensor ID;
      (e) batch ID;
      (f) sensor nA under calibration conditions;
      (j) temperature;
      (h) pressure.

9.3.7   Correct Use (Where Can It Go Wrong?)

Grab Sample Calibration

Unlike pH measurement, it is not possible to take grab samples (offline samples). With no aeration, no agitation in sample container and rapid outgassing under no pressure conditions and rapid atmospheric diffusion of gases into the sample, they would not be representative of media in the vessel.

What is Being Measured

The sensor measures the partial pressure of oxygen within the process. It is not possible to measure oxygen solubility or oxygen concentration (ppm, mg/L), directly. Some instrumentation provides measurement in ppm or mg/L. It should be appreciated that this is achieved by making a number of assumptions:

(i)   the temperature of the solution is known;
(ii) the pressure at the point of measurement, including any headspace pressure and hydrostatic head, is known;
(iii) the solubility of oxygen for the measured solution is known.

Given this information, the instrument applies correction factors to the partial pressure measurement based upon an 'ideal' solution.

9.3.8   Maintenance (Testing, Cleaning, Storage)

The following procedures could be included in a preventative maintenance programme:

(i)   Test the membrane: pressure test the cartridge for membrane integrity. Even the smallest leak in the membrane cartridge will cause catastrophic failure of the DO sensor. Before use, remove the membrane cartridge from the sensor and test for leaks, some manufacturers supply simple membrane testing kits to perform this procedure.
(ii) Inspect any o-ring seals: replace when necessary. Look for any damage to the o-ring seals, most manufacturers provide spare o-rings with each membrane cartridge, so these should be changed when the membrane is replaced.
(iii) Inspect the cathode and anode: clean when necessary. The condition of the cathode and the glass around it is critical to the correct operation of the sensor. A visual inspection should identify if the cathode is worn or cracked, or if it requires cleaning (see sections on cleaning and troubleshooting below).
(iv) Replenish the electrolyte. Refill with fresh electrolyte: follow the manufacturer's instructions.
(v)   Is the nano amp signal reasonable? Check the nA output (see section on checking the sensor above).
(vi) Store sensor with the membrane end capped with liquid filled boot.

9.3.9   Cleaning the Sensor

Different manufacturers make their own recommendations on cleaning. These should always be referred to first, before attempting any maintenance on the DO sensor. The following are general recommendations only.

The 'working' parts of the DO sensor are isolated from the process by a gas permeable membrane (Figure 9.10). If this membrane remains intact, the internals of the sensor will require very little maintenance. The silver (Ag) anode will darken in colour with use; this is silver chloride (AgCl) that is formed as part of the electrochemical reaction when oxygen is reduced at the cathode. This coating can be removed with a very light micron

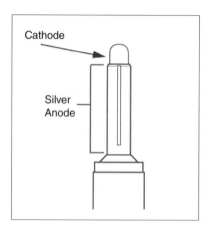

Figure 9.10   *Working parts of DO sensor*

polishing paper. Care should be taken not to abrade any part of the cathode or the surrounding glass during cleaning. The cathode can be cleaned using toothpaste by gently
cleaning the glass tip with a small amount of paste on some tissue or brush. Do not apply
any sideways force on the cathode assembly as it is very fragile. Finally rinse in purified
water and dry.

9.3.10   Testing the Sensor

One of the most important procedures when using a dissolved oxygen sensor is to make
a check on the sensor performance, prior to sterilisation. This is because it is possible to
calibrate a sensor that is faulty. All manufacturers of dissolved oxygen instrumentation
provide a very large adjustment for zero and span, to accommodate for significant variability in the output of different dissolved oxygen sensors. This means a sensor with a
high zero offset and/or a very low or very high output can be made to read zero and 100%.
It is only when the sensor is put to use, that it more closely resembles a random number
generator!

This test procedure should be followed, prior to sterilisation:

(i)   Polarise the sensor for 6 hours.
(ii)   Spot check nano amp output of the sensor for a reasonable range: 40–100 nano amps
       in air at 25 °C.
(iii) Spot check nano amp output of sensor at zero using nitrogen. Less than 1% of 100%
       value in 90 seconds, e.g. if the sensor gives 60 nA in air it should read less than
       0.6 nA in nitrogen.
(iv)   Measure the speed of response. The sensor output should fall from the 100% nA
       value to the zero nA value in less than 90 seconds.
(v)   Log this information with the sensor ID and date.

This data will be valuable in determining the performance of the DO sensor and its suitability for use. Interpretation of this data is covered in the next section.

9.3.11   Troubleshooting

To determine if a sensor is suitable for use, the test procedure described above can be applied. If the values fall outside of these tests then it is possible to use the information in Table 9.3 to troubleshoot the sensor.

Table 9.3 *Information for troubleshooting*

Effect	Cause	What to do (see section on repair)
High nA output, will not zero	Internal electrical leakage caused by a breakdown of insulation. This could be in the connector, the cable, or inside the sensor.	Disconnect the sensor from the cable, if the nA output drops to zero the sensor is defective. Reconnect the sensor, remove the membrane cartridge, rinse the cathode in deionised water and dry thoroughly. Test the sensor nA output which should be zero. If not the sensor is defective.
	Additional platinum exposed for reduction reaction, caused by a crack or chip in the glass surrounding the cathode	Visually inspect the cathode, if cracked the sensor is defective
	Membrane loose or stretched	Tighten membrane cartridge or replace.
	Insufficient polarisation time	Polarise for at least 6 hours. If using a battery polariser check the battery condition.
	Damaged membrane	Visually inspect the membrane and replace if necessary.
Low nA output	Coating on the membrane surface preventing oxygen from permeating through the membrane	Visually inspect the membrane and replace if necessary.
	Coating on the platinum surface	Clean
	Depleted electrolyte	Refill with fresh electrolyte. Check membrane is not ruptured and leaking electrolyte.
Slow response	Coating on the membrane surface preventing oxygen from permeating through the membrane	Visually inspect the membrane and replace if necessary.
	Crack or chip in the glass surrounding the cathode, leaving additional electrolyte around the cathode.	Visually inspect the cathode if cracked the sensor is defective.

Further Reading

Determination of pH Theory and Practice, R. G. Bates. John Wiley & Sons, Inc., New York, 1973.

pH Control, Gregory K. McMillan. Instrument Society of America, 1984.

Measurement of Dissolved Oxygen, Michael L. Hitchman – A Series of Monographs on Analytical Chemistry and its Applications, Volume 49. John Wiley & Sons, Inc., New York, 1978.

The Oxygen Electrode in Fermentation Systems, W. M. Krebs and I. A. Haddad. Developments in Industrial Microbiology, Volume 13. Society for Industrial Microbiology, Washington, DC, 1972.

Broadley-James Corporation website www.broadleyjames.com/dir-documents.html.

10

SCADA Systems for Bioreactors

Erik Kakes

10.1 Terminology

SCADA is an acronym for Supervisory Control And Data Acquisition.
DCS is a Distributed Control System
MES is a Manufacturing Execution System
TCP/IP is Transmission Control Protocol / Internet Protocol
RAID system is a Redundant Array of Independent Disks system
OPC is OLE for Process Control
OLE is Object Linking and Embedding
21CFR Part 11 is a Code of Federal Regulations that describes the use of Electronic
    Records and Electronic Signatures.
PLC is a Programmable Logic Controller
HMI is a Human Machine Interface
I/O system is Input / Output system
PID control is Proportional Integral and derivative control or three term control
ISA is Instrument Society of America
FDA is the US Food and Drug Administration

10.2 What is SCADA?

SCADA stands for Supervisory Control and Data Acquisition. SCADA systems are computer-based monitoring and control systems that centrally collect, display, and store information from remotely located data collection transducers and sensors to support the control of equipment, devices and automated functions.

Practical Fermentation Technology Edited by Brian McNeil and Linda M. Harvey
© 2008 John Wiley & Sons, Ltd

10.2.1   Why Use a SCADA System

Research Applications

Process controllers supply information on the current state of the process. To be able to optimize a process the historical process data need to be available for further analysis, and a recorder output was classically used for this purpose. This way of process analysis required a lot of accurate and tedious work. The computer has enabled easy data storage and data analysis by storing an enormous amount of data in standardized formats. These data can be read by data analysis software like Excel. Based on the outcome of these analyses new experiments can be designed. SCADA data are the best source for process development and process optimization.

A complete SCADA system for up to eight bioreactors in a (university) research setting will cost between Euro 2000 and Euro 4000.

Production

In the pharmaceutical industry the final product is just as important as the batch production data. These data can be stored in a reliable way using a SCADA system that complies with the rules in the pharmaceutical industry. The data will be stored and kept for inspection by authorities. Data security is of major importance in this application. SCADA systems can help in keeping the data in a secure format and in this way easing validation of the final product.

The investment needed for a validatable SCADA system for pharmaceutical industry ranges between Euro 2000 and Euro 20 000 per bioreactor system, depending on complexity and documentation needs.

In both cases (research and production) the supervisory control is an important part of the SCADA. This functionality allows procedures to be executed automatically without intervention of an operator. This is specifically useful when additions to the culture need to take place in the middle of the night, when a setpoint needs to be changed at the weekend, or when a complex sequence of actions needs to be executed in a short period. The SCADA system can be programmed to execute all these actions automatically and therefor perform these actions reliably at the right moment.

A lower cost option for the SCADA system is the data acquisition only version of a SCADA. This program has the data logging functionality of a SCADA system but lacks the supervisory control part. This is sufficient for a large number of bioreactor users. The cost of a data acquisition system only is between Euro 500 and Euro 1000 for a system that will store data for up to six bioreactors simultaneously.

10.2.2   Historical Development of SCADA

SCADA started in the 1960s. At this time mainframe computers were used for data storage from energy plants, chemical plants and other big industries (Figure 10.1).

At that time there were no standards for user interfaces, and there were no programming standards either. This resulted in a wide range of proprietary systems that had their own standard for communication with remote I/O systems. Choosing one supplier meant that all users were tied to one source for new developments and customization. Although this resulted in costly systems, this solution was still cheaper than sending operators around the plant to check visually and report the status of the processes manually.

Figure 10.1 *Computer control room in the 1960s*

When integrated circuits became available in the 1970s and 1980s the computers became smaller and more available. SCADA systems benefited from this development and were more widely used in the industry. The programming standards and user interfaces were still not standardized, thus, SCADA suppliers developed their own standards.

In the 1990s Microsoft Windows became the world standard operating system and the SCADA suppliers adapted to this standard in user interfaces. The architecture of the SCADA systems was still defined by the manufacturer. In the background the ISA was working on a standard to define batch control – a way to standardize the bits and pieces of batch control and how the various pieces should fit together. This standard is now known as the S88 standard. In the middle of the 1990s the OPC standard was defined and adapted by some SCADA suppliers, thus taking standardization one step further.

With the PC becoming cheaper the SCADA systems were increasingly moving from large industry into the pilot plants and laboratories of commercial companies, institutes and universities. The demand for data interchangeability grew, resulting in databases to be used for data storage replacing the proprietary file systems. The old way of running validated processes, where the operator usually just signed on the strip chart recorder printout, was disappearing, and this created a potential problem for the regulatory agencies and authorities. In the past all process changes were recorded manually in a logbook, and the operator signed all actions. By contrast, using the SCADA systems changes are made using a keyboard, and therefore, the operator could not sign for the changes. This is where the FDA defined the 21 CFR Part 11 standard for electronic signatures. This standard defines a way uniquely to identify an operator and limit access to the SCADA system.

10.2.3   SCADA versus DCS

A SCADA system is not the only solution for advanced supervisory control and data acquisition. The two main solutions for advanced supervisory control and data acquisition are SCADA and DCS.

10.2.4   DCS System

DCS is a distributed control system. The main difference between SCADA and DCS is that a DCS has one database for the complete system. A SCADA system has a separate database for the supervisory software and a separate database in the local controller. A

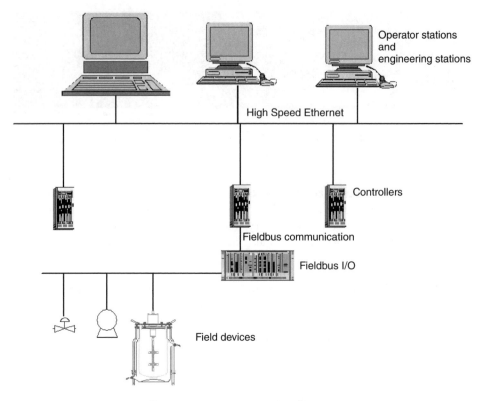

Figure 10.2 *DCS system architecture*

DCS is an integrated solution that usually comes from one supplier. Examples are Emmerson DeltaV, Honeywell PlantScape/Experion or the Yokogawa CENTUM CS3000 system. Of course there are others as well. Typical system architecture of a DCS is shown in Figure 10.2.

The operator station is used for operator access to the process and for process visualization.The engineering station is used for configuration of the processes and I/O as well as the regular operator functions. The database server collects all data and stores the process recipes. The local controller runs the process as defined in the operator/engineering station. The I/O system is controlled by the local controller and communicates through a bus system with the controller. The field devices are connected to the I/O although field devices might also have integrated fieldbus I/O transmitters.

The advantage of the integrated approach is that the complete control solution fits seamlessly together. The disadvantage is that there is only one provider for the solution. This reduces flexibility and very importantly, reduces the price pressure (there is no competition for the chosen solution) and is therefore usually a more expensive solution, especially for smaller installations (<10 bioreactor systems).

10.2.5 SCADA Architecture

A SCADA system (Figure 10.3) only replaces part of a DCS. The SCADA is the upper level of the system taking care of process visualization, data storage and supervisory control. There needs to be a local controller for the actual process loop control and local I/O comparable to the I/O level of a DCS. The level above the SCADA is usually a manufacturing execution system, which takes care of production scheduling, production execution and recipe management.

Typically industrial scale fermentation plants would operate at the three levels as shown in Figure 10.4, whilst academic research laboratories would typically operate only with local control and in some cases SCADA.

The local controller (PLC or other control system) has a local display for process visualization. The components in this system can be from different suppliers. The communication between the different components used to be critical, but with standardization of communication protocols this is not an issue nowadays. One advantage of this

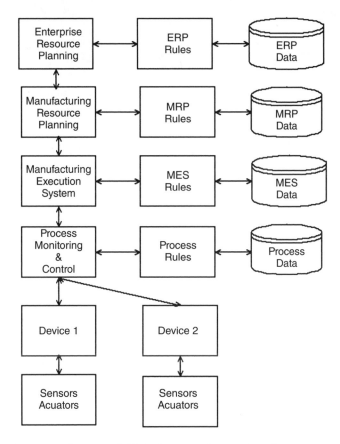

Figure 10.3 *Structure of a typical SCADA system for use in a fermentation pilot plant in industry or academia*

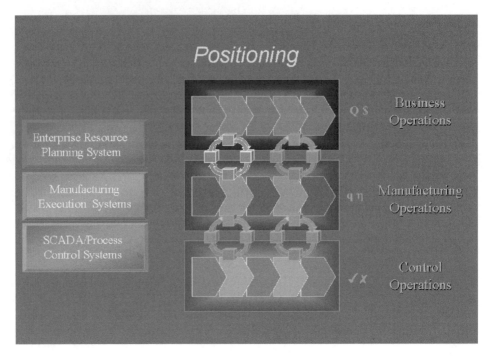

Figure 10.4   *The position of a SCADA system in a manufacturing environment*

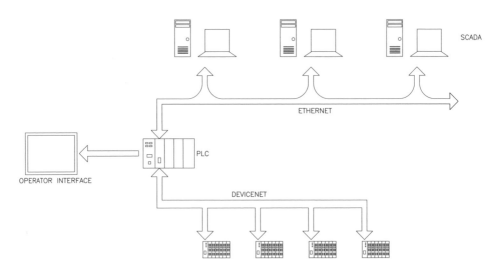

Figure 10.5   *SCADA system architecture*

architecture is that the local controller is running independently of the SCADA. The local controller can be of different brands just like the SCADA and the I/O modules. This demonstrates the major advantage of a SCADA system over the DCS system, namely the far greater flexibility of the SCADA architecture (Figure 10.5).

10.3   Communication Within a SCADA System

As discussed in the previous section there are different components that make up a SCADA based control system: (i) operator / engineer station; (ii) database server; (iii) PLC controller; (iv) I/O system, and (v) sensors and actuators. All these components need to communicate with each other. The SCADA collects data from the PLC and sends set-points and other parameters to the controller. The controller collects information from the inputs of the I/O system and sends commands to the outputs of I/O system to control the process. The data collected by the SCADA are transported to the database, and historical data is read from the database for data presentation.

Timing of the communication is of the highest importance in all steps of the SCADA. There are several communication standards used in the chain.

10.3.1   Communication Between the SCADA and the Database

Communication between SCADA and the database needs to be reliable but timing of the communication is not critical. The communication does not need to be deterministic since the data are time labeled by the SCADA system. (A deterministic communications protocol is one that provides ensured scheduleability of the actions in a real-time system. Deterministic communication is more complex and more costly than standard, off the shelf, TCP/IP communication and should therefore only be used when the application is time critical.)

Usually these systems run in a Windows environment. The database can run on the same computer as the SCADA or on a different computer. The advantage of a database in the SCADA computer is the simplicity and robustness of the design. These systems can be installed in any location and do not require any network knowledge. When the gathered data need to be available for multiple users, a centralized database can be used. In this set-up the data are gathered by the SCADA computer and send to the central database in the network. All authorized users (and software programs) can have access to the data stored in this database. This design requires a stable and reliable network. Using standard TCP/IP communication protocols there is no guarantee that the sent data have arrived in the database. When the network is down, the data will not be stored in the database and could be lost if no security measures are programmed into the SCADA computer. One of these security measures can be a temporary local data storage that will be transferred to the database when the network is back on line.

The size of the SCADA system is often the determining factor in the location of the database. A smaller SCADA (up to 10 bioreactors) usually has the data storage build into the system, while a SCADA with multiple operator stations for a larger number of controllers (usually more than 20 bioreactors) has a separate system for data storage. The choice of a separate computer for data storage (a so called database server) is based on security and reliability. The database server can be equipped with a RAID system for ultimate security. The RAID system is designed to write all data to two different hard disks. The second disk serves as a backup system for the process data in case the first hard disk is experiencing problems or crashes (Table 10.1).

Since the purpose of the RAID system is data security, the most commonly used RAID system in a local database is the RAID 1 system. For centralized network databases, the RAID 5 system is the best solution.

RAID-0 (also known as 'striping')

The advantage is that multiple disks are used in parallel for fast data transfer. Data (bits) is distributed over multiple disks. The disadvantage is that no disk will contain a complete data file and when one disk crashes, all data are lost.

RAID-1 (also known as 'mirroring')

The advantage is that one disk is a safety back-up for the other disk. All data written to disk one is continuously copied to disk two. When one of the two disks is experiencing a problem, the RAID system will identify this and notify the user. This is a simple and very cost effective way of protecting your valuable process data.

RAID 2, 3, 4 and 5 are variations of the RAID 1 and 2 system with added functionality as described in Table 10.1 and are not needed for the security that we are looking for in a local SCADA system. These standards are therefore not described in this chapter.

When the database is integrated into the SCADA station communication is done through the operating system. This is a fast way of communicating and timing is not an issue. However, when the database is in a separate server the data needs to be transported from one computer into the database over a network. The SCADA network is preferably separate from the company office network to improve performance, and to increase security of data. The communication speed of a 100Base-T Ethernet network is sufficient for

Table 10.1 RAID hard disk systems

RAID level	Number of disks	Positive	Negative
0	minimum 2	Fast data storage Simple set-up 100% error correction	No error correction One disk lost means all data lost
1	minimum 2	Extremely reliable Simple set-up	No fast data storage
2	minimum 3	Error correction for old hard disks without built-in error correction Fast data storage	No practical application with modern hard disks
3	minimum 3	Error correction Parity check is used to recover lost data Fast data retrieval	No simultaneous read/write actions.
4	minimum 3	Error correction Parity check is used to recover lost data Fast data storage and retrieval	Slow data storage
5	minimum 3	Error correction Parity check is used to recover lost data Best solution for network drives	Slower than RAID 0 and RAID 1 Expensive

SCADA applications. (100BASE-T is an implementation of Ethernet that allows stations to be attached via twisted pair cable. The name *100BASE-T* is derived from several aspects of the physical medium. The *100* refers to the transmission speed of 100 Mbit/s. The *BASE* is short for baseband. The *T* comes from twisted pair, which is the type of cable that is used).

The availability of the network determines the reliability of the SCADA system. Networks have become increasingly reliable, but still are a potential danger for communication in SCADA networks. For critical situations, back-up systems can be installed to guarantee network performance. If such a system is not in place, and the network is down a fall back system should be in place in the SCADA system.

One common option is to store the data in the SCADA computer. As soon as the database server is back online the SCADA system should send the missing data to the database server to complete the data file.

10.3.2   Communication Between the SCADA and the PLC

Communication between the SCADA and the PLC need to be deterministic (Table 10.2). The data are gathered by sensors, and sent to the I/O system. Data is transported into the PLC and the SCADA samples the PLC registers for new data. These data need to be time stamped, since they are process data. This time labeling can be done by the PLC, or can be done by the SCADA.

Classically, serial communication is used for the communication between the SCADA and the local controller. Serial communication standards used include RS-232, RS-422 and RS-485 (RS is Radio Standard). RS-232 is most commonly used. Most (older) computers used to have an RS-232 port so connecting devices to a PC was not an issue. The RS-232 standard could be used for distances up to 15 meters with a communication speed of up to 19 200 bit per second. RS-232 can only be used for communication between two instruments (one driver and one receiver). When multiple instruments need to communicate one could either use an RS-232 multiplexer, use RS-422 or switch to RS-485 communication. RS-422 overcomes the limit in distance between the instruments and the computer, and it can be used to connect up to ten receivers to one computer. RS-485 can connect up to 32 instruments with the same maximum distance as RS-422 of 1200 m.

Table 10.2 *The relative capabilities of different communication methods between PLC and SCADA systems*

	RS232	RS422	RS485
Max. number of drivers	1	1	32
Max. number of receivers	1	10	32
Modes of operation	half duplex full duplex	half duplex	half duplex
Network topology	point-to-point	multidrop	multipoint
Max distance (acc. Standard)	15 m	1200 m	1200 m
Max. speed at 12 m	20 kbs	10 mbs	35 Mbs
Max. speed at 1200 m	(1 kbs)	100 kbs	100 kbs

The last remaining drawback of the RS-communication protocols (the communication speed) can only be overcome by switching to a different communication technology, e.g. Ethernet.

Ethernet networks are widely used in office networks, but the control industry is somewhat conservative in switching to new technology. The risks of switching to new technology are a lot higher in the process industries. When a process control network goes down the results can be dramatic. However, with the mass usage of Ethernet systems, the technology has proven itself to be reliable.

Current Ethernet networks are getting better in terms of reliability of the communication timing, but industrial Ethernet is developing specifically for this purpose. Industrial Ethernet is an onward development of the standard Ethernet and is gaining market share in the fermentation industry.

10.3.3   Industrial Ethernet

Industrial Ethernet is the name given to the use of the Ethernet protocol in an industrial environment, for automation and production machine control.

Some of its advantages are:

- increased speed, up from 19.2 kbit/s with RS232 to 1 Gbit/s with IEEE 802 over Cat5e/ Cat6 cables or optical fiber;
- increased overall performance;
- increased distance;
- ability to use standard access points, routers, switches, hubs, cables and optical fibers, which are immensely cheaper than the equivalent serial-port devices;
- ability to have more than two nodes on link, which was possible with RS485 and RS 422 but not with RS232;
- peer-to-peer architectures may replace master–slave ones, thus reducing the complexity of the SCADA architecture. Every machine can be a master or a slave depending on the functions of that machine rather than the type of communication network that is used (as was the case with the RS communication);
- better interoperability.

The difficulties of using industrial Ethernet are:

- migrating existing serial communication systems to a new protocol (however many adapters are available);
- real-time uses may suffer for protocols using TCP. TCP guarantees the transport of data and makes sure that data packets arrive in the order they were sent, but the timing of the data is not guaranteed. This means that data might arrive faster on one occasion and slower the next time, which might cause problems for time dependent information. Process controllers, for example, require accurate timing on sending and receiving data from the process. The timing of the measured value is the basis for the proper functioning of a process controller.
- Managing a whole TCP/IP stack is more complex than just receiving serial data, but this is an internal problem for the communication system. The user will not see these differences.

10.3.4   Communication Between the PLC and the Field Devices

Field devices are all sensors and actuators that are used to control a process in a bio-reactor. A pump, a mass flow controller, an electrical heater, a pH sensor, and temperature sensors are examples of field devices.

I/O devices are the relay outputs that control the heater, the pump or analog outputs for mass flow control controller setting. I/O devices are also pH and dissolved oxygen sensor amplifiers.

Communication between PLC and the I/O devices must be very fast and very reliable. Most proprietary bioreactor controllers have the PLC and the controller inputs and outputs integrated in one box. The controller interfaces directly with the field devices like sensors and the actuators (pumps, heaters, coolers, flow devices, etc.). Communication between these devices is integrated in the controller, and is usually a supplier-designed communication bus system.

Most industrial controllers use fieldbus communication between the PLC and the external I/O devices. Fieldbus is a generic term for a communication standard that is used to replace the old analog (4–20 mA) communication standard in the industry. The fieldbus uses intelligent devices with its own communication system build into each device. The technology offers lower costs, increased speed, higher accuracy and simpler networking. Examples of this technology are CANopen, Profibus (Siemens), DeviceNet (Allen Bradley), and Foundation Fieldbus. Each standard has its own specific advantages.

10.4   Functions of a SCADA System

Functions of a SCADA system are:

online data collection;
off-line data collection;
online data storage;
off-line data storage;
data presentation;
calculations on measured data;
data import and export;
time and event based actions;
recipe definition, management and execution.

10.4.1   Online Data

On-line data is data that is continuously available for collection by the SCADA. This data can be collected from the local process controller, a chemical analyzer, off gas analyzers and so on. The SCADA takes the initiative to get the data, and the slave device sends the data to the SCADA on request. Examples of online data are setpoints and measured values of process temperature, pH value, dissolved oxygen concentration, liquid level, O_2 or CO_2 concentration in the fermenter exhaust gas, etc. The data collection interval (sample frequency) is selectable per measured parameter, and can be in the range of a measured point every second up to one measurement every day. Depending on the importance of a variable during a phase of the process the data will be stored more or less frequently. It is very important for lengthy fermentation processes or for multiple parallel fermentations, or those involving data rich measurements (e.g. the acquisition of IR spectra of fermentation fluids), to think carefully about the actual needs for frequent data collection and the purposes to which the data will be put relative to the volume of data that will have to be transferred and stored.

The online data are located in a local instrument. The SCADA will have to know where these data are located in the local device. A so called driver is the interface between the local instrument and the SCADA. The driver determines how the user has to address the data in the local instrument.

A typical dedicated driver window shows the available data and the address of these data in the local device (Figure 10.6). For every on-line measurement the address of the datum point needs to be defined before the process is started.

In the past all drivers were specific for the combination of controller and SCADA. In the 1990s the OPC standard was defined (Figure 10.7). This OLE for Process Control standard defines a unified interface between instruments and SCADA. The goal for OPC is to make it easier to let instruments exchange information.

An OPC communication system consists of an OPC client and an OPC server. The client is the part that is requesting the data while the OPC server is supplying the data. The server is getting the information from the local controller and the client supplies the data to the SCADA.

When an OPC client is looking for OPC servers on the network a window like the one in Figure 10.8 shows all available OPC servers on the network. The selected OPC server will than show the data that are available from the local device.

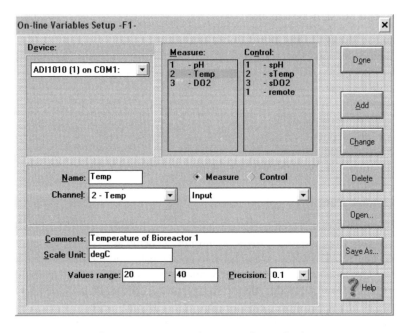

Figure 10.6   *Typical SCADA driver display*

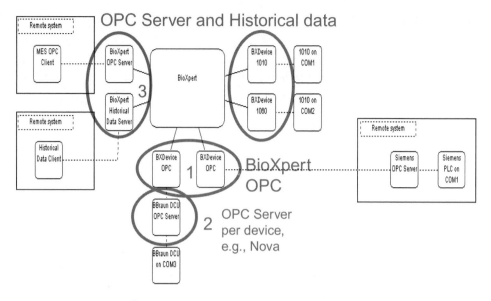

Figure 10.7 *OPC communication architecture*

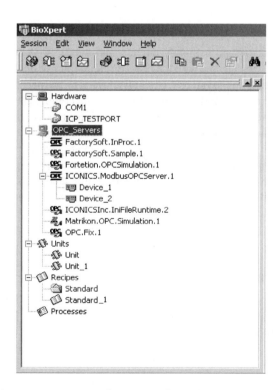

Figure 10.8 *A typical OPC interface of a SCADA system*

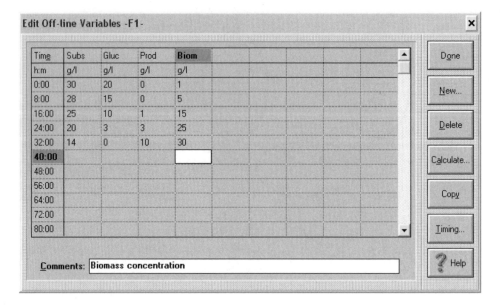

Figure 10.9   *Off-line data entry display*

10.4.2   Off-line Data

Off-line data are data that are manually entered into the SCADA system database. The initiative for entering the data lies with the operator. Examples of off-line data are biomass concentration (although this parameter can nowadays also be measured online), product concentration, substrate concentration, and other medium component concentrations. Off-line data can be entered at the time of sampling or minutes, hours or even days after the sample time. This places extra demands on the SCADA functionality for bioreactor processes. Usually in the process industry the off-line data are entered at the time or shortly after the time of sampling. In bioreactor processes the time between sampling and data entry is determined by the time it takes to get the sample analyzed. Biomass dry weight and other analysis might take days to determine. The database must have the ability manually to enter the data retrospectively. With every data point the date and time of the sample needs to be entered. After this the data can be presented in line charts (Figure 10.9).

10.4.3   Data Storage

Collected data can be stored in several ways. The simpler programs will have a proprietary file format. An example of such a program using a proprietary file format is BioXpert 2 by Applikon. The Industrial version of this software (BioXpert XP) uses a SQL-server for data storage. The industrial programs will use an external database for storage.

Proprietary file formats have the advantage that the program controls all elements of storing and retrieving data. The data are, however, not available for other software programs without a manual file export or without an additional driver program.

The external database can be any commercial database program, but an SQL (Structured Query Language) database has distinct advantages when moving in to data mining. The data are available for third party applications allowing further integration of the SCADA into the manufacturing plant management system. A central alarming system can read alarm data from a SQL database and use these for sending alarm messages to operators.

The main advantages of SQL databases include:

Speed: SQL databases are designed to handle large amounts of data. This is useful when retrieving large amounts of batch data.

Security: All data are stored in one place. The security built into the SQL database guards all data.

Compatibility: Through the well defined SQL standards, data from these databases are available for a wide range of supporting software programs.

When using a SQL database for data storage, the SCADA will not create one data file per process. All data from all processes go into one big SQL database. Separate batch data can be extracted from the SQL database using the SCADA or an other separate program.

10.4.4 Data Mining

Data mining has been defined as 'the non-trivial extraction of implicit, previously unknown, and potentially useful information from data' (from http://www.liacs.nl/home/kosters/AI/datam.html), and 'the science of extracting useful information from large data sets or databases'.

Data mining involves sorting large amounts of data and picking out relevant information and trying to determine patterns in the stored information. The large amount of data that can be derived for a single experiment using modern SCADA techniques, makes data mining a more and more common tool in fermentation technology. The technique is used to try to predict process events and to find relationships between parameters that did not seem related previously. Data mining is related to SCADA systems, but is a science on its own and will therefore not be covered in this chapter. Additional information can be found in Rommel and Schuppert (2004).

The amount of data determines the size of the data files or of the SQL database. These amounts of data can be huge, especially when working with perfusion cell culture processes for mammalian cells that can last for several months.

Typically in perfusion cultures the process parameters are stable. In this case the database will be filled with large amounts of data with little or no variation. This is an ideal situation for data reduction techniques.

10.4.5 Data Reduction

There are several ways to reduce the amount of process data stored in a database. Data reduction strategies include reducing the amount of data before it enters the database or reducing the data once it is in the database. Programs with proprietary file formats sometimes are limited in the number of datapoints that they can store. When this limit is reached, the amount of data in the database needs to be reduced to be able to keep on storing information from running processes. This data reduction needs to be fast since the database is running out of space and complex and time consuming algorithms cannot be used in this case. The most appropriate ways to do this is by averaging the measured datapoints and replacing every two points with the average value of the two. This immediately reduces the size of the database by 50%.

Although an SQL database is designed to handle a large amount of data, it can be useful to have some form of data reduction in these situations as well. Less data means faster access.

There are different ways of preventing excessive amounts of online data entering into the SQL database:

(i)   reduce the sample frequency of the measurements;
(ii) average the measured data;
(ii) store data only on deviation of the measurements.

Reducing the Sample Frequency of the Measurements

The easiest way to reduce the amount of data stored in the database is to reduce the sample frequency of the measured data. The sample frequency can be set per measured parameter, and the frequently changing parameters will be sampled more often than the stable parameters.

In the case of a bioprocess, the temperature is usually well controlled and stable, the pH and oxygen might be varying. Oxygen has a slow response due to the design of the Clark type electrode (See Chapter 9) but pH sensors respond instantaneously. In this situation the sample frequency for temperature could be one measurement per 5 minutes, the dissolved oxygen value could be stored every 2 minutes, and the pH value could be stored every 30 seconds. This avoids filling up the database with a large number of data points for temperature with identical values. The value for oxygen will not change fast but the trend will be recorded while the changes in the pH value will be stored as well. It is vital to think before opting for very frequent measurements. This should involve a real critical assessment of the time constants of the process (in simple terms, how quickly a fermentation process is proceeding), and how these change with time (very fast in exponential phase, much slower in stationary phase), and what use is likely to be made of the measurement. SCADA systems are tremendously powerful tools, but like all implements, thought is required in advance to get the best from them.

Averaging of the Measurements

An alternative to reducing the individual sample frequency per parameter is to store data on averaging that data. In this case, the expected changes per parameter are not ruling the storage frequencies, but it is a general time based storage reduction. A pre-defined number of measurements is collected at the same frequency; the average of these data is calculated and stored as one datum point (Figure 10.10). This way of reducing data does not make much sense since the alternative is storing process data at a lower frequency, which works just as well and does not require any calculations. It does, however, prevent the measuring of artefacts that can occur when the sample frequency is set to a too low a value.

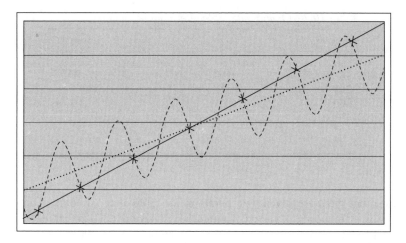

Figure 10.10   *The effect of averaging on measurement averaging and of too frequent measurement. The dotted line in this picture represents the average of the actual measurements (dashed line) while the line represents an artifact when sampling data at a too low frequency*

The averaging filter should only be used to reduce the amount of data in the database when the process is running and the amount of data starts to exceed the size of the database. If the average filter is not be used in this situation, the SCADA will stop storing data when the database is full.

Store Data on Deviation of the Measurements

This technique is very useful for fermentation processes in bioreactors. These processes are slow and the parameters do not change rapidly in many systems. When storing data on a time base, the database runs the risk of being filled with a lot of identical values for temperature, pH, dissolved oxygen and agitation speed. The deviation filter works by comparing the currently measured data with the last stored data point and comparing this with an operator-defined deviation interval. When the value is outside the deviation interval, the current and previous data point are stored and any new data are compared with the last stored data point.

In practice this looks appears as follows.

The setpoint for temperature of typical mammalian cell culture is 37.0 °C with a deviation of 0.1 °C

The measured value for temperature is 37.0 °C at time 0 (= T). When the next data point is measured one minute ($T + 1$) later and has a value of 37.0 °C, it is discarded. This is repeated until a temperature of 37.1 is measured at $T + x$. Now the temperature of 37.1 is stored at time $T + x$ and temperature 37.0 is stored at time $T + x − 1$ (when the sample frequency is 1 minute)

After 20 hours the data table could look like Table 10.3.

Table 10.3   Online data table after data reduction

Time (min)	Value (°C)
0	37.0
360	37.0
361	37.1
380	37.1
381	37.0
1137	37.0
1138	37.1

This means that there are seven data points in the table after 20 hours of temperature monitoring at a frequency of once per minute. Without the deviation filter there would have been 1200 data points in the table. The reduction is $(1-7/1200)*100\% = 99.4\%$.

The deviation interval is selectable per measurement and can be used for all measured online parameters. As shown in the example above, the amount of data stored can be greatly reduced, resulting in smaller data files and faster response when retrieving data from long-term cell culture processes.

10.4.6   Data Presentation

The SCADA operator can choose several different ways to view data from fermenters: (i) synoptic; (ii) trend chart (value versus time); (iii) scatter plot (value versus value); (iv) bar graph; (v) table display.

Synoptic

A synoptic display is a graphical presentation of the process with the actual measured data projected on the process presentation. This shows exactly what is occurring in the process at the time of viewing. This is a frequently used way of presenting data since the process installation is shown graphically with all major elements. The user can see very quickly whether or not the process is in a safe and controlled state. Alarms will be displayed in red and safe or expected situations will be green. Inactive I/O values will be printed in gray. From a synoptic display the user can have basic control over the process. Setpoints can be changed and control loops can be started or stopped.

Trend Graph

The trend graph (Figure 10.11) shows the measured data as a function of the (process) time. Displayed parameters can be changed during the process and the time axis can be changed to get a complete process overview, or to zoom into a specific part of the process. The trend chart gives a quick view of the trend of the measured signal and is one of the most commonly used process views in a SCADA system. It can usually be exported in a different graphical format to be used in presentations. No control over the process is possible from the trend graph.

The trend chart is often used to compare the current batch to the 'golden batch'. This is typically an ideal batch fermentation that meets the desired aims of the process

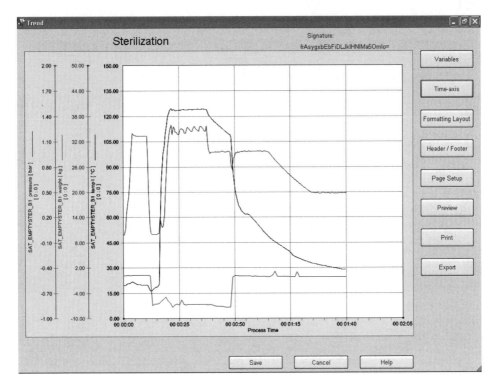

Figure 10.11   *Trendgraph with on-line data*

operators, and will usually be associated with closely defined limits of tolerance for key process parameters. For this purpose it must be possible to import data from previous cultures into the current graph. The trend of the important process parameters will be compared to the parameters in the golden batch allowing for an online prediction of the quality of the current batch.

Trend charts can display online data, off-line data and calculated parameters, a good example of the latter being the respiratory quotient (RQ) widely used in control of fermentation processes. The off-line data can be presented in different formats. Examples are constant mode, line mode and spline mode.

In *constant mode* the off-line data is displayed as a constant value until a new data point is entered (which will be the new constant value) (Figure 10.12). This is useful when the off-line parameter is used in calculations. The formula will be executed with the last entered value of the off-line parameter, so this parameter will be displayed as such (see Figure 10.12).

The *line mode* is the most frequently used (Figure 10.13). This way of displaying data uses linear interpolation between the data points to generate straight lines between the measured points. The linear interpolation will generate data points between the measured values. These data points are assumptions of a linear relation, and might not be correct in all cases (see Figure 10.10) but in most cases the linear relation gives a good representation of the real data.

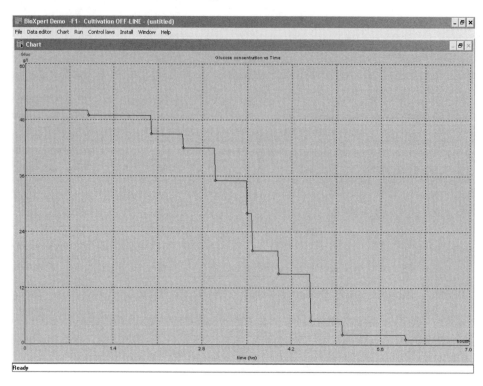

Figure 10.12   Trend graph in constant data mode

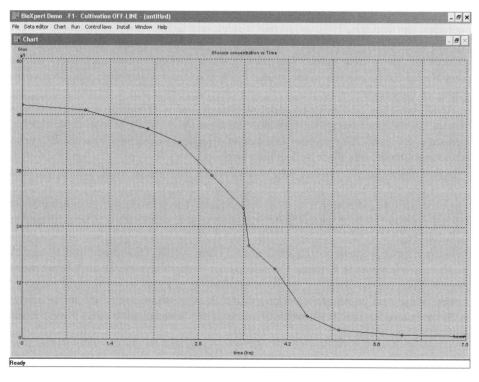

Figure 10.13   Trend graph in line data mode

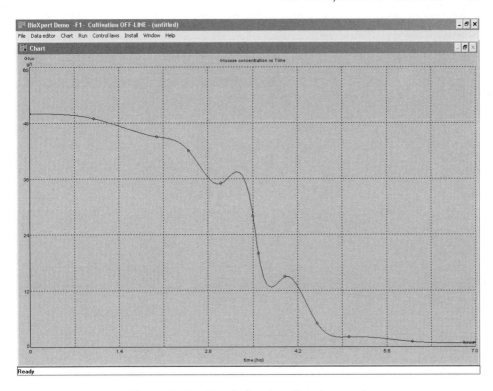

Figure 10.14   *Trend chart in spline data mode*

The *spline mode* draws a smooth curve through the measured points trying to approach the real values in the interpolated section (Figure 10.14). The fitting parameters can be adjusted to make the fit more realistic. The spline mode gives the nicest graphical presentation when displaying off-line data.

Scatter Plot

Where the trend charts show the trend of the measurement in time, there are also other uses for line charts. One of them is the scatter plot. In this chart the parameters are not plotted against the process time, but are plotted against another process variable. This allows the user to find a correlation between measured parameters or measured and calculated parameters. Examples of correlation plots are optical density (online parameter) versus biomass concentration (off-line parameter, Figure 10.15). When the correlation between the two parameters is determined, a calibration curve for the optical density sensor can be drawn up and the measurement can be used as a direct indication for the amount of biomass. Another example is agitation motor torque (power uptake, online) versus viscosity of the culture (off-line).

The scatter plot can be used for a variety of parameters and is the first small step into the data-mining field.

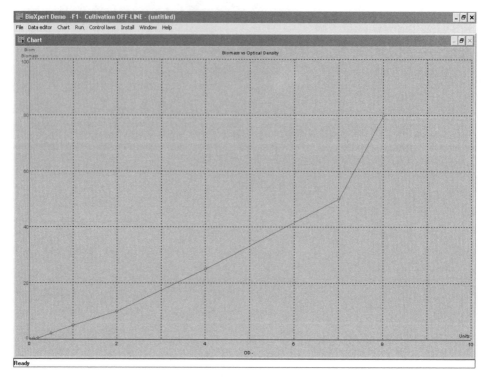

Figure 10.15 *Scatter plot of optical density versus biomass concentration*

Bar Graph

The bar graph display shows the status of the local control loops, and allows advanced control over the process. Setpoints and proportional, integral and derivative parameters can be changed, control loops can be (de-)activated and cascades of control outputs can be defined or changed. The bar graph usually shows the measured value, the setpoint and the controller output in the shape of a vertical bar. Multiple bar graphs are presented in one screen. Bar graphs are used for online parameters.

Table Display

The actual measured datapoints can be displayed with a table view (Figure 10.16). The table view can show the online, off-line and calculated data. The values in the table are user selectable and are displayed in combination with the measurement date and time.

10.4.7 Calculations on Measured Data

Not all parameters can be measured online or off-line. The RQ value, for example (respiration quotient), will need to be calculated from online variables. These calculations can be defined in the SCADA layer. Calculations can use online parameters and off-line parameters from one or many bioreactors or other external devices. Examples of these calculations include calculation of oxygen uptake rates (OUR), carbon dioxide evolution

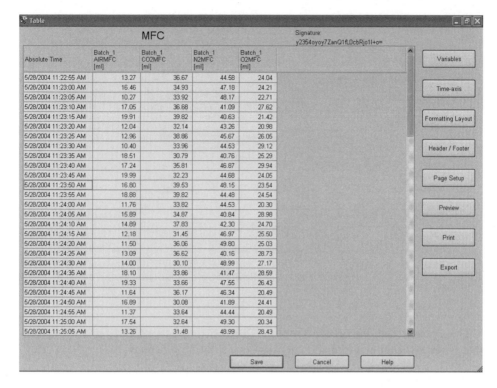

Figure 10.16 *Data table display*

rate (CER), specific growth rates, substrate consumption rates and so on (see Chapter 7 for details of calculations). The calculations require a calculation module where the user can define calculations and link these formulas to online or off-line data to make the calculations dynamic. Figure 10.17 shows an example of a calculation module.

The result of a formula can be used as an online variable for data presentation or for further supervisory control. Time and event based actions can be taken, based on calculated variables. For example, the above mentioned calculated RQ value can be used to control a feed pump for substrate.

Software (soft) sensors are based on these calculations. With a software sensor a parameter can be determined without measuring this parameter. Biomass dry weight can be determined by measuring the optical density and converting this into the biomass value using an appropriate formula. The software sensor can generate data that cannot be measured with normal sensors, but the error in the measurement is the sum of the errors of the 'real' measurements that the formula is based on, which is normally greater than the error in a 'normal' sensor.

10.5 Data Import and Export

Data import and export can be dynamic or static. The static version needs an operator action to do a one time data transfer from one program into the other.

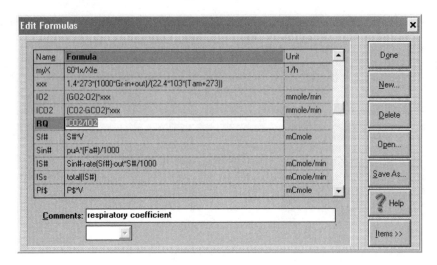

Figure 10.17   *Calculation/formula display*

The dynamic data import/export uses DDE/OLE (nowadays COM or DCOM) techniques for exchanging information between two software programs.

10.5.1   Static Data Export

Once the measured and calculated data is in the database of the SCADA, the demand to export these data to other software programs arises. The measured data can be exported to Microsoft Excel for further calculation and data manipulation, graphs can be exported to graphical editing software, and graphs and data can be used in Microsoft Powerpoint for presentation purposes. All these applications require their own file format. The SCADA program should be able to supply the data in several formats depending on the data sort. Graphs and charts should be available in BMP or JPG format, measured data can be exported in TXT (text), CSV (comma separated values) or SYLK (symbolic link) format. The different formats have different advantages. TXT can be imported into a variety of software programs, but this format requires more work before data can be interpreted in the other software program. The CSV format contains values separated by commas and includes a 'new line' command. This eases importing the values into Excel for example.

The SYLK format is an easy transfer to Excel. The files contain fieldnames and formatting information and do not require additional work to create a readable Excel spreadsheet.

When exporting data the user must be able to reduce the amount of data that will be exported. If this is not done the amount of data that will be sent to the next program might be too large to handle, or the amount of data is so big that the useful information cannot be separated from the rest. The data reduction can be done by limiting the parameters that will be exported to the export file, limiting the time period that will be exported, or by one of the data reduction techniques previously described.

10.5.2   Static Data Import

The import function is usually reserved for retrieving data from old batches stored on optical media or other back-up media. When batch data need to be compared to data from an older batch, these data can be imported into the SCADA. Depending on the SCADA application there will be different functions available for the imported data. In the best case scenario, these data can be used for presentation in the same graph as the actual batches, and this data can be used for online calculations. The calculations on old batch values can be used to base the setpoint value of a parameter of the current batch on the measured value of a 'golden batch' (see Chapter 11).

10.5.3   Dynamic Data Exchange

Dynamic data exchange uses the Windows information exchange layers to transfer data from one application into the other. This is useful when complex calculations cannot be done in the SCADA and these calculations are executed in programs like Excel. The dynamic data link transports a defined dynamic online measured variable to a location in the spreadsheet. The computing options of the spreadsheet are than used to execute more complex calculations. The spreadsheet can be used to collect data from multiple sources (SCADA and other software) and combine these data in one file. Using DCOM technology, the data source software and the spreadsheet can even be running on different machines.

10.6   Time and Event Based Actions

After the data storage and data presentation, one of the most frequently demanded functions of the SCADA is time and event based actions. Time and event based actions are those things that an operator usually does manually based on observations or on a recipe that is defined for the process. The problem with operator actions is that they are never exactly the same. Timing of actions is based on physical presence and observations of an operator. The SCADA, however, is always present and is continuously observing all parameters of the process, resulting in more reproducible actions based on time and events in the process. The SCADA program avoids the need, for example, for the operators to return to the laboratory in the middle of the night to switch on a feed pump after 30 hours of process time. The SCADA can do this automatically and without the presence of the operator.

Examples of time-event based actions are automatic sterilization of a bioreactor, automatic cooling of the bioreactor contents after the process is finished, automatic switching of manifold valves to direct the exhaust gas flow of a specific bioreactor to a gas analyzer and storing the data with the appropriate data file. The system for defining these time and event based actions differs from SCADA program to SCADA program. In some cases the programming is done by drawing lines between measurements and calculation modules (graphical programming) and other SCADA programs use a command line interpreter that allows programming in a BASIC-like programming language. An example is given below.

```
# scale totalflow,
TOTALFLOW=42.5,
# retrieve do-controller output,
OUTPUT=CODO2,
# if CO<-25 only use N2,
IF OUTPUT<-25,
N2MFC=TOTALFLOW,
O2MFC=0,
ELSE,
# if CO>100 only use O2,
IF OUTPUT>100,
O2MFC=TOTALFLOW,
N2MFC=0,
# in normal control range calculate a mixture of O2
and N2,
ELSE,
O2MFC=TOTALFLOW*(OUTPUT+25)/125,
N2MFC=TOTALFLOW*(1-(OUTPUT+25)/125),
ENDIF,
ENDIF,
```

10.6.1   Recipe Definition, Management and Execution

ISA S88 Standard

Before going in detail into the recipe handling, the S88 standard must be explained first. The standard defines a common set of models and terminology that can be used to describe and define batch manufacturing systems. Building blocks (Figure 10.18) of the model are

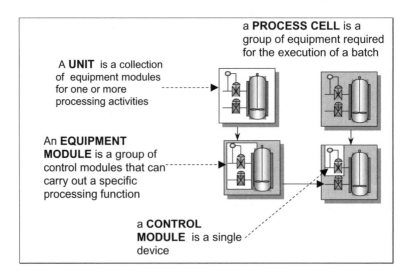

Figure 10.18   *ISA 88 building blocks*

equipment models and recipe models. The process is a combination of a unit and a recipe. The equipment is separated from the recipe because it should be possible to define a recipe without knowing the equipment it will run on. Of course, this requires a clear definition of the equipment modules. A uniform terminology is developed by the ISA that enables the uniform definition of the equipment independent of the recipe or the application.

The basic building blocks are process cell, units, equipment modules and control modules. All process hardware can be broken down into these blocks and this can be used to create a uniform process hardware description.

The recipe is built up to match the equipment definition. The general recipe is a company recipe that is valid for all sites of a specific company or distributed research group or network. It is the 'secret' recipe for product 'A'. The site recipe adds the local language and the local materials for the production of product 'A'. A master recipe is valid for the specific process cell and a control recipe is specific for the batch of the product 'A'.

The recipe handling in SCADA should be capable of working with master recipes and control recipes. The control recipes require strict version control and all recipe data should be stored with the batch data. When defining a batch the SCADA will make a copy of the master recipe and the operator can make specific changes to this recipe (before or during the batch) and this way turn it into a control recipe.

Creation of a product in an S88 compliant SCADA is simply selecting a unit, selecting a master recipe and combining the two to run a batch.

10.7   Validation

10.7.1   Electronic Signature

As indicated in the previous section, an operator can change a batch recipe to make a specific product. This may be acceptable for a lot of noncritical products, but the pharmaceutical industry is used to documenting all process data and changes that the operator makes to the process. The authorities require full traceability of the development, procedures and materials used in drug manufacturing. This places an extra demand on the SCADA software.

In the old days the operator would walk around with a notepad in which he would report all the actions and sign off on all the changes made to the process. Also the operator would sign off the strip chart recorder printouts and these paper process data would be kept as a batch record.

The SCADA replaces these paper process data and poses a problem for the authorities. There was no way to sign off on the digital data, and data manipulation after the batch is finished became easier. The FDA introduced the electronic signature to counter these potential problems. The electronic signature standard (21 CFR Part 11) defines the following:

(a) Validation of systems to ensure accuracy, reliability, consistent intended performance, and the ability to discern invalid or altered records.
(b) The ability to generate accurate and complete copies of records in both human readable and electronic form.

(c) Protection of records to enable their accurate and ready retrieval throughout the records' retention period.

(d) Limiting system access to authorized individuals.

(e) Use of secure, computer-generated, time-stamped, audit trails.

(f) Use of operational system checks to enforce permitted sequencing of steps and events.

(g) Use of authority checks to ensure that only authorized individuals can use the system, electronically sign a record, access the operation or computer system input or output device, alter a record, or perform the operation at hand.

(h) Use of device checks to determine the validity of the source of data input or operational instruction.

(i) Determination that persons who develop, maintain, or use electronic record/electronic signature systems have the education, training, and experience to perform their assigned task.

(j) The establishment of, and adherence to, written policies that hold individuals accountable and responsible for actions initiated under their electronic signatures.

(k) Use of appropriate controls over systems documentation.

In short, the 21 CFR Part 11 defines protection of the software against uncontrolled access and defines integrity of the data, readability of the data and the appropriate training of personnel.

10.7.2   GAMP

The other important part of validation in a SCADA system is GAMP compliance. GAMP was founded in 1991 by pharmaceutical experts in the UK who were interested in meeting evolving FDA expectations for GMP compliance of manufacturing and related systems.

Validation according to the FDA is: establishing *documented evidence* that provides a high degree of assurance that a specific process will consistently produce a product meeting its predetermined specifications and quality attributes.

The FDA *software* validation guideline states: 'The software development process should be sufficiently well planned, controlled, and documented to detect and correct unexpected results from software changes.'

GAMP describes five different categories for software and control systems:

- Category 1: Operating systems. Change control is needed for upgrades and retesting is needed to verify impact of new features.
- Category 2: Firmware. This applies to instrumentation and controller. The name, version number, configuration, calibration must be documented and verified during IQ. Functionality must be tested during OQ. Change control is needed for upgrades and retesting is needed to verify impact of new features. SOPs need to be in place and a training program must be in place. Supplier audits may be needed for highly critical or complex applications. Custom firmware is treated as a Category 5 product.
- Category 3: Standard software packages (statistical analysis software, HPLC software on PCs, etc.). Off-the-shelf programs not configurable. Name and version must be documented in IQ. User requirements must be documented, tested and reviewed during OQ. Supplier audits may be needed for critical applications.

- Category 4: Configurable software packages (SCADA, DCS, MES, LIMS, ERP, MRPII, etc., e.g. Applikon's BioXpert or Invensys Wonderware). When not mature packages, the product may be regarded at as Category 5. Supplier audit is usually required. Validation documentation should be present. Documentation should follow 'life-cycle' approach and should address the layers of the software and their category.
- Category 5: Custom software. Supplier audit is usually required. A validation plan is needed describing the life cycle and all validation documentation should be present.

As listed above a SCADA program is classified as Category 4 software and needs validation documentation according to the life-cycle approach.

The life cycle approach for software defines the development steps and controls for validatable software. Automated systems, especially software components, cannot be tested in the same way as a physical product.

Software is different, all software programs contain errors. How it is used will determine whether the errors become apparent or not. Software application complexity can mean that we could not test every permutation of inputs and use cases, therefore end-line testing (while still an important feature of a good quality system) cannot be relied upon on its own to ensure product quality.

To minimize the risk of unreliable software, the pharmaceutical industry has developed a standard for software development. All software for the pharmaceutical industry should be developed in compliance with this guideline.

Figure 10.19 shows the validation 'V' adapted to software development where the left process is for more complex software programs and the right track is for simple programs, configuration of SCADA programs and macros.

10.7.3 Validation Documentation

The vendor assessment is needed to review the vendor quality system and to review the level of documentation for the SCADA. A vendor audit can also be replaced by a document review. The development documentation of 'off-the-shelf' SCADA solutions is available from the supplier. An example of this documentation is given in Table 10.4.

Validation of software is needed for applications in the pharmaceutical industry. In a large number of other applications validation is not an issue. It is important to realize that validatable software is well documented, but is not necessarily better than nonvalidatable software.

10.8 What to Look for When Buying a SCADA System

As described before, there is a wide range of functions that SCADA system can perform. When selecting a SCADA system one should address the following questions:

(1) Will I need supervisory control or is data acquisition sufficient?
(2) What know-how is available in my company for SCADA software support?
(3) How many bioreactors am I planning to connect?
(4) Will I end up with a process where validation is required?
(5) Will the SCADA data be integrated with other (plant) automation systems?
(6) What is the budget that I have available for the SCADA system?

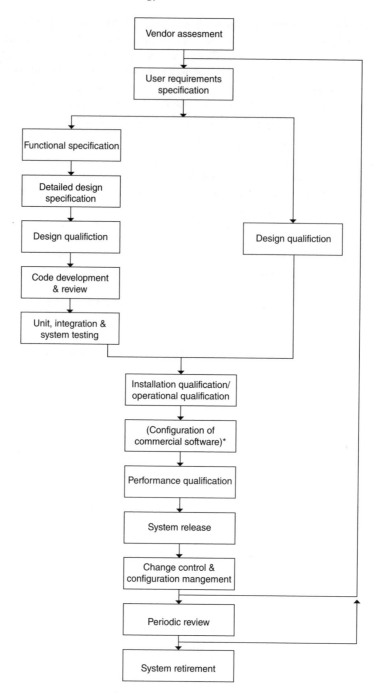

Figure 10.19   *Validation 'V' adapted to software development*

Table 10.4 *Example of documentation needed for software validation*

Phase	Document
CRD	CRD BioXpert
Definition	• High-Level system architecture • Operations concept • Release note definition • Requirement definition • Requirements definition phase plan • Reuse proposal
Analysis	• Release note analysis phase • Requirement analysis phase plan • Requirements analysis review • Requirements and functional specifications • Requirements • System and operations concept
Preliminary design	• Preliminary design phase plan • Preliminary design report • Release note preliminary design phase
Detailed design	• User interface architecture • BioXpert user interface • Object installation wizards • Hardware manager • Unit manager • Recipe manager • Process manager • Synoptic chart editor • Table chart editor • Trend chart editor • Import/export dll • Code editor dll • Data structures • Detailed design phase plan • Detailed design report • Requirements traceability • Service design • User interface design
Implementation	• Implementation phase plan • Test plan
System tests	Import/export wizard test results Add-on test plan Build 08 test reports (Part 1 and 2) Build 10 and 11 test reports (Part 1 and 2) Build 10 test reports Build 13 test reports Build 14 test reports Build 15 test reports

Table 10.4   *Continued*

Phase	Document
CRD	CRD BioXpert

	Build 16 test reports
	Build 17 test reports
	Build 18 test reports
	Build 19 test reports
	Build 21 test reports
	Build 22 test reports
	Build 23 test reports
	Build 25 test reports
	Build 26 test reports
	Build 27 test reports
	Test summary report build 13
	Test summary report build 14
	Test summary report build 15
	Test summary report build 16
	Test summary report build 17
	Test summary report build 18
	Test summary report build 19
	Test summary report build 22
	Test summary report build 23
	Test summary report build 25
	Test summary report build 26
	Test summary report build 27
	Test summary report build 23–27 overview
Acceptance	Acceptance test plan
	ADI 1010 driver verification
	Module hardware installation 1.01
	Module hardware installation
	Module test data presentation
	Module test data storage and archiving
	Module test help
	Module test information
	Module test networking
	Module test performance
	Module test process control
	Module test recipes
	Module test security
	PCI1711 driver test 0901
	PCI1711 driver test 0902
	Summary report build 29..40
	Summary report build 29..46
	Summary report build 29..48
	Summary report build 29..50
Additional	Device support for ADI1010 biocontroller
	Design
	Adda card driver design
	ADI 1010 Driver definition 1.2
	ADI 1010 Driver definition 1.1

Responses:

(1) A lot of users are only using the SCADA for data storage. There is no need to spend money on a part that will not be needed. Data acquisition systems are easier to operate than SCADA systems, since they have limited functionality.

(2) When no or limited local support is available, try to find a SCADA that will be supported by a local supplier. Software support is critical with SCADA systems.

(3) Limit the number of bioreactors that can be connected to your SCADA. A limited number of parallel processes reduces the complexity of the SCADA architecture and decreases the investment needed.

(4) When the final process needs to be validated, it is advisable to use the same SCADA from research to production. This increases the initial price for the research system, but reduces the complexity later on and ensures that all data from initial research to final production are available for inspection.

(5) When the SCADA will be integrated with other automation systems, the data structure and data storage system need to be designed to allow this data exchange with other systems. This adds to the initial complexity of the system, but eases the data exchange later on.

(6) Last but not least it is important to know what the budget is for the SCADA system. This might limit the available choices.

10.9 Summary

SCADA systems are available with a variety of functionality. The main functions are data storage, data presentation, alarm handling and supervisory control such as time and event based actions.

Communication between all parts of the bioreactor control system is of vital importance. This specific area has been changing rapidly in recent years. Instrument communication is getting faster, cheaper, more reliable and more transparent with the appearance of industrial standards (Ethernet hardware and OPC standards for communication).

The large amounts of stored process data can be reduced using clever data reduction techniques. Data can be presented in variety of ways. Examples are trend charts, bar graphs, synoptic and table displays.

Documentation is increasingly important with all software. In the pharmaceutical industry, the documentation is as important as the product itself. In this area of validation the standardization is enforced by the FDA and the ISPE, protecting the end user (consumer) from buying a faulty product (pharmaceutical). This standardization improves the quality of the SCADA product but also raises the price by increasing the development time needed.

Choosing the right SCADA system depends on the application. Validation, data interchangeability and communication with other devices are some of the criteria that will determine which SCADA is best fit for a specific application.

Further Reading

Practical Modern Scada Protocols. G. Clarke, D. Reynders. (2004) Newnes Computers/ Communications/Networking.

SCADA: Supervisory Control and Data Acquisition, third edition, S. A. Boyer (2004) ISA.

Practical SCADA for Industry IDC Technology. D. Bailey, E. Wright (2003) Newnes.

GAMP Good Practice Guide: Validation of Laboratory Computerized Systems (2005) International Society for Pharmaceutical Engineering.

ISPE GAMP Good Practice Guide: Global Information Systems Control and Compliance (2005) International Society for Pharmaceutical Engineering

S. Rommel, A. Schuppert (2004) Data mining for bioprocess optimization. *Engineering in Life Sciences* **4**, 266–270.

11

Using Basic Statistical Analyses in Fermentation

Stewart White and Bob Kinley

Variability is the killer. It's that simple. Whether running $150\,m^3$ industrial processes or starting off a PhD with a couple of 'control' runs. Until the variability in any process is understood then results and any conclusions drawn will be open to interpretation, scrutiny and worst of all, error. Somewhat worryingly, variation not only applies to existing processes, but to experiments not even thought of! In this chapter, we will discuss how one can use some basic statistical concepts and analyses to ensure that any differences seen in either the datasets from experiments or production processes are, in fact, real. We will discuss how to monitor variability, test for differences between datasets, examine correlations between datasets and finally walk through the theory of experimental design. These basic tools will enable interpretation of data and phrasing of conclusions in statistical terms. The statistical software used here was JMP, however, other packages are available such as Minitab.

Sources of variability include, but are not limited to: raw materials, operator error, engineering/ mechanical faults, documentation or calibration.

In order for us to proceed to combat process variability we must adopt a mindset that requires everything to be challenged – whether it's the analytical results, reagent shelf life, or, the reproducibility of a colleague's analytical technique! Consider that almost everything has an innate or inherent level of variability – and the level of variability will simply be acceptable or unacceptable in a specific context.

Process variability is typically cumulative. Each small change at every stage will be likely to increase, or at least contribute to, the overall variability of the process. Therefore, when analysing the process, it is useful to segment it into stages. For example, a process

Practical Fermentation Technology Edited by Brian McNeil and Linda M. Harvey
© 2008 John Wiley & Sons, Ltd

may typically begin with an inoculum work up procedure, for which there will be pre-set criteria such as cell concentration or pH of the medium, which will signify growth to adequate levels. These are two parameters that can easily be monitored to understand if any variability comes from this particular stage. The inoculum may then go onto a stirred tank reactor, or seed vessel, to generate further biomass, before being used to inoculate a final production vessel. Typically the work-up procedures will be simpler and less automated, than those for the production vessel, with fewer forward process criteria. Nevertheless, until there is some measure, and again understanding, of the variability within this stage, they must be considered as having the potential to cause wholesale variability. Later in the process, i.e. at the final production stage, product concentration and/or quality may be determined by an HPLC method. This method will typically require a standard to be run on a daily basis to account for change in personnel, reagents, etc. This standard itself will be scrutinised for variability, and so long as this falls within a particular range, then the standard will be deemed to be acceptable, as will results obtained from analyses performed during that day (under good manufacturing practice, such activities are required to maintain the procedure in a validated, or controlled, state). If however, no such control is in place or the standard fluctuated without being monitored, or was not observed on a daily basis, this would mean the estimate of product concentration, and by extension, the profitability of an industrial fermentation process, would be unpredictable. This is, of course, an unacceptable situation, and hardly the best basis on which to lay the foundations of a new biotech start up!

11.1 Measuring Variability Using Control and Capability: Control Charts

The control chart is a tool whose job is to identify those results in a dataset that do not 'belong' with the rest, so that the process can be improved by tracking down and fixing the cause(s) of the exceptional result(s). Control charts are now an industry standard, and are used with every process imaginable. It is possible to generate control charts from scratch using paper and pencil, however, there are some excellent software packages on the market such as JMP and Minitab, which will do it for you. Regardless of how the charts are produced, the trends/results figures will still require interpretation and phrasing in statistical terms.

Most data distributions are normal in nature (before presuming that a distribution is normal, it can be worthwhile to verify this using the preferred software package). Normal distributions can be fully described in terms of their mean and standard deviation: μ, the distribution's mean, determines its midpoint, while σ, the distribution's standard deviation, describes the spread of the data around the mean.

Figure 11.1 shows how different populations can have the same mean, but different standard deviations. This is a good example of how comparing simple numerical values of means of two processes can lead to significant errors.

Similarly, altering the mean and the standard deviation alters the midpoint and the spread. Figure 11.2 clearly displays two different populations. We will go on to discuss how to measure these differences, or indeed if there are any, later in the chapter.

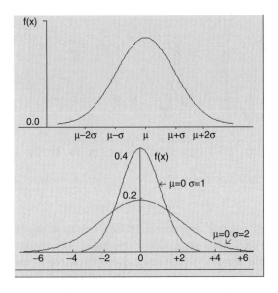

Figure 11.1   *Normal distribution of a dataset with mean μ and standard deviation σ, showing how the mean can remain constant, while the standard deviation can vary*

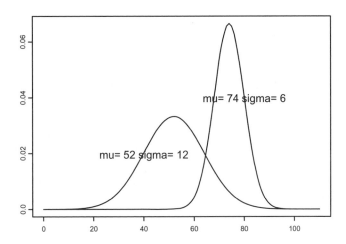

Figure 11.2   *Examples of how variations in both mean and standard deviations can generate results in very different datasets*

Although some argue that only normally distributed data should be placed on a control chart, many experts agree that there is more to be gained by charting than by not doing so, unless the distribution is extremely heavily skewed.

After ensuring that a dataset is not bizarrely nonnormal, and that it is chronological (or, one result follows another), one can begin to use control charts. Figure 11.3 shows how a few random observations from a normal distribution might look on a control chart.

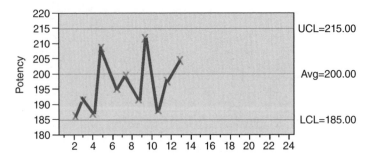

Figure 11.3 *The correspondence between a Normal distribution and a control chart*

Statistical theory tells us that approximately 99% of almost any distribution will fall is within three standard deviations either side of the mean. The mean in the control chart is 200.00, while the three standard deviations either side of the mean are called the upper and lower control limits, and are shown as 215.00 and 185.00. If points from our process fall within the control limits, and do not exhibit any patterning, then the fermentation is operating consistently and predictably (i.e., within statistical limits – sometimes called 'natural process limits') and is said to be 'in control'. Of course, labelling a process as 'out of control' is pointless unless some action is both necessary and it is possible to fix it. If the process is in control, then one can accept that the distribution of results is homogeneous and further statistical analyses can then be performed. However, we may find that the process is out of control. Examples of some processes that are out of statistical control are shown in Figure 11.4.

Test 1 in Figure 11.4 shows the simplest of out-of-control signals, that is one point out with the upper or lower control limits. Test 2 shows nine consecutive points in zone C (which may continue into zones B and A). This test shows that there has been a step change in the process that has been sustained. Test 5 demonstrates an out-of-control process where two points out of three are in zone A, or, outside the control limits. In all cases the analyses were performed using JMP software. These out-of-control signals are typically termed 'special cause' events, given that such excursions/patterns are unlikely to have occurred without some specific root cause, i.e. within the dataset of a process that is in statistical control. Examples of such special-causes events can include electrical failure causing loss of agitation and/or airflow, failing valves allowing incorrect addition of nitrogen/carbon/acid/alkali, or the usage of an incorrect raw material.

Control charts are invaluable as the daily, weekly or annual measure of any dataset's variability and process performance. With a greater understanding of where variability can come from, and what it looks like statistically, it is possible to measure and describe variability, and elucidate its effects in both continual improvement for existing processes, or indeed for laboratory experiments.

In Figure 11.4, we observed the effects of how processes can suffer from special cause variation. Naturally, variability will still exist in a process that is in statistical control, as illustrated in Figure 11.5. This natural background variation is known as 'common cause' variation, as distinct from the 'special cause variation', which gave rise to control-chart signals.

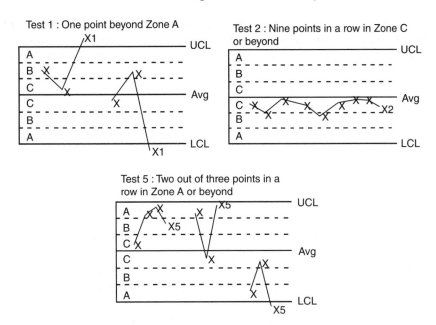

Figure 11.4   *Examples of control charts showing processes which are statistically 'out of control'*

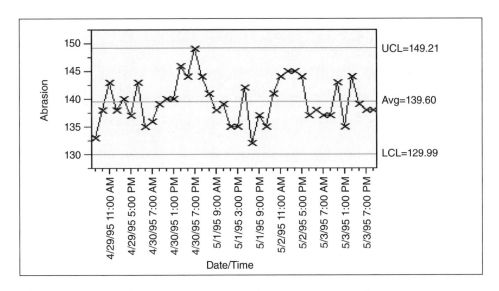

Figure 11.5   *Control chart of the parameter abrasion versus time, showing common cause variability*

When only 'common cause' variation is present we can say we are dealing with a distribution that is in statistical control. This means we can legitimately go on to apply standard statistical tests to both analyse the variability, and, to drive process improvements.

11.2   Process Capability

For many production processes, it is required that some prespecified criteria should be met, for instance the product concentration is required to be within a specific range. Such specifications are typically set using a balance of historical data and business need. 'Capability' is a measurement of how well a given process is meeting specification, and a popular 'capability index' is called 'Cpk'. For example, a CpK of 1.0 means that virtually all points lie within the appropriate release limits. That is, a process will have upper and lower release limits, which may or may not be within the statistical control limits. The control charts shown in Figure 11.6 demonstrate process capability. That is, a process with UCL and LCL 'inside' the specification limits will have a Cpk > 1, while process limits outside specification limits will give a less capable process with Cpk < 1. If process and control limits are the same then the Cpk = 1. Figure 11.6 demonstrates that although a process may be in statistical control, the dataset in question must be compared with a

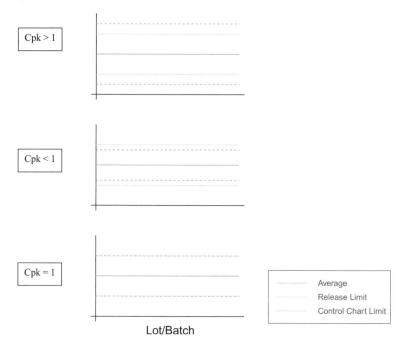

Figure 11.6 *The capability index, Cpk, described using control charts showing upper and lower control limits with reference to upper and lower specification, or release, limits. A Cpk > 1 means a process is capable, and performing within requirements. Cpk < 1 means that the process is not meeting the specification limits and is not capable*

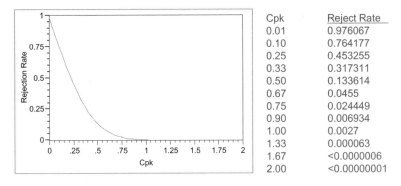

Cpk	Reject Rate
0.01	0.976067
0.10	0.764177
0.25	0.453255
0.33	0.317311
0.50	0.133614
0.67	0.0455
0.75	0.024449
0.90	0.006934
1.00	0.0027
1.33	0.000063
1.67	<0.0000006
2.00	<0.00000001

Figure 11.7 *The capability index, Cpk vs reject rate, with numerical reference*

target or specification as a reference to the meaningfulness of the data. Therefore, common cause variation within a process that is in statistical control, still has the potential to fail to meet the specification/ acceptance criterion.

In Figure 11.7 we can see the relationship between Cpk value and the reject rate, where a Cpk of 1 means that on average about three lots in every 1000 will be rejected, which is usually considered to be an acceptable limit of process performance in industrial fermentations. This means that all points of a particular dataset lie within both the UCL and LCL and the upper and lower specification limits, i.e. within three standard deviations (σ) of the mean. A Cpk value can be generated using a dataset of around 20 points, and is generally predictive.

The acceptability of any Cpk or rate of rejection will depend on the process itself. For example, a Cpk of 1 in industrial fermentations may be acceptable – but not in the aerospace sector where a Cpk of 1, would mean three flights in every thousand not reaching their destination, which would be unimaginable, let alone unacceptable. Instead of being a 3 σ process, with a Cpk of 1, flight fatalities are greater than a 6 σ process. For some processes, a target for Cpk of 1.33, or even 1.67 is used to provide additional assurance.

Figure 11.8 shows that a 6 σ process has an almost infinitesimal chance of any point from the dataset falling outside specification limits. 'Six sigma' is also a general term for what is now becoming a discipline in itself, i.e. 'six sigma' is a system to define, monitor, analyse, improve and control all kinds of processes. 'Six sigma' has become a central discipline within organisations, with different levels of attainment – green belts, black belts and master black belts. Organisations now commonly use all or some of the six sigma components to reduce wastage and variability and to drive continuous improvement. Importantly, much of the data gathering and interpretation utilise all (and more) of the statistical tools we are discussing here.

One more way to monitor and describe the variability in datasets from a process is the coefficient of variation (CoV). This is simply a percentage of the standard deviation/ mean. If the CoV is high, say 30%, then this indicates a process that has an unacceptably high level of variation. Typically in industrial fermentation processes, <5% is considered desirable. Again, this should be compared in context with the detailed process

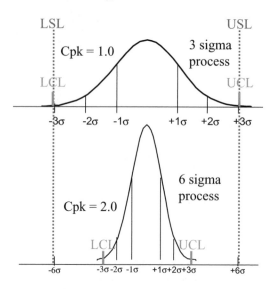

Figure 11.8 *Demonstration of a three-sigma and six-sigma process with reference to upper and lower control limits, and upper and lower specification limits*

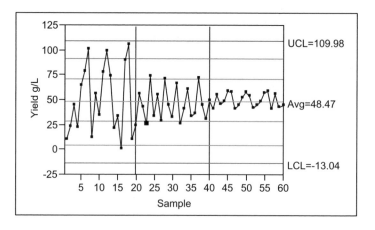

Figure 11.9 *A control chart of yield versus sample, demonstrating how reducing process variability brings the control limits closer to the mean and makes for a more robust process. Successive control limits correspond to step changes in the process, indicated by vertical lines on the x-axis*

requirements, e.g. the process specification/acceptance criteria, which may be target yield for each batch, or simply an acceptable level of reproducibility for control runs in the laboratory.

To end the discussion on control charts, Figure 11.9 shows the advantages in reducing variability. This is an example of how a process can be brought under control prior to actually making any changes or improvements.

11.3   *t*-Tests and the *P*-value

This is one of the most common statistical analyses, and it comes in two forms:

the *1-sample t-test*, where it tests whether the dataset is consistent with having come from a normal distribution with a particular specified ('hypothesised') mean, and the

two-sample test, where it tests whether the two datasets are consistent with having come from normal distributions with the same, unspecified, mean, for example the comparison of a standard process with a modified process.

The one-sample test works like this: if the difference between the specified mean and the observed dataset mean is more than can be accounted for by the variation between individuals in the dataset then a 'statistically significant' difference is said to exist.

Similarly, with the two-sample test, if the difference between the two observed dataset means is more than can be accounted for by the variation between individuals in the datasets then a 'statistically significant' difference is said to exist.

The 'differences' are measured by a *t*-statistic, which is calculated from the datasets, and the degree of 'consistency' is measured by the probability (or '*P*-value') of getting such a large *t*-statistic. When the *P*-value is small, the difference is significant.

In statistical convention, $P < 0.05$ is deemed to be the norm. Therefore, to have a *P*-value below this level indicates that a significant difference 'at the 0.05 or 5% level' is indicated.

If however, a *P* value is greater than 0.05, say 0.45 – then there is said not to be a statistically significant difference at the 5% level.

There are some caveats to this type of analysis, namely, the datasets must be normally distributed, and in the two-sample case one should test of for unequal variances – that is, there must be comparable variability within each dataset.

11.3.1   *t*-Tests

In order to determine the *P*-value, the *t*-statistic is compared with the distribution of *t*-values that can be expected for this size of dataset (*n*). These are extensively tabled, and are available in statistical software. If the probability of seeing a *t*-statistic of this size or greater is less than 0.05, then the difference is significant at the 5% level.

Figure 11.10 shows the distribution of *t*-statistics to be expected when $n = 9$, and shows that there is a 5% probability of getting a *t*-statistic either greater than 2.26 or less than −2.26. Therefore, if a dataset of this size generated a *t*-statistic beyond +/−2.26 we would say that the difference is statistically significant at the 5% level.

Table 11.1 shows the output from a JMP analysis where the *t* statistic (called the 'test statistic' above) is −2.68. This is beyond the 5% points (+/−2.26) so we know that the difference is significant at the 5% level. In fact JMP tells us that the probability of getting our actual *t*-statistic (2.68) is .0251, or 2.51%, and this is often referred to as the 'significance-level' of this particular test.

Although somewhat heavy on jargon, it is crucial to understand statistical theory along the way, as this allows us to understand how significantly different datasets are from each other using the *P*-value as a guide.

Warning note: It is important *not* to assume that because a difference is statistically significant, that there is a difference that is of practical importance. For example, the

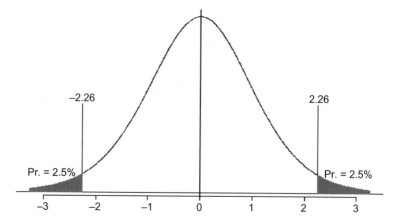

Figure 11.10   *Normal distribution of t-values for t(9), showing a t-value of +/–2.26, where P = 0.05*

Table 11.1   *One-sample t-test comparing hypothesised mean of 12.1 with dataset mean of 11.77*

Test Mean=value	
Hypothesized Value	12.1
Actual Estimate	11.77
df	9
Std Dev	0.38887
	t-Test
Test Statistic	-2.6835
Prob > |t|	0.0251
Prob > t	0.9875
Prob < t	0.0125

one-sample test in Table 11.2 shows the significance level $P = 0.0112$, meaning that the actual mean is statistically significantly different from the hypothesised mean.

However, we must also view the practical significance in context. Note that the confidence interval (an estimate of a plausible range for the true, but unknown, mean) is very narrow, and the observed mean is very close to the hypothesised mean, so we have to ask whether such a small difference matters in the context of the process.

Always remember that the statistical analysis is only part of the interpretation of an experiment – it is crucial always to try to view results in context (Table 11.2).

Two Sample t-tests

When comparing a standard process with a process we believe has changed, we can compare datasets using a two-sample *t*-test. The JMP output below illustrates how we

Table 11.2 *Output from analysis with a hypothesised mean 100 against the actual estimate of 100.3. Although statistically significant, with P = 0.0112, there may be limited benefit to this conclusion as the difference between actual and hypothesised means is so small*

Moments		Test mean=value	
Mean	100.32107	Hypothesized Value	100
Std Dev	2.8209014	Actual Estimate	100.321
Std Err Mean	0.1261545	df	499
upper 95% Mean	100.56893	Std Dev	2.8209
lower 95% Mean	100.07321		*t* Test
N	500	Test Statistic	2.5450
		Prob > |t|	0.0112
		Prob > *t*	0.0056
		Prob < *t*	0.9944

can use the two-sample test to compare datasets, that is, to test the whether the mean of one dataset is the same as that of the other.

The main piece of information in Figure 11.11 is the *P*-value, which is 0.2536, showing that means of the two datasets are not significantly different. That is, the two sample means are no more different from each other than could be accounted for by the random variation between individuals. Note also that the lower and upper 95% confidence interval contains 0. This is equivalent to saying that the difference betwen the two dataset means is not significantly different from zero.

When we need to compare more than two datasets, we use the 'analysis of variance'. Figure 11.12 shows such an analysis for the effect of a possible drug on blood pressure.

Like the *t*-test, this procedure also works by comparing the variation between dataset-means, with the variation between individuals and, in this case generates an *F*-statistic (which is a generalisation of the *t*-statistic), and from the *F*-statistic, we use the *F*-distribution to obtain the *P*-value or significance level.

The *P*-value from the analyses in Figure 11.12 is 0.0754, which is not significant at the 0.05 or 5% level, but is significant at the 10% level. In an experimental setting one might be satisfied that a significance at this level is sufficient evidence for action.

From this, and from the graphical representation, we can see that that drug B may have some benefit and warrants further investigation. The diamonds and circles are simple graphical representations used by JMP to allow easier assessment of the data – e.g. the further apart the circles, the bigger the difference between datasets.

This chapter is designed to provide a taster of some commonly used techniques to better monitor and describe fermentation processes. Even the most defined and established fermentation processes still have processing issues, or there is a business driver to increase productivity/product quality. These techniques will be at the heart of any continual improvement strategy in industry. Understanding these techniques will allow extraction of as much information as possible from each experiment and allow description of their effects in a universally accepted form.

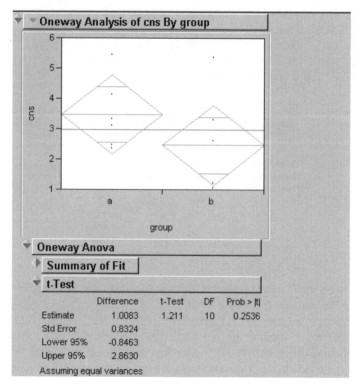

Figure 11.11 t-Test of sample datasets a and b, showing that these data sets over lap. P value = 0.2536, meaning that the difference is not statistically significant

11.4   Correlation and Regression

There are occasions where the variable of interest changes with the level of some other continuous variable. Using correlation and regression analyses we can test if such relationships are in fact correct. Figure 11.13 shows both good and poor correlations in *x* by *y* graphs. Correlation is described by the 'correlation coefficient', a dimensionless descriptor, always between −1 and 1.

Such simple assessments carry the risk of the correlation coefficient being affected by an outlying point which can have a tremendous impact. Or indeed, the relationship may not be linear, but parabolic in nature. Therefore, values must be interpreted with care.

In addition to correlation, we can use regression analyses to assess the best fit of a line using the equation $y = \beta_0 + \beta_1 x$. The ideal line of best fit will have the sum of the squares of the distances from *x* to the line of fit as small as possible. That is, 0. The breakdown of regression analysis is shown in Figure 11.14.

We can see the equation, $y = \beta_0 + \beta_1 x$, fitted by JMP in Figure 11.15 (underneath 'linear fit'). The *R* square value is 0.503, which is the square of the correlation coefficient, *r*.

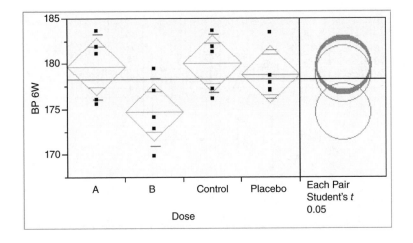

Analysis of Variance

Source	DF	Sum of Squares	Mean Square	F Ratio	Prob >F
Dose	3	92.21660	30.7389	2.7713	0.0754
Error	16	177.46774	11.0917		
C. Total	19	269.68433			

Figure 11.12 *Analysis of variance comparing the effects of drugs A and B on blood pressure versus a control group and a placebo*

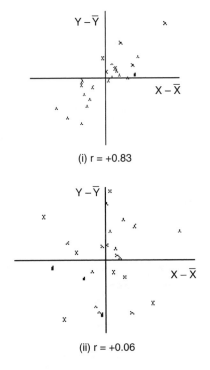

Figure 11.13 x *by* y *plots showing good correlation between variables (i), and poor correlation (ii), as described by the correlation coefficient* r

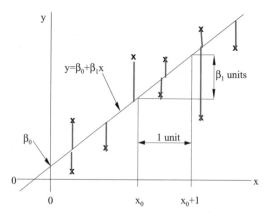

Figure 11.14   *Regression analysis of x and y, outlining the components of equation* $y = \beta_0 + \beta_1 x$

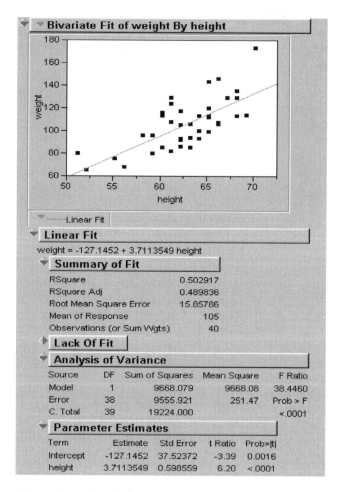

Figure 11.15   *Example of x by y regression analyses and statistical output*

If R square is 1, the straight line model fits perfectly, if it is 0 then there no evidence connecting the change in one variable with another. Importantly, the R square value tells us what proportion of variation is explained by the fitted line.

The statistical test again works by comparing the amount of variability in y, that is 'explained' by the linear dependence on x, with the variation between individual ys and the fitted line. As in the analysis of variance, the test is carried out by generating an F-statistic (or 'F ratio') and applying an F-test.

In the above example, we see a large F ratio and hence negligible P value 0.001, therefore the regression is significant.

In summary, to test the quality of a perceived linear 'fit', or relationship between two variables, we must assess this relationship in terms of r^2, and P-value, but whatever else we do, we must always plot the data.

11.5 Design of Experiments

In an ideal world we would be able to run vast experiments, where time and resources were limitless, and we could test every theory or effect in one vast experiment. Unfortunately, time and resources are usually limited and we need to find a way to test our theories by experimentation, but still in a significant manner and, with as much clarity as is possible. As difficult as this may be, it can be done by using experimental design (or design of experiments, DOE). As in previous sections we will attempt to touch on the benefits of DOE, some of the statistical theory behind the tools, which will hopefully allow others to apply this thinking to their own work. This approach has been widely deployed to accelerate process development in fermentation using micro- and minireactor systems, as described in Chapters 2 and 8.

When we plan an experiment, we need to understand what data to collect, how much data, and under what conditions. We also need to plan how to analyse, interpret and present the results of the change.

One of the most powerful and economical tools of experimental design is the factorial design, where we may have any number of factors, at any number of levels. However, a widely-used, powerful and flexible factorial design is the two-level factorial design, specifically fractional factorial design. Factors may be continuous (low to high concentrations) or be categorical (yes/ no).

A very powerful and important factorial design is where we have k factors, each at two levels i.e. the $2 \times 2 \times \ldots \times 2$ or 2^k factorial design. The advantages of 2^k factorial designs are:

- they need only a few runs per factor;
- they can indicate major trends very efficiently;
- they can be added to in order to form composite designs for more detailed study;
- a fraction of a 2^k design can be run, to screen many factors quickly;
- they can be used as building blocks, so that experimentation can be iterative.

However, the experiments must be correctly structured to be accurate and effective. For example, an incomplete experiment (lost data or early termination) will destroy the structure.

11.5.1   Fractional Factorial Designs

If we have more than a few factors, i.e. if k is large, say 8, then 2^k will be of the order of 256 experimental runs. However, it is possible to perform a 'fraction' (say 1/2 or 1/8 or 1/32) of these runs in such a way that we only sacrifice a limited amount of information. Although possible, we must be extremely careful what fraction we use as we must maintain a suitable experimental structure. Importantly, fractional factorial design will allow us dramatically to reduce the amount of work we need to do. If the fraction is denoted by $\frac{1}{2}^p$ where p is an integer, these are often called 2^{k-p} Factorial designs, e.g. half of a 2^6 design would be called a 2^{6-1} design.

11.5.2   Replication

Even if a factor has little or *no* effect, we won't get identical results at the different levels, as there will be other sources of variation within the design. Therefore, we need to understand the level of random variation to expect and compare this with the size of the apparent effect, in which case a single 'replicate' experiment may be sufficient. Otherwise there may be a need for more replications of the experiment in order to obtain an estimate of error variability, and thus reassure ourselves that, on average, the difference in response is too great to be explained by random variation. If k is small, e.g. 3, one could afford to do all of the possible $2^3 = 8$ runs, but if variability is high, and it may be necessary to replicate the experiment n times, giving $n*2^k$ results altogether.

Using the example below, it will be demonstrated how to perform a fractional factorial experiment.

Example

In a certain chemical reaction, the response is % purity, and there are three factors that are believed to be important in controlling purity, namely:

catalyst,   with two levels: absent or present   [categorical]
pH,   with two levels: 8 or 10   [continuous]
buffer,   with two levels: type1 or type2   [categorical]

After inputting this data into JMP software, the options for the experimental design are as presented in Figure 11.16.

Figure 11.16 shows that we have selected performing the minimum number of runs possible, i.e. four, and the design is shown in Table 11.3 below, with Y being the response, in this case purity. Notice that, as we have three factors, a full factorial design would have had $2^3 = 8$ experimental runs. So what we have here is a half-fraction of the full design, i.e. a 2^{3-1} design:

By analysing this using the effect screening tool, the analysis suggests that by increasing pH, the purity is reduced by 8.5, the absence of the catalyst reduces purity. The buffer reduces purity by 4.1 if type 1 is used, while type 2 increases purity by 4.1.

Table 11.3 shows that the parameter estimates output contained no standard error estimate, and no t-tests of the significance of the fitted parameters. Also, there are no effects tests, in terms of F ratio or P value. This is because there were only four results, therefore there were only three degrees of freedom, with one degree of freedom 'used up' in esti-

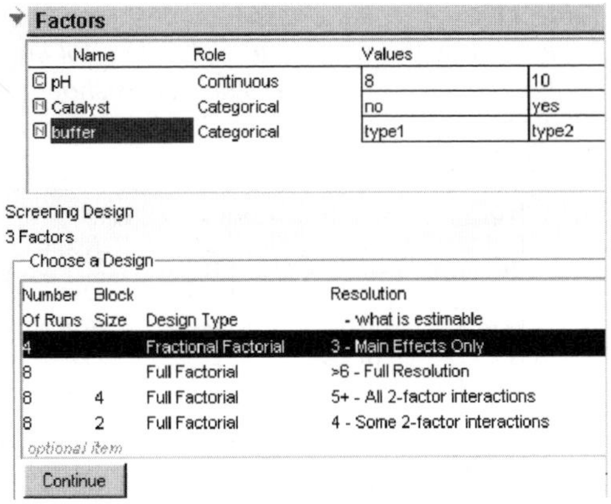

Figure 11.16 *JMP output of suggested number of experimental runs for a 2³ factorial design*

Table 11.3 *Design and effect screening analysis of a 2³⁻¹ fractional factorial design, showing parameter estimates and effect tests*

	Pattern	pH	Catalyst	buffer	purity
1	--+	8	no	type2	-44.81
2	-+-	8	yes	type1	-34.5
3	+--	10	no	type1	-70.01
4	+++	10	yes	type2	-43.3

Parameter Estimates

Term	Estimate	Std Error	t Ratio	Prob>|t|
Intercept	-48.155	.	.	.
pH(8,10)	-8.5	.	.	.
Catalyst[no]	-9.255	.	.	.
buffer[type1]	-4.1	.	.	.

Effect Tests

Source	Nparm	DF	Sum of Squares	F Ratio	Prob > F
pH(8,10)	1	1	289.00000	.	.
Catalyst	1	1	342.62010	.	.
buffer	1	1	67.24000	.	.

mating each of the three effects (parameters). Therefore, no degrees of freedom remain in order to estimate the error. With no estimate of error in our test, can we conclude that these answers are correct?

Fortunately, we can check our results by replication – this is the complete repetition of the whole experiment (it is not simply repeated readings or measurements). Replication will allow us to get an estimate of experimental error, which we can then use to decide

Table 11.4 *A repeat of experiment shown in Table 11.3, with rows five to eight identical in design. Parameter estimates and effects show all effects are significant at the 5% level (P < 0.05)*

	Pattern	pH	Catalyst	buffer	purity
1	--+	8	no	type2	-44.81
2	-+-	8	yes	type1	-34.5
3	+--	10	no	type1	-70.01
4	+++	10	yes	type2	-43.3
5	--+	8	no	type2	-42
6	-+-	8	yes	type1	-38
7	+--	10	no	type1	-60
8	+++	10	yes	type2	-40

Parameter Estimates

Term	Estimate	Std Error	t Ratio	Prob>|t|
Intercept	-46.5775	1.43198	-32.53	<.0001
pH(8,10)	-6.75	1.43198	-4.71	0.0092
Catalyst[no]	-7.6275	1.43198	-5.33	0.0060
buffer[type1]	-4.05	1.43198	-2.83	0.0474

Effect Tests

Source	Nparm	DF	Sum of Squares	F Ratio	Prob > F
pH(8,10)	1	1	364.50000	22.2195	0.0092
Catalyst	1	1	465.43005	28.3721	0.0060
buffer	1	1	131.22000	7.9990	0.0474

whether observed differences in data represent a real difference in the underlying structure or model. Also, the effects of factors upon the response can be determined with greater precision.

By having more data, as shown in Table 11.4, we are now able to obtain a statistical interpretation, providing more insight into the significance of the effects.

Another major consideration in performing analyses such as these is the statistical power of the tests. By understanding power, we can assess whether or not the test is appropriate for the level of effect we are looking for, i.e. have we done enough runs to detect the difference we are looking for. Table 11.5 shows that the two replicates (eight runs) we have so far give a power of only about 19%. If we want a good chance (usually >80%) of detecting a catalyst effect of 2.0 for example, we will need a total of 36 runs, i.e. an additional 28 runs = another seven replicates!

Returning to our purity example, it is clear that it would have been wise to carry out the full 2^3 experimental design. Table 11.6 shows rows five to eight as identical, except that the buffer type has swapped around, now allowing us to consider combinations of all three factors. This shows that performing the smallest number of runs can be a false economy. This design would have allowed us to assess any interactions.

Table 11.5 Power data demonstrating the number of runs required versus the statistical power, to detect a difference (delta) of 2

Power				
Alpha	Sigma	Delta	Number	Power
0.0500	4.05025	2	8	0.1916
0.0500	4.05025	2	12	0.3258
0.0500	4.05025	2	16	0.4432
0.0500	4.05025	2	20	0.5458
0.0500	4.05025	2	24	0.6338
0.0500	4.05025	2	28	0.7079
0.0500	4.05025	2	32	0.7693
0.0500	4.05025	2	36	0.8193
0.0500	4.05025	2	40	0.8597

Table 11.6 Full 2^3 Factorial experimental design of three factors, pH, catalyst and buffer, at two levels

	Pattern	pH	Catalyst	buffer	Y
1	--+	8	no	type2	-44.81
2	-+-	8	yes	type1	-34.5
3	+--	10	no	type1	-70.01
4	+++	10	yes	type2	-43.3
5	---	8	no	type1	•
6	-++	8	yes	type2	•
7	+-+	10	no	type2	•
8	++-	10	yes	type1	•

11.5.3 2^{k-p} Factorial Designs – Interaction

So far we've dealt with factors that are independent. When factor A and factor B act independently, the effect of factor A is the same, no matter the level of factor B. However, when factor A and factor B are not independent, but interact, the effect of factor A does depend on the level of factor B. In a given experiment, reaction time will depend on the amount of a reagent present, but the reaction time in response to a particular amount of reagent may differ depending on the presence or absence of a catalyst. Figure 11.17 is an example of an interaction plot, i.e. the greater the interaction, the less parallel the lines. We can incorporate the search for interactions into our experimental design.

Using some new data for our purity experiment in Table 11.7, we can use this example to carry out a screening effect analysis, which will include two factor interactions.

The data in Table 11.7 shows that there is no significant change in the parameter estimates and effect tests, apart from the overall average, and perhaps pH. Again, this is because our estimate of error is based on only one degree of freedom. That is, there are eight results, so there are seven degrees of freedom, six have been 'used up' in estimating the three individual effects and the three interactions, so there is only one left for error. This means that *any* effect will be difficult to detect.

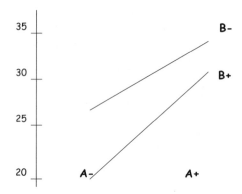

Figure 11.17   *Schematic representation of an interaction effect between two parameters A and B. Parallel lines would indicate no interaction*

Table 11.7   *A 2^3 Full factorial design, with resulting analyses showing effects of two-factor interactions*

	pH	catalyst	buffer	purity
1	8	no	type1	-44.81
2	8	yes	type2	-12
3	10	no	type2	-134.5
4	10	yes	type1	-168
5	8	no	type2	-70.01
6	8	yes	type1	-40
7	10	no	type1	-143.3
8	10	yes	type2	-188

Parameter Estimates

Term	Estimate	Std Error	t Ratio	Prob>|t|
Intercept	-100.0775	10.25	-9.76	0.0650
pH(8,10)	-58.3725	10.25	-5.69	0.1107
catalyst[no]	1.9225	10.25	0.19	0.8820
buffer[type1]	1.05	10.25	0.10	0.9350
pH(8,10)*catalyst[no]	17.6275	10.25	1.72	0.3353
pH(8,10)*buffer[type1]	1.75	10.25	0.17	0.8923
catalyst[no]*buffer[type1]	3.05	10.25	0.30	0.8159

Effect Tests

Source	Nparm	DF	Sum of Squares	F Ratio	Prob >F
pH(8,10)	1	1	27258.790	32.4316	0.1107
catalyst	1	1	29.568	0.0352	0.8820
buffer	1	1	8.820	0.0105	0.9350
pH(8,10)*catalyst	1	1	2485.830	2.9576	0.3353
pH(8,10)*buffer	1	1	24.500	0.0291	0.8923
catalyst*buffer	1	1	74.420	0.0885	0.8159

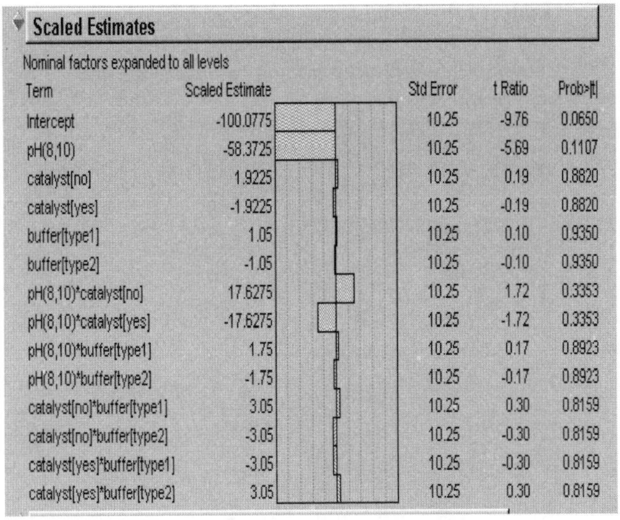

Scaled Estimates

Nominal factors expanded to all levels

Term	Scaled Estimate		Std Error	t Ratio	Prob>|t|
Intercept	-100.0775		10.25	-9.76	0.0650
pH(8,10)	-58.3725		10.25	-5.69	0.1107
catalyst[no]	1.9225		10.25	0.19	0.8820
catalyst[yes]	-1.9225		10.25	-0.19	0.8820
buffer[type1]	1.05		10.25	0.10	0.9350
buffer[type2]	-1.05		10.25	-0.10	0.9350
pH(8,10)*catalyst[no]	17.6275		10.25	1.72	0.3353
pH(8,10)*catalyst[yes]	-17.6275		10.25	-1.72	0.3353
pH(8,10)*buffer[type1]	1.75		10.25	0.17	0.8923
pH(8,10)*buffer[type2]	-1.75		10.25	-0.17	0.8923
catalyst[no]*buffer[type1]	3.05		10.25	0.30	0.8159
catalyst[no]*buffer[type2]	-3.05		10.25	-0.30	0.8159
catalyst[yes]*buffer[type1]	-3.05		10.25	-0.30	0.8159
catalyst[yes]*buffer[type2]	3.05		10.25	0.30	0.8159

Figure 11.18 Scaled estimates graphical output incorporating the P-value (prob > [t]). The larger the horizontal bars, the greater the effect of a particular factor or interaction

Although the basis of whether or not a particular experimental factor has an effect is based on F value and the P value, there are some useful graphical outputs, used in JMP, that allow an 'at a glance' assessment, that is, the scaled estimates, shown below in Figure 11.18. Here we can see that pH seems to have the greatest effect, while the interaction between the pH and catalysts is also directionally significant, if not statistically at the 5% level.

How can we improve these tentative conclusions from this screening process? Experimentation is an iterative process, and one outstanding feature of the fractional factorial design is the ease with which we can add additional runs in a structured but flexible way. For example, we noted above that we could not determine the significance or otherwise of effects in the model, due to its having only one degree of freedom for error. One way around this would be to replicate the whole experiment. However, we can also start by adding additional runs at one of the treatment combinations. Suppose we choose to do two more runs with all three factors at the high level, which results in Table 11.8 below.

For the cost of two extra runs, we have established that the effects we are seeing are highly unlikely to be due to chance, i.e. P value is 0.0127, while the effects that we thought were most important in the previous table do turn out to be statistically significant, i.e. pH and the pH * catalyst interaction.

Adding the two extra runs has now given us yet another opportunity. Recall that with the initial eight runs we had problems because our estimate of error is based on only one degree of freedom. This was not strictly true. That left over one degree of freedom actually represented the one term we hadn't asked JMP to fit, namely the three-way interaction between pH*catalyst*buffer. What happens here is that when there is no 'pure error' in

Table 11.8 *2³ full factorial design effects with two way interactions, and resulting analyses showing the benefit of adding two replicates of one of the combinations*

pH	catalyst	buffer	Purity
8	no	type1	-44.81
8	yes	type2	-12
10	no	type2	-134.5
10	yes	type1	-168
8	no	type2	-70.01
8	yes	type1	-40
10	no	type1	-143.3
10	yes	type2	-188
10	yes	type2	-185
10	yes	type2	-183

Analysis of Variance

Source	DF	Sum of Squares	Mean Square	F Ratio
Model	6	41122.297	6853.72	23.6088
Error	3	870.909	290.30	Prob > F
C. Total	9	41993.206		0.0127

Effect Tests

Source	Nparm	DF	Sum of Squares	F Ratio	Prob > F
pH(8,10)	1	1	30571.235	105.3080	0.0020
catalyst	1	1	54.591	0.1880	0.6938
buffer	1	1	23.043	0.0794	0.7965
pH(8,10)*catalyst	1	1	2913.529	10.0362	0.0506
pH(8,10)*buffer	1	1	47.291	0.1629	0.7135
catalyst*buffer	1	1	54.203	0.1867	0.6948

Effect Details

the design (as with that provided by the extra runs just added) the procedure is to treat the interaction terms that weren't put into the model *as if* they were pure error.

Now that the two new runs have been added, we have two degrees of freedom for pure error, so we can now go back to fit the model, add the three-way interaction into the model, and re-run. Table 11.9 shows that the three way interaction is significant after all, with a *P* value of 0.0073. This has the effect of reducing the mean square error, thus increasing the significance of the terms already known about, and revealing that the catalyst*buffer interaction may be important too.

This iterative approach is very typical of experimental design. This example demonstrates the need for proper experimental design, while showing that with some forethought

Table 11.9   *A 2³ Full factorial design effects having two-way and three-way interactions, with resulting analyses showing that pH alone has a significant effect, with both pH * catalyst and pH * catalyst * buffer interaction also being of importance*

Source	Nparm	DF	Sum of Squares	F Ratio	Prob > F
pH(8,10)	1	1	29398.210	4641.823	0.0002
catalyst	1	1	22.040	3.4800	0.2031
buffer	1	1	4.482	0.7078	0.4887
pH(8,10)*catalyst	1	1	2610.224	412.1407	0.0024
pH(8,10)*buffer	1	1	17.515	2.7656	0.2382
catalyst*buffer	1	1	99.901	15.7738	0.0579
pH(8,10)*catalyst*buffer	1	1	858.242	135.5120	0.0073

and understanding how results can be built upon to form an accurate statistical picture of the experimental results.

11.5.4   Blocking

Sometimes there will be physical limitations that prevent the performing of a whole experiment with all other influences held constant, e.g. limited size of trial raw material. In these cases we have to resort to carrying out the experimental runs in 'blocks', so that we keep a suitable structure within the blocks. In a sense, the block becomes an additional 'nuisance factor' and if its only possible to do, say, four runs under the same conditions, then we say that we have block size of four. Again, it is possible to allocate experimental 'runs' to blocks in such a way that a limited amount of information is lost. Using the same 'purity' example we will consider blocking, in this case we are blocking by assay occasion – as all tests were not performed at the same time under the same conditions. (This is a also a good way to consider operator error). Table 11.10 below shows how JMP has generated the design, with the corresponding results for the four assay occasions.

11.6   Conclusion

In this chapter we have sought to highlight and exemplify the benefits of statistical analyses to understand, and plan for, variability in the fermentation process. Using control charts, it is possible to monitor and improve any aspect of the process, allowing one to focus resources on areas that are showing highest variability. In addition, we have described how to examine the relationships between datasets using *t*-tests, correlation and regression. By beginning with the end in mind, i.e. low variability, we can assess this at the start of the experimental process using experimental design. Our aim has been to provide a taster of the benefits of applying basic statistical tools to understand and control fermentation processes. Armed with this information and additional reading list, it is now up to the individual to assess which tools are best for the job at hand, and select a suitable software package, which will enable these analyses to be performed to greatest effect.

Table 11.10 *2^3 full factorial design effects with two way interactions factoring in assay occasion as distinct 'blocks'. By factoring in assay occasion the analysis shows that the main effects are still significant. However, the benefits of blocking are clear in that assay occasion has been identified as statistically significant at the P = 0.0079 level, which would warrant further investigation as part of this experimental scheme*

	pH	catalyst	buffer	assay.occ	purity
1	10	yes	type2	1	56.15
2	8	no	type2	1	62.94
3	8	yes	type1	1	116.11
4	10	no	type1	1	71.8
5	10	no	type2	2	48.99
6	10	yes	type1	2	97.79
7	8	no	type1	2	102.72
8	8	yes	type2	2	96.7
9	10	yes	type2	3	65.03
10	10	no	type1	3	76.33
11	8	yes	type1	3	121.8
12	8	no	type2	3	76.08
13	10	no	type2	4	50.07
14	8	no	type1	4	104.58
15	10	yes	type1	4	93.56
16	8	yes	type2	4	91.12

▼ Effect Tests

Source	Nparm	DF	Sum of Squares	F Ratio	Prob > F
pH(8,10)	1	1	2817.7518	391.2970	<.0001
catalyst	1	1	1309.5352	181.8532	<.0001
buffer	1	1	3528.6570	490.0194	<.0001
assay.occ	3	3	232.3083	10.7534	0.0079
pH(8,10)*catalyst	1	1	12.3728	1.7182	0.2379
pH(8,10)*buffer	1	1	0.0473	0.0066	0.9380
catalyst*buffer	1	1	0.5293	0.0735	0.7954

References and Further Reading

Wheeler, D. (1993) *Understanding Variation: The Key to Managing Chaos*. SPC Press, Knoxville, TN.

Wheeler, D.J. and Chambers, D.S. (1992) *Understanding Statistical Process Control*. SPC Press, Knoxville, TN.

Box, G.E.P., Hunter, J.S., and Hunter, W.G. (2005) *Statistics for Experimenters*, second edition. John Wiley & Sons, Inc., New York.

Clarke, G.M. and Cooke, D. (2004) *A Basic Course in Statistics*, fifth edition. Edward Arnold, London.

Sall, J., Creighton, L., and Lehmann, A. (2001) *JMP Start Statistics*. SAS Institute, Blemont, CA.

12

The Fermenter in Research and Development

Ger T. Fleming and John W. Patching

Symbols Used in This Chapter

μ: specific growth rate (units: time^{-1})

μ_{max}: maximum (nutrient sufficient) growth rate (units: time^{-1})

μ_{dif} : growth rate difference between mutant and nonmutant (units: time^{-1})

D: dilution rate (units: time^{-1})

X: biomass concentration (units: biomass.volume^{-1})

s: nutrient concentration (units: substrate.volume^{-1})

S_0: initial (influent) nutrient concentration (units: substrate.volume^{-1})

k_s: saturation constant (units: substrate.volume^{-1})

Y: growth yield

q: specific rate of substrate utilisation (units: substrate.biomass^{-1}.time^{-1})

n: total number of organisms in culture

m: total number of mutants in culture

r: mutation rate (units: number of mutants arising in the culture. time^{-1})

The study of the growth of bacterial cultures does not constitute a specialised subject or a branch of research: it is the basic method of microbiology. J. Monod (1949) quoted in Kovarova-Kovar and Egli (1998)

Practical Fermentation Technology Edited by Brian McNeil and Linda M. Harvey
© 2008 John Wiley & Sons, Ltd

12.1   Introduction

Culturing microorganisms is an essential part of experimental microbiology, either as an integral part of experiments or as a source of biomass for subsequent experiment and analysis. In this chapter we initially describe the use of fermenters as experimental tools in the laboratory. The advantages and disadvantages of batch and continuous systems are discussed together with practical information on their design and operation. Sterility, asepsis and containment are dealt with as separate topics, since they appear to be the cause of most of the problems experienced by those unfamiliar with laboratory fermenters, especially when they are used in the continuous culture mode. Finally, we discuss the theory and practice of using laboratory fermenters to obtain microbial strains with useful characteristics, either by enrichment culture or by the processes of mutation and selection, which together may be referred to as evolution.

12.2   Batch Cultures

Herbert has stated that 'it is virtually meaningless to speak of the chemical composition of a microorganism without at the same time specifying the environmental conditions producing it' (Herbert 1961). This statement is equally relevant to phenotype in general, yet it is clear that many researchers pay insufficient attention to growing their cultures under controlled and reproducible conditions. 'Overnight batch cultures' of an organism, unless further defined, will result in wide variations in biomass, chemical composition and physiology.

Batch culture is, fundamentally, a poor experimental tool. Constant conditions cannot be maintained since biomass, substrate and product concentrations must change with time. It is insufficient to rely on a comprehensive description of the environment at any point in a batch culture to define the conditions influencing phenotype, since this phenotype will also be influenced by earlier and different conditions (an effect referred to as hysteresis). The researcher also has little effective control over much of the process. Growth at a constant cell-specific rate is only possible in the exponential phase of culture, when conditions are nutrient sufficient and growth is not inhibited by the accumulation of metabolites. This growth rate (μ_{max}) can only be changed by altering factors such as the type of substrate in the medium or the temperature. Such factors will also have other direct effects on the cell's phenotype. Growth rates less than μ_{max} will be found in the deceleration phase of growth as the result of nutrient limitation or product build-up, but this is a dynamic phase when relationships (between growth rate and the concentration of limiting nutrient, for example) are difficult to determine against a background of changing biomass, substrate and metabolite concentration and growth rate. Batch cultures can only be operated effectively when the medium contains nutrients at levels above those that would limit growth rate. If such nutrient-sufficient conditions are not present at the beginning of the culture, the dynamic phases of lag (if present) and acceleration will immediately be followed by the equally dynamic phase of deceleration. Such cultures will be of short duration and will only yield a small amount of cells for further study. It is generally impractical to use batch cultures to study microbial growth on substrates that can prove toxic at higher concentrations (e.g., phenol). Here, growth limiting concentrations may be directly adjacent to those concentrations that inhibit growth, making a nutrient sufficient state impossible to achieve. Fed-batch cultures may be used to extend the nutrient-limited deceleration phase or to work with poisonous substrates. Conditions can be arranged so that cell concentration remains constant during the feeding phase, but observations must still be made against a background of changing growth rate and substrate level in the culture.

In spite of the forgoing it must be conceded that most experimental microbiologists employ batch (or sequential batch cultures) rather than continuous cultures in their studies because of the perception (not always justified) that these are easier to set up and run than are continuous systems. The minimum acceptable standard for batch cultures intended to produce material for phenotype studies would seem to be shake flasks in a controlled temperature environment. Growth on the surface of plates and in stationary liquid cultures cannot be recommended because of their high degree of spatial variability. Shake-flask cultures provide the operator with some degree of control over factors such as temperature, aeration/agitation and pH (by the use of well-buffered media). For a greater level of control and monitoring, a laboratory fermenter should be used.

Laboratory fermenters with vessels of <5 litre working volume have several advantages some of which are discussed in Chapter 2. It is worthwhile reviewing some of these in this context. Such vessels can be sterilised in an autoclave and can therefore be constructed of borosilicate glass with single or double stainless steel end plates, enabling easy viewing of the contents (See Figure 2.1 (a) in Chapter 2). They are also more economical on medium usage. Using small fermenters (<1 litre working volume) will, however, limit the size and number of samples that can be taken for process monitoring, Working with a low culture mass can also lead to problems with the accurate control of process parameters, especially temperature. Research fermenters of 10 litre and greater working volume will usually require *in situ* steam sterilisation (for reasons discussed in Chapter 2). These will need to withstand the pressure differential of sterilisation and will thus be constructed mainly of stainless steel. They are relatively expensive to purchase and run, though they may be used when larger quantities of biomass or product are required for further research.

At its most basic, a fermenter for research will provide an environment with constant agitation, aeration and temperature, with monitoring by RPM gauge, rotameter and thermometer (electrical or mechanical). Constant temperatures in small fermenters are often maintained by the circulation of water through coils or jackets or, more simply, by placing the fermenters in constant temperature rooms or water baths. Temperature, pH, oxygen partial pressure and foam are the parameters most often measured and controlled in laboratory fermenters (Table 12.1). It should be noted that the control of oxygen partial pressure by adjusting the agitator speed will also alter the mixing of the culture. To avoid this, control may be accomplished by bleeding nitrogen or oxygen into the air input by means of electrically controlled valves. Mechanical antifoam systems such as spinning discs or cones immediately above the culture surface are only practicable in larger laboratory fermenters. Beyond these basics, an increasing complexity of measurement and control is available including substrate- and product-specific electrodes, though these sometimes lack the reliability and robustness required for routine use. In most well instrumented laboratory fermenters intended for physiological studies, a computer will almost inevitably form an essential component of measurement and control systems, facilitating

Table 12.1   *Measurement and control systems most commonly provided on laboratory fermenters*

Parameter	Measurement	Control
Temperature	Platinum resistance thermometer or thermistor	Water jacket or coils, electrical heating element
pH	Autoclavable pH electrode	Acid, alkali or nutrient addition
Oxygen partial pressure	Polarographic oxygen electrode	Aeration rate, agitation speed, changing composition of gas input
Foam	Conductance or capacitance probe	Antifoam addition, mechanical systems

the logging and manipulation of process data and the use of complex control algorithms. Pumps will be needed for the addition of pH and antifoam control agents, and for nutrient or medium addition for fed-batch and continuous cultures. Because of the small size of laboratory fermenters, low and precise pumping rates may be needed. Though 'aseptic' piston-based precision metering pumps are available, they suffer from problems of sticking valves and contamination problems. Peristaltic pumps are readily available and are recommended. If the pumping rate required falls below the minimum reliable pump rotation speed or the diameter of the tubing available, a simple electrical timer can be used to operate the pump on an on/off cycle. We have found that, in the case of nutrient addition, if this cycle is of duration significantly less than the culture generation time there should be no significant difference in culture behaviour to that found with continuous addition. We routinely use an on/off cycle of a duration less than or equal to one-third of the generation time.

The size and nature of the inoculum is a critical factor in determining the course of a batch fermentation. Its physiological state will influence the initial stages of the culture. An inoculum of inactive cells (from a stationary-phase culture, for example) will require a period of adjustment before growth can commence. The existence or length of the lag phase of growth may depend on the presence or absence of inducers in the medium used for the growth of the inoculum. As with large-scale fermentations, the use of an active, adjusted inoculum will minimise the time the culture spends at a low density and activity, when it is most susceptible to the effects of contamination. Unless the inoculum consists of washed cells, carry-over of nutrients or waste products from the inoculum culture will also have an influence. When examining the effects of growth factors, etc., the use of thoroughly washed cells is essential. Even then, it may take several serial batch cultures for intracellular pools of growth factors to reach limiting levels.

For reproducible results, inocula must be standardised with respect to the quantity of cells, their physiological state and the nature and quantity of the substrate in which they are suspended. Low carbon dioxide partial pressures are believed to inhibit the growth of inocula in aerobic cultures. The source of this carbon dioxide is the inoculum itself. It is therefore advisable to delay culture aeration until the end of the lag phase.

Because of possible problems in achieving a constant inoculum and operating conditions in a laboratory fermenter, cells cannot be said to be standardised if they are harvested on the basis of time since inoculation. A '24h' culture could be in the exponential or stationary phase of growth. It is better to standardise on cells from a specific phase of growth and use regular optical density measurements during culture to determine the correct time for harvesting.

12.3   Continuous Cultures

As discussed in Chapter 4, continuous cultures have many advantages for the experimental microbiologist. Chemostats are the type of continuous culture typically used, though there are other types, some of which will be mentioned subsequently, which are useful for specific studies. The reader is referred to Chapter 5 for practical details, and to Chapter 7 for a full description of the theory of chemostat culture.

Fundamentally, a chemostat consists of a perfectly mixed (in theory) culture, which is fed at a constant rate with sterile medium and from which culture is removed, or overflows, at the same rate. A critical factor in controlling the behaviour of a chemostat is the dilution rate (D). D is defined as the rate of medium addition divided by the culture volume. It has units of time^{-1}, as does the instantaneous growth rate constant (μ: the cell-specific rate of growth). If a chemostat is operating at a steady state the rate that cells are produced is balanced by the rate that they are lost from the culture through the overflow and the density of the culture remains constant:

$$\mu = D \tag{12.1}$$

Clearly there is a limit to the value of D (which is under the operator's control) if a steady state is to be achieved. This value is known as D_{crit}. Unless nutrient levels in the unused medium are low enough to inhibit growth, D_{crit} will equal μ_{max}, the growth rate under nutrient sufficient conditions. This growth rate would be observed in the exponential phase of a batch culture. Attempting to use values of $D > D_{crit}$ will result in the culture washing out of the system. If the dilution rate of an actively growing culture is set at a value $<D_{crit}$, the culture will adjust itself to achieve a steady state, which is typically self regulating. When in steady state, growth will be under nutrient limitation. There are several equations available that describe the link between nutrient concentration and growth rate. In practice, the Monod equation is generally used when working with chemostats:

$$\mu = \mu_{max}\left(\frac{s}{k_s + s}\right) \tag{12.2}$$

where μ is the growth rate, s is the nutrient concentration and k_s is the nutrient concentration at which the growth rate is half that achieved under nutrient-sufficient conditions (μ_{max}). In a steady-state chemostat, μ may be replaced by D in this equation. Media are generally formulated so that all nutrients are present at concentrations much greater than their respective k_s values. Nutrient levels in an equilibrated chemostat culture will be much less than those in the unused medium. One of the nutrients will be present at the correct level to provide a growth rate equivalent to D. This is known as the limiting nutrient and all other nutrients in the culture will be at relatively higher levels. The limiting nutrient may be defined as the nutrient whose decreasing concentration first limits growth

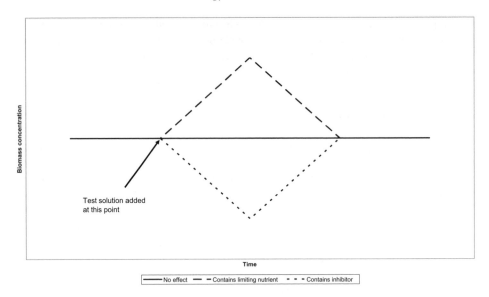

Figure 12.1 *Identifying the limiting nutrient in a chemostat by the method of Goldberg and Er-el (1981). If a solution containing a limiting nutrient is added, the release of limitation will cause a temporary increase in biomass concentration (see text for more details). Reprinted from Process Biochemistry 16, Goldberg and Er-el. The chemostat – an efficient technique for medium optimization, pp. 2–7, Copyright Elsevier 1981*

in a batch culture and may have a major effect on the cell's composition and physiology. Because of this the nature of the limitation in an experimental chemostat culture is generally stated: cultures are said to be 'carbon-limited', 'nitrogen limited', etc. In order to achieve a specific limitation, media may be formulated with a low concentration of one nutrient (typically one-tenth of the concentration used in a batch culture medium). It is advisable, however, to check directly which nutrient is limiting in a steady-state culture whenever possible. The method devised by Goldberg and Er-el (Goldberg and Er-el 1981) is recommended (Figure 12.1). Solutions of individual medium ingredients or groups of ingredients are introduced directly into the steady-state culture. If the addition contains the limiting nutrient, the limitation is temporarily removed and the culture density increases and then drops back to the original value as the nutrient addition is washed out and equilibrium is re-established. This method may also be used to identify medium ingredients with the potential of limiting growth by inhibition. As well as its specific use in defining the limiting nutrient in chemostat cultures, this method has a wider application in the formulation of media for industrial batch fermentations. By identifying the nutrient that will in effect 'run out' first, the medium may be 'balanced' so that no substrate is present in more than minimal amounts at harvest.

When a chemostat is at a steady state the concentration of a substrate in the influent medium (S_0) will be equivalent to the concentration in the culture (s), plus a concentration equivalent to that which has been used by the cells. The latter is derived from the biomass concentration (X) by means of a yield coefficient (Y), which is the ratio of substrate used to biomass produced. Thus:

$$S_0 = s + X/Y \tag{12.3}$$

There are two major advantages in using chemostats for physiological studies. Firstly they are usually intended to be operated at a steady state. Unlike batch cultures, substrate, biomass and metabolite levels are not constantly changing with time and growth at a steady rate is not restricted to a limited period, such as an exponential growth phase. It may be argued that the forgoing statements are untrue, since mutation and selection (to be discussed subsequently) are continuously occurring within the culture population. As a result, biomass and substrate levels will change with time, but the rate and magnitude of change will be far less than that of batch cultures. In practice these events do not generally hinder the use of chemostats for physiological studies, provided that cultures are not maintained for periods greater than 3 to 6 days.

The second advantage of using chemostats is the facility to manipulate culture-related variables independently of each other. Whilst the relationship between dilution (growth) rate and the concentration of the limiting nutrient within a steady state culture will be fixed and defined by the Monod equation (Equation 12.2), the operator has independent control over the levels of non-limiting nutrients as well as other non-nutrient substances (inducers, inhibitors, etc.) though care must be taken to avoid concentrations that may cause them to take over the growth regulating function of the limiting nutrient. Levels of both metabolisable and non-metabolisable medium ingredients will be maintained at a steady state in equilibrated cultures. By changing the level of the limiting nutrient in the influent medium (S_0: see Equation 12.3), low density and high density cultures may be studied under otherwise identical conditions, providing the function of limitation is not passed on to some other medium ingredient.

Chemostats may be used to dissect the direct effects of physical factors, such as temperature, on cellular structure and function from those caused indirectly via their effect on growth rate. Such studies have yielded results that appear to be at variance with those obtained in batch cultures. For example, exponential phase cells from batch cultures grown at different temperatures may have the same ribosomal content, whereas the ribosomal content of cells from chemostat cultures (at constant D) may be higher at lower temperatures. These anomalies may be resolved by considering the relative growth rate (μ/μ_{max}) rather than the absolute growth rate of the cultures. Though temperature will influence the exponential growth rate of batch cultures, the relative growth rate will remain constant (and equal to 1). Temperature will influence μ_{max} and thus relative growth rates in chemostat cultures at constant μ (=D). Ribosomal content is thus proportional to relative growth rate in both batch and continuous cultures. These findings are summarised in Table 12.2. The technique of using relative growth rate (and other dimensionless growth related variables) was developed by the MRE Porton group in the 1970s (Herbert 1976; Tempest 1976) and its use is recommended when analysing the results of chemostat-based studies on cell physiology. For example, changing the type of limiting nutrient in a carbon-limited chemostat operating at constant D may cause effects related to a change in μ_{max}.

The ability to control growth rate by manipulating D is a major advantage for the users of chemostats. Monod was only able to develop his formula relating substrate concentration and growth rate (Equation 12.2) by the technically difficult method of studying transient states in the deceleration phase of batch cultures. By obtaining values for s (the

Table 12.2 *The effect of temperature on ribosomal content and growth rates: – no effect: + varies with temperature (for more details, see text. Derived from Tempest, 1976: The concept of 'relative growth rate'; its theoretical basis and practical application. In* Continuous Culture 6; Applications and New Fields *[eds A. Dean, D. Ellwood, C. Evans and J. Melling], pp. 349–352. Ellis Horwood, Chichester, UK)*

	Batch culture (exponential phase)	Chemostat culture (constant D)
Ribosome content	–	+
μ	+	–
μ_{max}	+	+
μ/μ_{max}	–	+

concentration of the limiting nutrient in the culture) at different values of D it is possible to derive values for k_s and to examine alternatives to the Monod relationship. The measurement of s can present problems. The limiting nutrient will be present at a very low concentration and will be used rapidly by the cells in a sample after it has been taken. Practical solutions to this problem are discussed later. Buttons (1969) has also suggested an alternative approach. By operating chemostats at different D values and with different concentrations of limiting nutrient in the influent medium (S_0), a series of graphs can be obtained, each one of which shows the relationship between S_0 and X (the concentration of biomass in the culture). By extending the graph, the value of S_0 when X is zero may be derived (Figure 12.2). This value will be equal to s (see Equation 12.3). Unfortunately we have found that this approach does not always work in practice. One problem may be variation in yield coefficient, which can occur with growth rate.

The variation of yield coefficient with growth rate, which can only be studied by means of chemostats, has yielded vital information on the influence of growth rate and its limitation on cell composition and function. For example, fast growing cells will contain a higher proportion of ribosomes, and slow growing cells may incorporate any excess carbon compounds in the medium as storage material. Biomass yield with respect to the limiting nutrient may thus increase with decreasing growth rate in nitrogen-, phosphorus- and (especially) magnesium-limited cultures.

At low growth rates, there may be a strong correlation between the growth rate of heterotrophs and biomass yield with respect to the carbon and energy source. Because of this, carbon-limited continuous cultures may be unstable at low D values. Though changes occur slowly, the biomass may oscillate and the culture may ultimately wash out. Cells require energy for processes other than those directly involved in the production of biomass. This energy, referred to as maintenance energy, will become a significant part of the energy budget of slow growing cells and can cause these effects. Pirt (Pirt, 1982) has shown how continuous cultures are uniquely suited to the quantification and study of maintenance energy requirements. In the absence of a maintenance energy requirement, the rate of energy source utilisation (q) will be directly proportional to the growth rate (μ) with a zero rate of growth corresponding to a zero rate of energy source utilisation (line a, Figure 12.3). If the cell has a maintenance energy requirement that does not vary with growth rate, the relationship shown by line a will be offset (Figure 12.3, line b). The

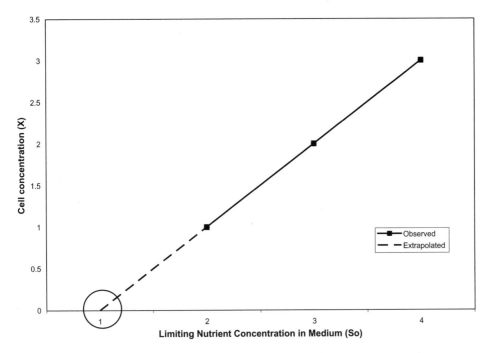

Figure 12.2 *Obtaining the limiting nutrient concentration (s) in a chemostat culture running at a fixed dilution rate (D) by the method of Buttons (1969). Steady state biomass concentrations (X) are plotted against different concentrations of limiting nutrient in the unused medium (S₀). By extrapolating the graph, the point where S₀ = s (circled) may be found. From J. Gen. Microbiol. (1969) 58, pp. 15–21. (Reproduced by permission of the Society for General Microbiology)*

offset shows this maintenance energy requirement as the intercept with the q axis when $\mu = 0$. Pirt has also identified a component of maintenance energy requirement that is inversely proportional to growth rate. In this case the relationship between q and μ would be that shown as line c in Figure 12.3. Using data obtained from steady state chemostat cultures of *Klebsiella aerogenes*, Pirt was able to plot the cell specific rate of oxygen uptake (effectively the rate of energy utilisation: q) against D and show that energy source (glucose) limited cultures only showed a fixed maintenance energy requirement, whereas cultures under other limitations (ammonia, sulfate or phosphate) showed both growth-rate-independent and growth-rate-dependent maintenance energy requirements.

Chemostats are uniquely suited for studies on the effects of nutrient limitation on cell composition and physiology. When grown under a specific nutrient limitation, cells will maximise the efficiency of their uptake and assimilation of the nutrient whilst controlling the uptake rates of the nonlimiting nutrients, so as to avoid toxic build-ups of intermediate metabolites. Chemostat studies have shown the existence of dual high and low affinity pathways for handling some nutrients. As an example, *Klebsiella aerogenes* uses a low affinity system for nitrogen assimilation under nitrogen-sufficient conditions, but employs

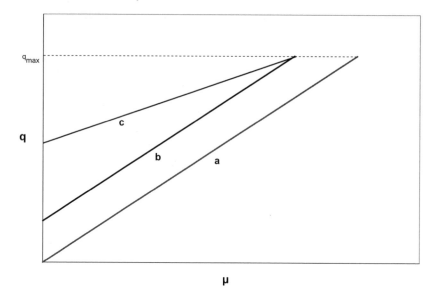

Figure 12.3 *Effects of maintenance energy requirement on the relationship between the specific rate of energy source utilisation (q) and the specific growth rate (μ): (a) no maintenance energy requirement; (b) constant maintenance energy requirement; (c) constant and growth-rate-dependent maintenance energy requirements. Based on Figure 1,* Arch. Microbiol *(1982)* **133**, *pp. 300–302*

a high efficiency assimilation system under ammonia limitation. As with other dual pathways, the high affinity system requires more energy for its operation.

Different nutrients will have specific effects on cell composition and physiology as they adjust to achieve the most effective use of the limiting substrate. As an example, the magnesium content of cells grown under magnesium limitation is proportional to their growth rate and is closely related to their RNA (ribosomal) content. The magnesium content of the cell envelope may be reduced and cells may be more resistant to lysis by EDTA or polymixin. Magnesium limitation may also result in increased respiration with a corresponding decrease in biomass yield with respect to oxygen.

In theory, all that is needed to change from batch culture to chemostat culture is to supply a fermenter with a medium feed and overflow. In practice, experimental fermenters used for continuous culture tend to be smaller (often 500 mL to 1 litre) than those used for batch cultures in order to avoid the expense of large quantities of medium. Reservoirs will also be needed for the sterile medium and to receive the culture overflow. These usually consist of glass or rigid polypropylene vessels, which must be vented by means of reliable in-line sterilising air filters. Unless a study calls for changes in medium composition or concentration, reservoirs should be a large as practicable to reduce the frequency of changeovers, with the attached risks of contamination. In our experience, 20 litres is the largest size of reservoir that may be autoclave sterilised and handled safely in the laboratory. They are enclosed in metal baskets and processed using a vertical autoclave and a chain hoist. Because the cultures are maintained at a steady state, control

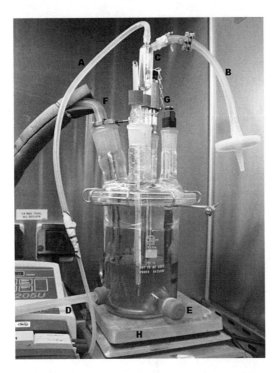

Figure 12.4   *A small experimental chemostat (500-mL working volume). A: medium feed (from peristaltic pump); B: air supply via on-line filter (to be collected to supply); C: medium break; D: effluent line carrying air exhaust and culture from overflow; E: port for sampling by means of hypodermic needle and syringe; F: water supply and return for constant temperature coil; G: pH and oxygen electrodes (not connected); H: magnetic stirrer bed*

of process parameters becomes easier, though there is the added parameter of flow through the system to monitor. This is most easily done by replacing the waste reservoir with a cotton-wool-plugged measuring cylinder for a fixed time. Alternatively, the body of a burette (plugged with a sterilising air filter at the top) may be attached to the medium feed upstream of the pump by a T junction. Tubing clips can then be used to run the medium feed from the burette for a fixed time instead of from the reservoir. Figure 12.4 shows a simple type of experimental chemostat used by us. It consists of a modified all-glass reaction vessel, which is constantly aerated and provided with constant agitation by a magnetic stirrer bar. Its temperature is maintained by water at a constant temperature circulated through a jacket or an internal metal coil. We find such fermenters adequate for measuring basic growth parameters and carrying out competition and selection experiments. They may be considered as the continuous culture equivalent of the shake flask.

The need to feed medium and to allow the overflow or removal of culture are fundamental requirements of continuous cultures. In slow growing cultures it is possible for grow-back to occur from the culture to the medium reservoir or from a contaminated waste reservoir into the culture. Typically the paths of medium/culture and the aeration supply and exhaust are combined to minimise grow-back. Figure 12.5 shows a medium

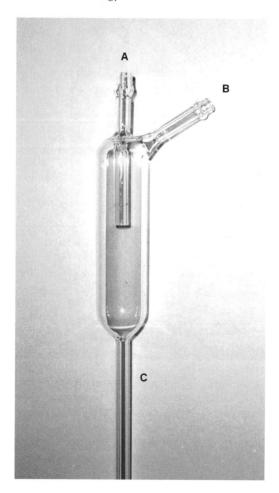

Figure 12.5 *A medium break for continuous cultures. A: medium feed; B: air feed;
C air/medium feed to culture vessel (through head plate)*

break where medium and air supply are combined before entering the culture vessel. The
medium supply breaks up into drops during its passage though the medium break, thus
providing a barrier to grow-back. By preventing or restricting direct exhaust from the
culture headspace, air can be forced through the overflow together with culture, minimising the possibility of grow-back and also helping to maintain a constant culture volume.
Figure 12.6 shows some alternative forms of culture overflow. We routinely use an internal weir (Type A) that exits the vessel through a rubber seal, thus allowing the use of
overflows of different lengths to change the working volume of the vessel. It is important
to measure the working volume of small chemostats accurately. To do this, the culture
vessel is weighed empty, and then again after the vessel has been filled with water by
the medium feed pump until it issues from the overflow. It is important to aerate and
agitate the vessel in the same way that is to be employed during culture whilst doing this.

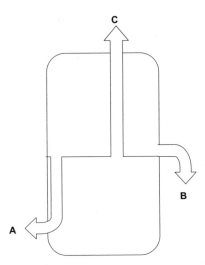

Figure 12.6 *Diagram showing alternative forms of chemostat overflow. A: internal weir; B: side arm C: through head plate*

Figure 12.6 also shows two other types of culture overflow that we have used. The side arm (Type B), though apparently simple is not recommended. Adjusting the culture volume is impossible with such overflows and we have found that they have a tendency to introduce fluctuations in working volume during operation. It is also worth noting that incorporating glass overflows directly into the fabric of glass vessels can cause stresses that result in vessels cracking during autoclaving. Others have actively pumped out culture when using overflows that exit through the head plate (Type C), but we have found that if the headspace is sealed (apart from the overflow), the exhausting air will maintain a constant working volume without the need for a pump on the effluent line.

The inoculum for a chemostat must be active enough to permit the successful initiation of the culture but apart from this (and excluding enrichment cultures), its properties will have a minimal influence on the culture once it is established in steady state. Because of the effects of competition and selection, it is of vital importance, however, to ensure that it is free from contamination and, for pure culture studies, to ensure that it is monoclonal. If the organism under study can grow as a batch culture in the medium to be used, we use an overnight batch culture in the same medium as an inoculum. This is added (as a 5% inoculum) to a chemostat vessel that has been filled to 50% of its working volume with medium, and allowed to grow as a batch culture overnight. The culture vessel is then topped up to its full working volume with medium and allowed to continue as a batch culture for a short period during which the culture's optical density is monitored to ensure that active growth is occurring. These measurements may also be used to determine μ_{max} for the culture. Medium supply may then be established so as to give the required dilution rate, and the culture allowed to equilibrate. Some organisms are unable to grow as batch cultures in the influent medium, either because it contains a substrate that is present at an inhibitory level, or because the organism is an oligotroph. Under such circumstances, an inoculum similar in size to the culture working volume must be used and medium supply commenced immediately without attempting an initial batch culture. Regular measurements of culture density can be used to monitor the process of equilibration. If a culture shows signs of washing out, it may be necessary to decrease the dilution rate for a while. If the medium contains an inhibitor, it may also be necessary to decrease its concentration temporarily.

When starting a culture or after a change in operating parameters, it is generally assumed that a steady state will have been achieved after the passage of three working volumes of medium through the system. Thus a 1 litre chemostat operating at a dilution rate of $0.05\,h^{-1}$ would require at least 60 hours to reach a steady state. Steady state is usually confirmed by a constant OD (optical density) reading for three generation times. Changes during equilibration are slow at low dilution rates. It is best to confirm a constant OD for five generations in such cultures before declaring that they are at a steady state. Simple theory would indicate that chemostats may be run indefinitely, but in practice, mutation and selection will occur in long term cultures, causing them to evolve and change their characteristics. These matters will be discussed in more detail later. Dykhuisen and Hartl (Dykhuizen and Hartl 1983) have suggested that working with chemostat cultures of 100-h duration or less should avoid significant changes. In practice, we have found that cultures growing at low relative growth rates (e.g., $0.05\,\mu_{max}$) will not change significantly in 150 h, whereas those growing near the maximum practical value of D $(0.5\,\mu_{max})$ may show changes in 75 h or less. With mutator or recombinant-proficient strains, significant changes may occur with 20–30 h.

Samples for monitoring or further study can be taken from the culture vessel. We have found that removal of up to 50% of the working volume may be carried out with minimal disturbance to the equilibrium of the culture. If the medium feed is maintained, the system will operate as a fed-batch culture with similar characteristics to the continuous culture and will be at or near equilibrium once the working volume has been re-established. Considering the low volume of most chemostats, this is fortunate, though operators should check the recovery time of their cultures for themselves. Once a sample is withdrawn, its conditions change from an equilibrium state to those of a batch culture. The limiting nutrient will typically be present at a very low concentration and will thus be used up rapidly, as will any other substrates that are present in low concentrations. In order to avoid this it may be necessary to use a syringe to remove the sample from the vessel rapidly and then immediately separate cells and medium by the use of a Swinney filter, or alternatively add a metabolic inhibitor to the sample. If such changes are not critical for subsequent studies, the contents of the reservoir receiving the overflow may be used. The properties of the culture at the point of overflow are identical to those of the bulk of the culture. These properties will change whilst the sample is collecting in the reservoir, but such changes can be lessened by chilling the reservoir and its contents.

In practice, chemostats are usually run at dilution rates between 0.05 and 0.5 of μ_{max}. At values higher than this, small changes in D, such as those caused by fluctuations in pump rates, etc., will cause large changes in cell density and nutrient concentrations. As a result the system will become unstable and will eventually wash out. Cultures grown at very low dilution rates are slow to reach steady state. If the limiting nutrient is the carbon and energy source, they may also be unstable because of maintenance energy requirements. It is possible to force stability and shorten the time taken to reach equilibrium by the use of a measurement and control loop: The speed of medium addition (and hence *D*) is controlled so as to maintain either constant biomass or growth. Both these variables may be measured directly or indirectly. At steady state, the basic equations describing growth in chemostats will still apply (see Chapter 7), though the role of some variables (such as growth rate and nutrient level) in cause and effect will be reversed. At present the main use of auxostats in the laboratory has been for the study of cultures growing at rates approaching μ_{max}, but they can also be used to achieve steady state operation where the equilibrium is inherently unstable. This is the case where growth is controlled by the inhibiting properties of a medium ingredient which is also used for growth. The first auxostats used biomass concentration, measured by optical methods, to control pump speed and were known as turbidistats. These are not easy to operate because of problems of optical fouling. Various state variables have been used to control medium feed, including pH, rates of acid or carbon dioxide formation and dissolved oxygen. Control by measuring nutrient levels in the culture has been hindered by a lack of reliable sensors with fast reaction times. The use of auxostats for strain selection purposes will be discussed later.

Whilst chemostat/auxostat systems are the major forms of continuous culture employed for physiological studies, continuous cultures with cell recycle or retention have proved useful for studying the behaviour of populations growing at very slow rates under severe nutrient limitation. In practice, systems set up as chemostats, but with a facility to remove cells from the effluent culture and return them to the culture, are the easiest to operate in the laboratory. Tangential flow filtration is the best way to achieve separation of cells from medium. By employing such a system with 100% recycle of cells, it is possible to subject a population to conditions that progress from nutrient sufficiency through chronic starvation to severe chronic starvation with doubling times progressing from <14 h to >100 h.

12.4   Sterility, Asepsis and Containment

Most of those who use fermenters for research and development would regard aseptic operation as essential if defined and reproducible experiments are to be carried out. In practice, low level contamination towards the end of a batch culture will probably have a minimal effect on the experimental outcome, unless the contaminant has a massive growth advantage over the bulk of the culture. Such cultures could not be used as inocula, or for further growth studies, however. In addition, it should be noted that infection, at any stage, with a bacteriophage may completely destroy a culture. Fortunately such occurrences are rare unless there is a significant source of bacteriophage in the laboratory environment: once such contamination occurs it can be difficult to eradicate. Maintaining aseptic operation is a major challenge to those operating continuous cultures. In addition to the requirements of batch cultures, the operator must ensure the continuous aseptic addition of medium and removal of culture over a long period (as compared with a batch culture) in the knowledge that, in theory, contamination by even a single cell of a fitter strain could result in the complete take-over of the culture by that strain.

In essence, aseptic culture involves removing all contaminants from apparatus and materials by sterilisation and then ensuring that none are introduced during inoculation, operation, sampling and (in some cases) harvesting. Because the high costs of running a large culture for production purposes, remedial measures may be applied to minimise or reverse the effects of contamination, once it has occurred. Such measures are rarely employed with small research cultures because of the lower costs involved and the importance of reliable results. Whatever the size or nature of a culture, prevention is far better than cure when dealing with contamination.

An additional requirement for those growing pathogens and some genetically engineered microorganisms (GM organisms) will be the need to avoid their escape into the environment. Effective, validated, systems of containment may be mandatory under such circumstances. Some aspects of containment are mentioned here, such as the use of absolute filters on all air ingress/egress points and magnetic couplings for agitators. Additionally, it may be necessary to place the fermenter in an effective barrier isolator and to operate it with a negative pressure in the culture vessel, though this is only practicable with larger experimental fermenters.

Critical points where contamination of a culture may occur are listed in Table 12.3 Reliable asepsis and containment require a combination of properly designed equipment and effective procedures. With larger fermenters, which are sterilised *in situ*, the problems of sterility and asepsis will be handled in the same way, whether the fermenter is to be used for research or production purposes. Those using such fermenters are therefore referred to the relevant sections of Chapters 2 and 8. The rest of this section will concentrate on small research fermenters (both batch and continuous), which are sterilised by autoclaving.

As well as achieving effective sterilisation of the culture environment and everything that will come into contact with it (including pH control agents/chemical antifoam additions, etc.), care should be taken to avoid the possibility of contamination during set-up (post-sterilisation and pre-inoculation). For this reason, autoclavable fermenters are typically sterilised after filling with medium. Heat-labile medium components will need to be sterilised separately by filtration. Some medium ingredients (for example, reducing sugars and ammonium compounds) may react to form undesirable compounds or precipitates if autoclaved together. This will also require the separate sterilisation of one or more of the offending ingredients. Medium ingredients that have been sterilised separately will need to be added aseptically to the fermenter. The syringe and septum technique (discussed later) is well suited to this purpose, unless dealing with large volumes. Providing it can be handled safely, sterilising fermenters with their pH and antifoam control systems (reservoirs, contents connecting tubing and pumphead or peristaltic tubing) attached reduces the chances of contamination during set-up.

Table 12.3 *Critical areas to be considered to assure the aseptic operation of small experimental fermenters*

- Sterilising the apparatus, medium and any additions that may be made during culture (pH control agents, antifoam, etc.)
- Inoculation, sampling and harvesting (when an aseptic harvest is needed).
- The agitation system
- The supply of sterile air to the culture and its exhaust
- Nutrient supply for fed-batch and continuous cultures, and culture removal for continuous cultures.

Table 12.4   *Typical procedures for setting up a small research fermenter for autoclaving*

Fermenter vessel	Detach agitator drive motor. Fill with medium. Sterilise solutions of thermolabile/incompatible ingredients separately and add aseptically after autoclaving
Air inlet	Assemble to dirty side of in-line filter and close with tubing clamp. Cap end of tube with foil
Air outlet	Leave open. Cap end of tube with foil. If outlet filter is to be used, foil wrap and autoclave separately
pH electrode	Check reference electrolyte level. Leave connection open circuit and protect from moisture
Oxygen electrode	Short out connection and protect from moisture
Other probes	Protect connection from moisture Connection open circuit or short circuit according to manufacturer's instructions
pH/antifoam reservoirs (if attached)	Isolate using tubing clamps. Leave air filters on reservoirs open and foil wrap them
Tubing connections for nutrient feeds, etc.	Close with tubing clamps if there is any possibility of medium loss during autoclaving, otherwise leave open. Cover connectors with foil
Sampling system	Isolate from fermenter vessel with tubing clamp

Table 12.4 lists the typical steps taken when preparing a small fermenter and its ancillaries for autoclaving. It is important to prepare the system so as to avoid the formation of air pockets during autoclaving, as these would result in a failure in sterilisation, and also to avoid the development of pressure differences that could damage electrodes or result in the expulsion of medium from the fermenter.

Steam sterilisation requires the application of moist heat at a specific temperature for a specific time. Established time/temperature combinations (e.g., 121 °C for 15 minutes) may be used or they may be specifically calculated from the estimated bioburden of the system before sterilisation and the sterility assurance level to be achieved. In theory, the oily nature of some antifoams would make any contaminants they contain more resistant to steam sterilisation, but this does not seem to cause problems in practice. Autoclaves are usually allowed to steam with their exhaust valve open until air has been driven from the system. This process is usually referred to as venting, and is absolutely essential to achievement of sterility, as the temperature of an air/steam mixture is lower than that of steam at the same temperature, the density of air is different from that of steam, which may lead to zoning in the autoclave, and hence cool spots. Because of the thermal capacity of a fermenter and its contents, it may be necessary to leave the autoclave longer at this stage to allow the medium to heat up. Ideally, dummy runs, with a temperature probe inserted in the medium (or an equivalent amount of water), should be used to determine the heating and cooling characteristics of the system and a suitable sterilisation cycle calculated using Richard's rapid method (Richards 1968) or some other algorithm that allows for the death of microorganisms during the heating and cooling portions of the sterilisation cycle. No attempt should be made to speed the cooling phase by opening the chamber exhaust valve or other means, as this will cause damage to the equipment and loss of medium. It is important, however, to allow air to enter the chamber once the temperature has dropped to 100 °C to avoid the development of a vacuum that may cause the same kinds of damage as rapid cooling.

Once the fermenter and ancillaries have been sterilised, asepsis must be maintained by avoiding contamination during set-up and operation. As with larger fermenters, it is good practice to confine penetrations through the vessel wall (for electrodes, stirrer shafts, supply and sample tubes, etc.) to the head-space area, where they will not be immersed in medium though, because of the lower hydrostatic pressures encountered, vessel penetrations though side arms or the fermenter base rarely give rise to the problems that may be found with larger systems.

The need to transfer agitation drive from the outside to the inside of the fermenter vessel is a potential problem area. Systems using a double shaft seals with sterile water passed through the intermediate space are available for small laboratory fermenters, but their effective use is limited to industrial laboratories where sterile condensate will be available on-tap from the plant utilities. Magnetic drives (discussed in Chapter 2) provide the assurance of an absolute barrier between the culture and the outside environment. We have used a system consisting of a standard laboratory magnetic stirrer and a plastic coated follower bar to agitate our small chemostat vessels, but such simple systems are limited in the torque that they can transmit. If the speed of agitation is increased too rapidly or to too high a level, the bar will cease to follow the transmitting magnet. Highly efficient magnetic drives are now available commercially. These transmit the drive from the stirrer motor to a conventional agitation system of shaft and impellers within the fermenter. Their use is mandatory when dealing with pathogens or GM organisms.

As with larger systems, the air supply for a laboratory-scale fermenter will need to be sterilised by means of a suitable in-line filter. High efficiency depth filters are preferred, since they give a lower pressure drop and are less susceptible to clogging than are absolute filters, although because of the relatively low volumes of air needed for laboratory cultures it is also feasible to use the latter (preferably equipped with a depth prefilter) if asepsis or containment are critical. In our experience, filters in the exhaust gas line are prone to clogging and often give problems by restricting air flow. With normal aeration rates, asepsis may be maintained without the use of in-line filters in the air exhaust if it is passed through a metre or two of tubing, the end of which may be placed in a container of disinfectant. Such an approach cannot be used when working with pathogens or GM organisms when containment by an absolute filter on the exhaust line is mandatory.

As air bubbles through a culture it will pick up moisture equivalent to the saturated vapor pressure at the temperature of the culture. This pickup of water may be considerable when operating fermenters at temperatures significantly above ambient (when culturing thermophiles, for example), resulting in a decreasing culture volume and increasing concentration. The moisture in the exhaust gas stream will condense as it cools and may clog in-line filters. To avoid these problems, the air exhaust from batch fermenters may be passed through a reflux condenser mounted directly on the head plate so that moisture in the exhaust stream condenses and is returned to the culture. This approach is not feasible in continuous systems, where air and culture leave through the same overflow, but we have not encountered problems with clogging air filters on spent medium reservoirs, possibly because the reservoirs themselves act as effective condensers. Loss of water by evaporation from continuous cultures is only a problem in the minority of cases where the loss is significant in comparison with the rate of medium addition to the culture. The solution in such cases is to bubble the sterile air supply through sterile water at the culture temperature immediately before supplying it to the fermenter.

Several systems may be employed for aseptic additions (inocula, nutrients, antifoam, pH regulating agents, etc.) to small fermenters, depending on the volume to be added and whether the addition is 'once off' or not. Small (<50 mL) 'once off' additions are best added by inoculation through a rubber septum mounted on the head plate of the fermenter, using a sterile disposable syringe and needle (gauge 19 or larger). In view of the cost of large disposable syringes, it is worth knowing that those made of polypropylene may be washed, wrapped in foil and resterilised for further use. Disposable sterile in-line syringe filters (0.2 μm pore size) may be used to assure the sterility of nutrients at the point of addition. The rubber septum should be swabbed with 90% ethanol or (preferably) isopropanol before and after use and kept covered with foil or a cap when not needed. This type of system may also be used for small (<1 L) 'one off' additions from aspirators, using either gravity feed or a peristaltic pump.

Larger volume additions will need to be pumped in from a separate vessel. Suitable vessels may be made from glass or autoclavable plastic (e.g., polypropylene). Borosilicate glass vessels are inert. The use of glass also makes it easier to observe the contents to check that ingredients have dissolved properly, etc. We sometimes place a magnetic follower in our glass vessels to facilitate mixing during batching and after autoclaving. For safety reasons and for ease of handling, glass vessels $\geq$5 L should be enclosed in a metal cage equipped with a carrying handle. To avoid problems from plasticizers leaching from plastic vessels, new vessels should be filled with a 0.5% (w/v) solution of sodium bicarbonate and left overnight. They should then be rinsed, filled with deionised water and given a further overnight soaking before use. The poor thermal conductivity of plastic should be remembered when autoclaving: a filled 10-litre plastic aspirator will need to be steamed for 30 minutes before the sterilisation cycle is commenced. We routinely use silicone rubber tubing for connections between vessels and fermenters because of its ease of autoclaving and inertness. Cable ties or similar restraints should be used to ensure that tubes do not become detached from their connections during autoclaving or handling. Small aspirators intended for antifoam or acid/alkali addition may have a side connection at their base to which the tubing supplying the fermenter may be attached, but it is usually cheaper to use vessels equipped only with a top opening and arrange for the supply tube to reach its bottom. When using flexible tube for this purpose, the end will need to be weighted. Vessels must be equipped with a suitable in-line filter to allow air to replace the contents as they are withdrawn. We routinely wrap the filter in foil when autoclaving to avoid water penetration. Even so, such filters should be checked for clogging immediately after autoclaving. It is then relatively easy to transfer the vessel to a laminar flow hood and replace the filter.

Problems of making aseptic connections may be avoided if the fermenter is autoclaved with the addition vessel attached. The connection must incorporate peristaltic pump tubing or a pumphead and the contents of the vessel isolated from the fermenter by means of a tubing clamp. This approach is usually not feasible, especially in the case of continuous cultures when several changes of the medium supply vessel may be needed during the course of an experiment. Unless medium composition is to be changed, it is best to avoid excessive changeovers during a continuous culture by using the using the largest medium supply vessels that may be handled safely (typically 20 L). The connection between the vessel and the fermenter may be made upstream or downstream of the pump. When using peristaltic pumps for continuous cultures, it is preferable to make the

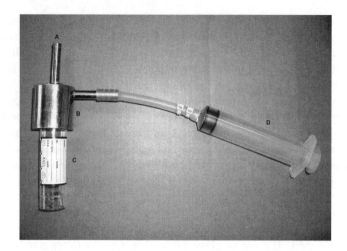

Figure 12.7   *Sampling system suitable for small laboratory fermenters. A: connection to sample line (through head plate); B: metal fitting holding sample container; C: sample container; D: syringe (50 mL)*

connection downstream, so that the pump tube is replaced each time and problems of tube failure are avoided.

Many forms of aseptic connector are available. We have successfully used both plastic luer-lock and commercially made stainless connectors where sealing is achieved by a rubber ring, and the connector parts are locked together by a screwed sleeve. Whatever form of connection is used, it is vital that it incorporates a mechanism to lock the two parts together positively, otherwise connectors may work loose causing leakage and contamination. Effective aseptic technique is essential when connecting or disconnecting vessels from a fermenter. Connectors should be foil wrapped before autoclaving to avoid post-sterilisation contamination. We routinely immerse all connectors in alcohol for 30 minutes before making a connection or change over. Excess alcohol may be shaken or (in the case of stainless fittings) flamed off, though the latter shortens the life of the rubber seals.

Withdrawal of small samples (<50 mL) may be carried out using a sterile syringe and needle as employed for small 'once-off' additions, if the fermenter is equipped with a side port terminated with a rubber septum. If there is no side port, or larger samples are needed, they are withdrawn through a tube that penetrates through the head-plate using a system of the type shown in Figure 12.7. The metal fitting (B) has an internal screw thread that facilitates the attachment of the 30-mL Universal containers (C), which we normally use when sampling from small chemostats. Larger containers may also be used, if suitably threaded. The sample flows into the container through the central arm of the metal fitting, whilst the side arm allows for air to be withdrawn or forced into the head-space of the sample container, using the permanently attached syringe (D). We would typically use a 50 mL polypropylene 'disposable' syringe for this purpose. The system is attached to the sampling tube by silicone rubber tubing. During autoclaving and fermenter operation (unless sampling), this tubing is closed by a clamp, a sample vessel is in place, the syringe plunger is in the withdrawn position and the sampling system is full of air. Before sampling the syringe plunger is depressed so as to force any culture contained in the tube (and thus not representative of the culture), out and into the bulk of the culture. If necessary, the sample vessel which is in place may then be replaced by a more suitable sterile vessel. The flow of sample may be initiated by withdrawing the syringe plunger. This may be sufficient to fill a small sample container. When using larger containers, it will be necessary to loosen the screw thread slightly to allow air to escape and then maintain the sample flow, either by gravity (operating the system as a siphon) or by forcing the sample out of the culture vessel by restricting the fermenter's air exhaust. Once the sample has been obtained, the syringe plunger is depressed sufficiently to stop the flow and force culture in the sample tube back into the fermenter, the sample tube is clamped and the full sample vessel replaced by a sterile vessel.

The overflow from continuous cultures will need to be connected to a suitable effluent reservoir. All the points that have been made in connection with medium supply vessels and their attachment will apply to these reservoirs, The air/effluent culture tube should enter through the top of the vessel and penetrate a few cm below the tube connected to the in-line air filter, to avoid culture entering the filter.

12.5   Mutation, Selection and Enrichment in Fermenters

Enrichment cultures are used to isolate organisms with interesting or desirable character-istics and are widely used in the areas of industrial microbiology and microbial ecology. The process works by arranging conditions so that clones with the desired characteristics are favored and out-compete or outgrow other components of the community and thus become 'enriched' in the culture. Such strains may be defined as 'fitter strains'. The term 'fitter strain' will be used throughout this section to refer to strains that can out-compete other disadvantaged strains under the selection conditions that apply at the time. Enrich-ment cultures require a mixture of clones with varying characteristics. At its most extreme this can consist of a mixture of genera and species in an inoculum from a natural source. Varying characteristics may also be introduced by mutation (natural or induced) in a so called 'pure culture'. Such mutants may be present in the culture inoculum or arise during the culture.

Batch and chemostat culture enrichments are compared in Table 12.5. In batch culture enrichments, the fitter strains will be those which show the fastest nutrient-sufficient growth rate (μ_{max}) under the culture conditions, since nutrient limitation may only be found in the deceleration and stationary phases of such cultures. Disadvantaged strains in batch cultures will grow more slowly, or not at all if they are unable to use the substrates supplied. These less fit strains will not be removed from a single batch culture, though serial batch cultures may be used to dilute them out. Batch culture enrichments are of further use in enriching for consortia that are predominantly enriched in a sequential manner. Selection for fitness is not based in this case on the ability to grow at a specific growth rate (as in the chemostat) but on the ability to utilize substrates, e.g. recalcitrant compounds. Growth rates in these instances may indeed be very slow. If a recalcitrant compound is used as the limiting carbon source in the chemostat, this may eventually result in the enrichment for consortia that can utilize this source.

Table 12.5 *A comparison of the characteristics of batch and chemostat enrichment cultures*

Batch	Chemostat
Fitter strains selected on the basis of μ_{max}	Fitter strains typically selected on the basis of competition for limiting nutrient at a fixed growth rate
Disadvantaged strains not lost from single batch cultures but may be diluted out by using serial batch cultures	Disadvantaged strains constantly lost through overflow
Single batch enrichments of limited duration – length of enrichment process may be extended by using serial batch cultures	No limit (in theory) to the length of a chemostat enrichment culture
Components of consortia enriched sequentially	Enrichment may produce consortia

In chemostats, competition is always based on the ability of competing strains to utilize a specific limiting nutrient. The competitor that is better able to capture and utilize the limiting nutrient at the particular dilution rate will always grow faster. The system will re-equilibrate to a lower concentration of limiting nutrient, and less efficient utilisers will be lost from the system. As a result fitter strains will progressively take over chemostat cultures even if they initially represent a small fraction of the population. The rate of this takeover will be proportional to the difference in the growth rates (μ_{dif}) of the fitter and less-fit clones but it will eventually occur, even if the difference in growth rates is very small. Chemostats are thus highly efficient in selecting fitter strains. It is unfortunate, however that such strains are often the equivalent of garden weeds to those attempting to study or utilize industrial strains that have been developed for efficient product formation, rather than growth.

Mutations will arise in any culture whether 'pure' or mixed, even in the absence of obvious mutagens. Many mutations are lethal or result in clones that are less fit, but occasionally, mutants will occur that have the potential to out-compete the rest of the population and thus influence population structure. The combined processes of mutation and selection, which will inevitably occur in growing cultures, may be referred to as evolution. Unlike evolution in higher organisms, which has a time scale of decades, centuries or even longer, the time scale of microbial evolution in cultures may be measured in days or weeks. Microbial cultures are thus valuable experimental tools for those studying mutation and evolution. Culture evolution, if carried out under suitable selection conditions, can result in the development or improvement of strains used for bioremediation or industrial processes. The principle of directed evolution suggests that microorganisms will adapt to an environmental or stress pressure (such as a metabolic inhibitor) applied to the culture. Evolution is primarily driven by two components. Actively dividing cells will produce mutants at random. A small proportion of mutants will show adaptation to the applied stress pressure. Secondly those that are fitter or better adapted to the stress pressure will out-compete the non-adapted or less adapted clones. Environmental adaptations, including strains producing novel enzymes that can use exotic five-carbon sugars or adaptation to a low phosphate chemostat environment by yeast, have been reported.

We would encourage the reader to read an excellent paper by Zelder and Hauer (2000) on this topic.

Classical strain development has used batch, serial batch or fed batch cultures coupled with mutagenesis and random screening for improved phenotypes. These procedures, whilst time intensive and often cumbersome, are relatively easy to perform and have long yielded strains with greatly enhanced levels of primary and secondary metabolite titres, biomass production and tolerance for inhibitors. They often require extensive screening protocols and are sometimes not reproducible because of changing environmental conditions. Generally, the larger a wild-type population, the greater will be the probability of a mutation event. Mutations in the early stages of a batch culture (particularly with low levels of inocula) are therefore less likely, but they can occur. Such 'jackpot mutations' may result in a relatively large proportion of the population being mutants. It is difficult to differentiate such events from the influence of mutation rates and the competitive advantage or disadvantage of the mutant. In the absence of a 'jackpot mutation' a single batch culture does not provide sufficient time for effective mutation and selection to occur. Serial batch cultures may be used to study growing populations over suitable time scales, but there are also disadvantages to their use for such studies (lack of experimental flexibility and steady state conditions, for example). They are of use when selecting mutants that may be slow growing, even under nutrient sufficient conditions. We have successfully used a semicontinuous fed batch culture technique to enrich for tributyl tin (TBT)-tolerant microorganisms from the marine environment. This type of enrichment would be impracticable using standard chemostat culture since the organisms selected were so slow growing that they would be lost from the chemostat.

Chemostats are ideally suited to study the kinetics of mutation, selection and evolution. Chemostat culture offers the possibility of establishing a highly selective and constant environment. The rate of mutation can be accurately quantified and competition/selection outcomes can be investigated over longer periods of time ($>1000\,h$) than those possible with batch cultures. Dykhuisen and coworkers have used such systems extensively for the study of evolution (Dykhuizen and Hartl 1983; Dykhuizen 1990). The practical power of the chemostat over batch culture as a method for enriching for useful mutants may be considered by the following example. It would require $100\,L$ of cells from a batch culture to produce a single mutant with a specific attribute under normal conditions of mutation frequency and culture density. It should also be pointed out that these mutants must be selected from 10^{14}–10^{15} cells, which is practically impossible from a screening aspect. In contrast, a chemostat containing 5×10^8 cells/mL and running at a dilution rare of $1\,h^{-1}$, would only need to be operated for 20 hours to yield the mutant of interest. At least 2.5×10^8 mutants with the required phenotype would be generated within 1000 hours of fermentation.

One of the first uses of the chemostat was to study the kinetics of mutation in growing populations (Novick and Szilard, 1950). The proportion of a strain carrying a specific mutation in a chemostat population will be governed by three factors: the frequency of the mutation event; the frequency of back mutation, and the difference between the growth rate of the mutant and the wild type. In most studies, the effect of back mutation may be ignored, since it is assumed that forward and back mutation rates are similar and experiments are usually terminated whilst mutants still only represent a small fraction of the total population. This also means that the size of the total population is not significantly different from the size of the wild-type population. Thus, when a mutant has a growth rate disadvantage in the culture, its appearance may be represented as follows:

$$dm/dt = nr - m\mu_{dif} \tag{12.4}$$

where dm/dt is the rate of change of mutant numbers in the culture, r the mutation rate (the proportion of the wild-type population mutating per unit time), μ_{dif} the growth rate difference between the mutant and wild-type, and n and m the sizes of the total and mutant populations respectively. Mutants of this type will establish themselves as a constant proportion of the culture, even if they are unable to grow (Figure 12.8, line C). In a similar fashion, the kinetics of mutant appearance when the mutant has a growth rate advantage are as follows:

$$dm/dt = nr + m\mu_{dif} \tag{12.5}$$

It follows that the number of such mutants will increase exponentially and have the potential to take over the culture (Figure 12.8, line A), though it must be remembered that μ_{dif} will typically be very small and take over a slow process.

If a mutation has no selective advantage or disadvantage in the culture environment, the kinetics simplify to:

$$dm/dt = nr \tag{12.6}$$

Since x should be constant in a chemostat, plotting the numbers of such mutants against time produces a straight line whose slope is proportional to r, the mutation rate (Figure 12.8, line B). It is difficult to prove conclusively that a mutation is truly neutral, but studies on mutation kinetics have employed the change from phage sensitivity to resistance as a change that is easily scored and that is essentially selection neutral in the absence of phage. Studies on the appearance of phage T5 resistance in populations of sensitive *Escherichia coli* in chemostats were able to show the relationship between growth rate and mutation rate that occurs when mutation takes place at the time of gene replication (Novick and Szilard, 1950). Other studies (Kubitschek, 1966) were able to show the lag between the addition of a mutagen and a corresponding increase in the rate of appearance

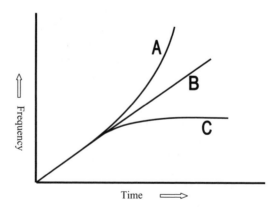

Figure 12.8 *Mutant accumulation in chemostat culture without selection pressure. A: mutant with selective advantage; B: neutral mutant; C: mutant with selective disadvantage*

of expressed mutants (the phenotypic lag), and to analyse the processes contributing to this.

Considering a specific mutation and ignoring all other mutation events that may occur will result in a gross over-simplification of the true situation. In long term cultures, the simple increase in numbers of the neutral mutant (as shown in Figure 12.8, line B) may be interrupted by the appearance of other mutants carrying 'fitness' mutations that will tend to take over the culture. Such mutants are more likely to occur in the clones without the neutral mutation because they will form the overwhelming bulk of the culture. This leads to a fall in the proportion of the population with the neutral marker. Once the fitter clone has taken over, it, in turn will give rise to neutral mutants. A sawtooth pattern in the appearance of the neutral mutation thus ensues (Figure 12.9). This phenomenon is termed 'periodic selection'. Adaptive shifts between competing subpopulations with and without the neutral maker have a number of consequences (Berg, 1995). The frequency of cells with the neutral mutation will be relatively small compared with those without, but when fitter mutants do occur in this sub population, cells containing the neutral mutation will be favoured and it can become dominant in the population through 'hitchhiking' with the fitter clone. This phenomenon, known as 'piggy-backing' will apply to all genes present in the favoured clone. Fitter strains can be generated concurrently from advantaged and less fit clones (which are in the process of being displaced from the chemostat). Competition thus ensues between subpopulations carrying an array of beneficial mutations. This is referred to as 'clonal interference' and theoretically results in a chemostat

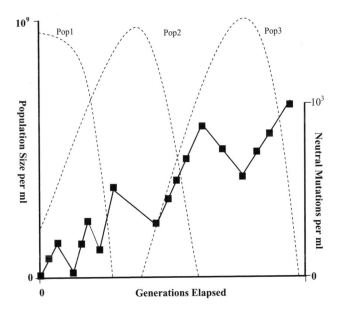

Figure 12.9 *Periodic selection against a neutral mutant in chemostat culture. The solid dark line represents the number of neutral mutations generated in the chemostat. The dashed lines represent the populations (Pop1, Pop2, Pop3, etc.) containing advantageous mutations that sequentially replace each other in the chemostat*

population of larger fitness where adaptations are fixed within the genotype for a significant period of time. In practice, the genetic diversity is increased and maintained in the reactor for longer than is theoretically predicted. The reader is referred to a paper by de Visser and Rozen (2006), which discusses this phenomenon in more detail.

Mutants that evolve later in chemostat cultures are expected to be of higher fitness than those selected earlier in the culture's history. In simple terms, mutant 2 is generated from cell line 1, and displaces it from the culture. Mutant 3 displaces 2, 4 displaces 3 and so forth. The profile of fitness is thus: 4>3>2>1. It might be assumed from this that later clones will always outcompete earlier ones. The mechanisms by which each succeeding takeover occurs may differ however. Thus there is no guarantee that clone 4 will be able to outcompete clone 1 or 2. We isolated mutants from a magnesium-limited chemostat culture *Pseudomonas aeruginosa* grown in magnesium-limited chemostats. Competition studies carried out under the same conditions showed that mutants isolated towards the end of the culture could be displaced by those first isolated.

A brief summary of the types of improvements noted after selection events is shown in Table 12.6. Selection for increased fitness in chemostats is ultimately dependent on the relative growth rates of the parent and mutant strain under the culture conditions. Growth rates in turn are governed by the affinity of the cell for the limiting nutrient. Fitness characteristics are thus generally manifested as increase in growth rate or an elevated affinity for the limiting nutrient. An increase in the proportion of 'mutator' strains is often noted in long term continuous cultures. The mutator characteristic causes an increased frequency of mutation and is thus more likely to be 'piggy-backed' into dominance by more directly advantageous mutations. Mutator strains are believed to offer a particular selective advantage when there is a rapid change in environmental conditions, such as sudden temperature or pH shocks to culture.

Visible growth will normally develop on vessel walls after 100–150 h of chemostat operation. Bacteria attached to the chemostat walls (as a biofilm) are normally less adapted to the chemostat environment than are their planktonic counterparts in the liquid culture. Attachment to the vessel walls may provide a mechanism that protects these dis-

Table 12.6 *Changes in culture characteristics during long-term chemostat cultures*

Characteristic	Advantage
Lowered substrate affinity constant (K_s) for limiting nutrient	Competition for limiting nutrient
Higher nutrient-sufficient growth rate (μ_{max})	Competition for limiting nutrient
Filamentation	Higher surface to volume ratio
Inducible enzymes involved in handling the limiting nutrient become constitutive	Limiting nutrient concentration may equilibrate at sub-induction levels
Multiple copies of genes involved in metabolism of limiting nutrient	Greater metabolic efficiency for critical pathways
Changes in uptake systems	More efficient uptake at low ambient levels
Auxotrophy	Uncertain: may confer energetic advantage
Clumping and wall biofilm growth ('sticky' strains)	Avoid wash-out
Mutator strains	Will produce fitter mutants at greater frequency

advantaged cells from washout. We have attempted to generate benzalkonium chloride (BKC) resistant mutants of *Pseudomonas aeruginosa* by applying increasing biocide selection pressure to chemostat cultures. Visible wall growth appeared after 25 generations ($D = 0.4\,h^{-1}$) and increased in intensity until 36 generations of growth. At this juncture, the liquid contents of the chemostat were transferred to a new vessel. The rate of adaptation to biocide tolerance in the culture increased. The wall population in the old vessel was found to consist of slow-growing phenotypes that had MICs for BKC similar to that of the original non-adapted strain. Wall growth severely affected the rate of adaptation because less fit populations adhered to the vessel walls. Furthermore, slough-off of such cells into the bulk medium resulted in a larger frequency of nonadapted phenotypes in the reactor.

In the absence of a specific selection agent, evolution in chemostats (as has already been noted) will result in a population that is highly efficient in competing in a nutrient-limited environment. Unfortunately, such populations may be seriously disadvantaged in the nutrient-sufficient environment of a batch culture. For example, growth of *Escherichia coli* in long term lactose-limited chemostats may evolve a population that has adapted to the limitation by dramatically overproducing β-galactosidase. This is an unnecessary metabolic burden in a nutrient sufficient situation. Thus, when such a population is used as an inoculum for a batch culture, the overproducing strains will be rapidly outcompeted by revertants (Pavlasova *et al.* 1986). This does not mean that strains evolved by chemostat culture will always be unstable, especially if a selection pressure other than nutrient limitation is applied.

As a practical example of directed evolution in chemostat culture, we were able to dramatically improve the stability of a plasmid carrying an industrially useful gene (a thermostable amylase) that had been inserted into *Bacillus subtilis*. The pUB110-based plasmid in question was segregationally unstable and this instability was exacerbated by the faster growth rate of plasmid-less cells compared with their plasmid-containing counterparts. The plasmid also encoded for chloramphenicol and kanamycin resistance markers. Selection was based on long term starch-limited cultures. A simple enrichment strategy based on a heuristic model of selection was devised. The model 'considers a population in which each cell represents a clone. In the absence of selection pressure, one daughter of each cell division is lost from the chemostat vessel; the other will remain to continue the cell line. In a system where selection pressure is applied against plasmid-free cells, the retained offspring of stable recombinant clones will always be resistant to the selective pressure, and thus able to continue the clone. With increasing degrees of instability, the probability of one or both of the daughters being plasmid-free increases. There will thus be an increasing probability that the retained daughter will be plasmid free, disadvantaged and unable to continue the cell line. Hence, as the culture proceeds we expected the spectrum of stabilities of recombinant clones within the system' (Fleming *et al.* 1988). Since chloramphenicol resistance was plasmid encoded, increasing concentrations of the antibiotic were added to the chemostat. Variant strains were isolated after 160 generations, which stably maintained the plasmid in non-selective batch and chemostat cultures. These variants showed growth rates that were similar to, or greater than, the plasmid-less cells. Though the frequency at which cells without plasmid were generated from the stable mutants was similar to the original strain (no change in plasmid copy number), plasmid-less cells were nevertheless competitively disadvantaged and the improved strains appeared phenotypically stable. In other words, the plasmid that previously decreased the growth rate of the unstable archetype no longer imposed a metabolic burden in the stable host organism.

Care must be taken when adding an inhibitor to a chemostat as a selective agent. One approach is to add a constant concentration to the medium reservoir. This is suitable if the agent in question is not metabolised or inactivated by the microbial population. If this does occur, it may be necessary to increase the concentration of the agent from time to time so as to maintain selection pressure. When improving plasmid stability in *Bacillus subtilis* we found it necessary incrementally to increase the concentrations of chloramphenicol in the medium feed because the antibiotic was metabolised by a plasmid-encoded chloramphenicol acetyl transferase. Furthermore, as the stability increased the adapted variants were better able to inactivate the selection pressure. However, an over-enthusiastic rate of addition of an inhibitor will result in washout of culture. We have found that the initial application of a selection agent should be in the region of one-half to two-thirds of the agents MIC as measured using the inoculum population. As feeding of an inhibitor is commenced, the culture density will decrease. It should be allowed to increase to 75% of its original value before increasing the inhibitor concentration is attempted. The degree of increase will depend on the individual selection regime and the potency of the inhibitor. Another method of judging a culture's ability to tolerate an increase in inhibitor concentration is to plate out from the chemostat to plates containing 2-, 4- and 6-times the concentration of inhibitor, and on plates that are selector free. The level of inhibitor can be increased if the proportions of colonies on the 4 times concentration plate are at least 10% of those on the plate without inhibitor. We have found this particularly useful when carrying out enrichment protocols at very low dilution rates. Optical density measurements should be taken of the culture at least three times daily. If excessive washout appears to be occurring after the addition of too high a concentration of an inhibitor, half the working volume of the culture should be aseptically removed and replaced with inhibitor-free medium. The culture should then be allowed to grow as a batch culture for 6 hours. After this, feeding may be recommenced with the inhibitor level at half that which caused washout. Finally, isolates should be obtained from the chemostat on a daily basis. These may then be used to inoculate a fresh chemostat should culture retrieval fail. In this manner culture history is not lost.

The probability of culture washout can be reduced using modified chemostat configurations. The turbidostat and auxostat (see Chapter 4) use optical density and pH feed back loops respectively to control the density of culture. In these systems, fall in optical density or perturbation in pH resulting from partial washout is corrected using feed-back loops to the media pumps. Brown and Oliver (1982) developed the BOICS ('Brown and Oliver interactive-continuous selection') continuous culture system for the selection of yeast mutants that were ultratolerant (12% w/v) to ethanol. BOICS employed a feed-back loop where the carbon dioxide output from the chemostat culture was monitored and this regulated the input of the inhibitor to the medium feed. These have been widely used to study R and K-type selection in chemostat-like ecosystems. Gradostat systems consist of a number of bi-directional linked chemostats that are fed with media/inhibitors from either end of the array. These and modified configurations of the gradostat have some potential for studying the kinetics of adaptation towards inhibitor tolerance along distinct concentration gradients of the selective agent.

12.6   Conclusions

For those working in microbial research laboratories, fermenters are important experimental tools. As with all such tools, however, the value and significance of the results obtained will rely on the appropriate use of the correct tool. Chemostat systems have two major advantages as experimental tools: the ability to maintain a steady state and the degree of independent control that the experimenter can exert on culture variables. The steady state environment is, however, a nutrient-limited one. The chemostat is also a powerful selection agent for fitter strains. Whilst the changes caused by this even when dealing with 'pure' cultures may cause problems to those engaged in studies on cell physiology and composition, this selective power may be put to good use by those wishing to isolate strains with improved characteristics or those who wish to study the mechanisms of mutation, selection and evolution.

The content of this chapter reflects our interests and we have tried to pass on some practical advice on small fermenter operation that is based on our experience in the laboratory. References cited in the text will provide a fuller coverage of many of the topics mentioned. Short reviews by Hoskisson and Hobbs (2005) and Denamur and Matic (2006) provide up-to-date information for those interested in the use of continuous cultures for strain improvement. Inevitably some topics, such as the use of fermenters by the microbial ecologist, have only received a cursory treatment in this chapter. The following recommendations for further reading may help to address this imbalance:

- A critical evaluation of the basis of the kinetics of microbial growth is provided by Kovárová-Kovar and Egli (1998). This review also deals with subjects of interest to the microbial ecologist: situations envolving mixed substrates and mixed cultures.
- Also of interest to the microbial ecologist is the earlier review by Gottschal (1990), which deals with the use of chemostats and other forms of continuous culture for ecological studies.
- The review by Feldgarten *et al.* (2003) deals with the relevance of selection cultures to the ecologist.

References and Further Reading

Berg, O. G. (1995). Periodic selection and hitchhiking in a bacterial population. *Journal of Theoretical Biology* **173**: 307–320.

Brown, S. W. and Oliver, S. G. (1982). Isolation of ethanol-tolerant mutants of yeast by continuous selection. *European Journal of Applied Microbiology and Biotechnology* **16**: 119–122.

Buttons, D. (1969). 'Thymine limited steady state growth of the yeast *Cryptoccus albidus. Journal of General Microbiology* **58**: 15–21.

de Visser, J. and Rozen, D. E. (2006). Clonal interference and the periodic selection of new beneficial mutations in *Escherichia coli. Genetics* **172**: 2093–2100.

Denamur, E. and Matic, I. (2006). Evolution of mutation rates in bacteria. *Molecular Microbiology* **60**: 820–827.

Dykhuizen, D. E. (1990). Experimental studies of natural-selection in bacteria. *Annual Review of Ecology and Systematics* **21**: 373–398.

Dykhuizen, D. E. and Hartl, D. L. (1983). Selection in chemostats. *Microbiological Reviews* **47**: 150–168.

Feldgarden, M., Stoebel, D. M., *et al.* (2003). Size doesn't matter: microbial selection experiments address ecological phenomena. *Ecology* **84**: 1679–1687.

Fleming, G., Dawson, M. T., *et al.* (1988). The isolation of strains of *Bacillus subtilis* showing improved plasmid stability characteristics by means of selective chemostat culture. *Journal of General Microbiology* **134**: 2095–2101.

Goldberg, I. and Er-el, Z. (1981). The chemostat – an efficient technique for medium optimization. *Process Biochemistry* **16**: 2–8.

Gottschal, J. C. (1990). Different types of continuous culture in ecological studies. *Methods in Microbiology* **22**: 87–124.

Herbert, D. (1961). The chemical composition of micro-organisms as a function of their environment. *Microbial Reaction to Environment: SGM Symposium 11*. (G. Meynell and H. Gooder, eds) Cambridge University Press, Cambridge, pp. 391–416.

Herbert, D. (1976). Expression of bacterial growth equations in dimensionless form. *Continuous Culture 6; Applications and New Fields*. (A. Dean, D. Ellwood, C. Evans and J. Melling, eds) Ellis Horwood, Chichester, UK, pp. 353–356.

Hoskisson, P. A. and Hobbs, G. (2005). Continuous culture – making a comeback? *Microbiology-SGM* **151**: 3153–3159.

Kovarova-Kovar, K. and Egli, T. (1998). Growth kinetics of suspended microbial cells: from single-substrate-controlled growth to mixed-substrate kinetics. *Microbiology and Molecular Biology Reviews* **62**: 646–666.

Kubitschek, H. (1966). Mutation without segregation in bacteria with reduced dark repair ability. *Proceedings of the National Academy of Sciences of the United States of America* **55**: 269–274.

Novick, A. and Szilard, L. (1950). Experiments with the chemostat on spontaneous mutations of bacteria. *Proceedings of the National Academy of Sciences of the United States of America* **36**: 708–719.

Pavlasova, E., Stejskalova, E., *et al.* (1986). Stability and storage of *Escherichia coli* mutants hyperproducing beta-galactosidase. *Biotechnology Letters* **8**: 475–478.

Pirt, S. J. (1982). Maintenance energy – a general-model for energy-limited and energy-sufficient growth. *Archives of Microbiology* **133**: 300–302.

Richards, J. (1968). *Introduction to Industrial Sterilisation*. Academic Press, London.

Tempest, D. (1976). The concept of 'relative growth rate'; its theoretical basis and practical application. *Continuous Culture 6; Applications and New Fields*. (A. Dean, D. Ellwood, C. Evans and J. Melling, eds). Ellis Horwood, Chichester, UK, pp. 349–352.

Zelder, O. and Hauer, B. (2000). Environmentally directed mutations and their impact on industrial biotransformation and fermentation processes. *Current Opinion in Microbiology* **3**: 248–251.

Index

Note: Figures and Tables are indicated by *italic page numbers*

Practical Fermentation Technology Edited by Brian McNeil and Linda M. Harvey
© 2008 John Wiley & Sons, Ltd